THE ISLE OF MAN

THE ISLE OF MAN
Celebrating a Sense of Place

edited by

Vaughan Robinson and Danny McCarroll

LIVERPOOL UNIVERSITY PRESS

First published 1990 by
Liverpool University Press
PO Box 147, Liverpool L69 3BX

Copyright © 1990 Liverpool University Press

All rights reserved. No part of this book may be reproduced, stored in a retrieval system, or transmitted, in any form or by any means, electronic, mechanical, photocopying, recording, or otherwise without the prior written permission of the publishers.

British Library Cataloguing in Publication Data
The Isle of Man: celebrating a sense of place.
   1. Isle of Man
   I. Robinson, Vaughan   II. McCarroll, Danny
   942.790858
   ISBN 0–85323–036–6 *Cloth*
        0–85323–296–2 *Paper*

Set in 10/12 pt Linotron 202 Palatino by
Wilmaset, Birkenhead, Wirral
Printed and bound by
Courier International, Colchester, Essex

For Brian Hardcastle, a true geographer (VR)
To my parents, for their continuing support (DMcC)

CONTENTS

List of Plates ix
List of Figures xiii
List of Tables xv
Contributors xvii
Preface xix
Acknowledgements xxi

CHAPTER 1   A return to landscape and place: rationale for a new geography
Vaughan Robinson and Danny McCarroll   1

PART ONE: THE PHYSICAL ENVIRONMENT

CHAPTER 2   The Manx landscape
Roger Dackombe and Danny McCarroll   10

CHAPTER 3   Solid geology
Roger Dackombe   18

CHAPTER 4   The Quaternary Ice Age in the Isle of Man: an historical perspective
Danny McCarroll   40

CHAPTER 5   Lateglacial and Postglacial environmental history
Danny McCarroll, Larch Garrad and Roger Dackombe   55

CHAPTER 6   The Isle of Man's unstable coast
Colin Rouse   77

CHAPTER 7   Nature conservation
Larch Garrad   84

PART TWO: HISTORY, CONSTITUTION AND POPULATION

CHAPTER 8   History
David Freke   103

CHAPTER 9   The Isle of Man constitution
Robert Quayle   123

CHAPTER 10   Social demography
Vaughan Robinson   133

PART THREE: THE ISLAND'S INFRASTRUCTURE

CHAPTER 11	Social infrastructure *Richard Prentice*	163
CHAPTER 12	Economic infrastructure *Vaughan Robinson, Richard Prentice and* *Gwyneth Davies*	177

PART FOUR: THE ECONOMY

CHAPTER 13	Agriculture, forestry and fishing *Gwyneth Davies*	207
CHAPTER 14	Extractive and manufacturing industries *Vaughan Robinson*	219
CHAPTER 15	Producer services *W. Dawson*	238
CHAPTER 16	Tourism *Richard Prentice*	248
CHAPTER 17	Geography of the Isle of Man into the 1990s *Danny McCarroll and Vaughan Robinson*	271

Bibliography — 274
Geographical Index — 284
Subject Index — 287

LIST OF PLATES

1.1	An evocative view of the steam locomotive 'Loch'	3
1.2	View of Victoria Street in Douglas in 1890	4
1.3	The village of Cregneash near Port St Mary, now restored as an open-air museum of Manx folk-life	5
2.1	The general accordance of summit heights in the Manx uplands	12
2.2	The large channel which separates Bradda Hill from the southern upland massif, now occupied by a very small underfit stream	13
2.3	Glen Maye, typical of many steeply incised stream valleys bordering the upland massifs	14
3.1	The Carboniferous basal conglomerate lying unconformably on the Manx Group at The Arches, Langness	22
3.2	Pillow lavas in the Scarlett Volcanic Formation	35
3.3	Columnar jointing on the Stack	36
4.1	The upper part of Laxey Glen displays some semblance of the U-shape typically produced by glacial erosion	44
4.2	The Bride Hills of the northern plain	45
5.1	Small organic-filled depressions on the surface of the gravel fan issuing from Glen Dhoo	60
5.2	The gravel fans of the west coast, seen incised at the top of a coastal section	61
5.3	The virtually complete skeleton of a Great Deer excavated at Close y Garey, near St Johns, in 1887	75
6.1	Huge joint-bounded blocks of Manx slate at the Chasms, south-west of Port St Mary	79
6.2	The interbedding of clay-rich and more permeable sediments in the drift cliffs of the northern plain	80
6.3	Breached sea defence wall at Gansey, Port St Mary	80
6.4	Property lost to the sea in Kirk Michael parish	81
7.1	The spectacular cliffs of the south-west coast	86
7.2	The Calf of Man, now a bird sanctuary	89
7.3	Ballaugh Curraghs	91
7.4	Laggagh Mooar (the big miry place) by John Miller Nicholson, 31 December 1883	93
8.1	Meayl Circle overlooking Port Erin	104
8.2	A view of St Patrick's Isle at Peel	112
8.3	Bishop's Court, built by Bishop Simon in the 1230s	116
8.4	A period view of Castle Rushen	117
8.5	The fourteenth-century Monk's Bridge over the Silver Burn adjacent to Rushen Abbey	118

8.6	Castle Rushen, taken from the main square in Castletown	119
8.7	A period wood-cut of Castle Rushen at Castletown	120
8.8	A period view along the earlier promenades of Douglas	121
9.1	The building, adjacent to Castle Rushen in Castletown, which was the home of the House of Keys between 1710 and 1874	124
9.2	Tynwald Hill at St Johns	125
10.1	Still one of the most characterful parts of Douglas, the North Quay	134
10.2	Traditional cottages as occupied by the early farmer-fishermen on the Island	136
10.3	North Quay in Douglas	138
10.4	A general view of Port Erin showing the way that the nineteenth-century hotels clustered along the bay front	139
10.5	Cronkbourne Village on the outskirts of Douglas	142
10.6	The Saddle Mews retirement village, a recent development on the outskirts of Douglas	158
11.1	Bishop's Court Glen	171
11.2	A waterwheel, part of the illusion the Edwardians sought when promenading in the Island's glens	171
11.3	Water-powered roundabout at Silverdale	172
12.1	Mona's Isle III at Douglas in 1890	178
12.2	Panorama of Douglas taken from the Head	179
12.3	Ramsey Harbour in 1902	180
12.4	Many people's first introduction to the Isle of Man, the unusual ferry terminal	181
12.5	Part of the modern fleet operated by Manx Airlines	185
12.6	The new Manx Telecom satellite station at Port-e-Chee outside Douglas	188
12.7	Two of the Island's generating stations found on the same site at Pulrose, near Douglas	190
12.8	Tramcar of the Snaefell Mountain Railway	196
12.9	Preserved vehicles at Douglas represent the Island's former transport companies	199
13.1	Sail-powered fishing smacks in Ramsey harbour, 1931	215
14.1	A period view of Lady Isabella at Laxey	221
14.2	A 1900 view of Laxey lead mines, showing the industry in its twilight years	222
14.3	Snaefell lead mines in 1898	223
14.4	Corlett's flour mill at Laxey	227
14.5	The first buildings erected in the much-delayed Freeport, sited next to Ronaldsway airport	234
14.6	Airship Industries' manufacturing plant at Jurby	235
15.1	A new office block in Douglas to cater for the growth in offshore services	245
15.2	A new housing development at Port-e-Chee near Douglas	246

16.1	Port Soderick in 1898	249
16.2	A period view of Douglas Head	250
16.3	Interior of a cottage at the Cregneash village museum of Manx folk-life	254
16.4	A modern view of Lady Isabella, the huge waterwheel at Laxey	255
16.5	The Island's railways are still represented by the line from Douglas to Port Erin	256
16.6	A night view of Douglas promenade	266

LIST OF FIGURES

1.1	The location of the Isle of Man	7
2.1	The Manx landscape divided into eight physiographic regions	11
2.2	Drainage diversions between Douglas and Ronaldsway	15
3.1	Outline solid geology of the Isle of Man	19
3.2	The geology of the northern Irish Sea basin	21
3.3	The stratigraphy of the Manx Group	23
3.4	The evolution of the structure of the Manx Group	27
3.5	The stratigraphy of the carboniferous limestones and the volcanic rocks of the Isle of Man	33
3.6	The location of the Manx mines and their relationship to the structure and lithology of the Manx Group	39
4.1	Limits of glaciation in the Irish Sea Basin	41
4.2	Oxygen isotope and palaeomagnetic record of the last 1–6 million years in core V28–238 from the equatorial pacific	42
4.3	Generalised Quaternary stratigraphy of the Isle of Man	42
4.4	Glacial landforms and deposits of the Isle of Man	43
4.5	Schematic diagram to represent cleavage of basal, debris-rich ice around the flanks of the Isle of Man whilst the uplands are overridden by clean ice devoid of erratics	48
4.6	Idealised cross-section of the Quaternary stratigraphy of the Isle of Man	50
4.7	The modelled surface topography and flowlines of the Late Devensian ice sheet	52
4.8	Simplified schematic model for deposition of 'push-ridge-subaqueous-outwash' depositional system exposed on the east coast of Isle of Man	53
5.1	Water depth, in metres, in the northern Irish Sea	55
5.2	Lateglacial and Postglacial landforms and deposits of the Isle of Man	57
5.3	Lateglacial organic deposits exposed on the west coast of the Isle of Man	58
5.4	Pollen diagrams obtained from organic depressions north of Kirk Michael	62
5.5	Present-day distribution of beetle species found in Lateglacial deposits of the Isle of Man	64
5.6	Diagrammatic illustration of the variations in mean July temperature in lowland Britain during the Lateglacial	65
5.7	Part of the sequence exposed on the coast near Phurt	67
5.8	Distribution of Manx Mesolithic sites	70
5.9	Ronaldsway Neolithic sites in the Isle of Man	72

6.1	Coastal erosion and accretion in the north of the Isle of Man	77
7.1	Location of sites designated as of special ecological importance in the *Isle of Man Development Plan*	86
7.2	The main features of the Calf of Man	90
7.3	Main features of the Ballaugh Curraghs	92
8.1	Selected prehistoric sites	105
8.2	Selected early Christian, Norse and Medieval sites	110
10.1	The distribution of members of the North American Manx Association, 1979	148
10.2	The distribution of the Manx-born in England and Wales, 1951	149
10.3	The changing distribution of the population in the Isle of Man, 1976–86	157
10.4	The distribution of New Residents, 1981	159
11.1	The importance of residents' opinions on the Glens as an essential part of the Island's heritage	173
11.2	The importance of residents' opinions on a national nature conservation plan	174
11.3	Walks promoted by the Manx Conservation Council	175
12.1	The location of key economic infrastructure, 1988	182
12.2	The railway network: as proposed and as built	193
12.3	The road network in 1789	197
12.4	The road network in 1989	197
12.5	Blair's motor coach tours, 1928	198
12.6	Stage bus routes operated outside Douglas, 1983	199
12.7	Stage bus routes operated outside Douglas, 1934	199
13.1	Changes in farm size, 1910–87	208
13.2	Changes in acreage under cereal crops, 1905–87	210
13.3	Changes in cattle and sheep numbers, 1945–87	213
13.4	Herring, scallops and Queen Scallops landed, 1955–85	216
14.1	Location of industrial enterprises in the period prior to the Second World War	225
14.2	Location of industrial enterprises, 1985	236
14.3	Location of main industrial estates	237
16.1	Advertised flights and sailings to the Isle of Man, 1987	259
16.2	The spatial distribution of different grades of tourist accommodation, 1987	263

LIST OF TABLES

3.1	The timing of geological events represented in the Isle of Man	20
4.1	Formal stratigraphic classification of the Quaternary deposits of the Isle of Man	51
5.1	Comparison of the Palaeoenvironment indicated by studies of pollen and plant remains from Lateglacial organic deposits in the Isle of Man	66
6.1	'Milestones' in the investigation of coastal erosion, IOM	78
6.2	Rates of coastal accretion or erosion	81
6.3	Amount of coastal accretion or erosion, 1869–1976	82
7.1	Basis for the listing of sites of special ecological importance	85
7.2	Habitat types represented by the listed areas	86
10.1	The Manx population in the period 1831–91	141
10.2	The Manx population in the period 1896–1945	143
10.3	The Manx population in the period 1945–61	144
10.4	Components of Manx population change, 1961–86	150
10.5	The Manx population in the period 1961–86	150
10.6	Changing fertility and mortality in the Isle of Man, 1951–86, per 1,000	153
10.7	The age structure of the Isle of Man's population, 1986	154
10.8	Changing employment by industrial sector of Manx residents, 1951–81	154
10.9	Social class of economically active residents of the Isle of Man, 1981	155
10.10	Changing population distribution, 1976–86	156
11.1	New housing construction, 1972–85	166
11.2	Visitors recorded at the Manx Museum and branch museums	176
12.1	Numbers of motor vehicles licensed on the Isle of Man	197
12.2	Numbers of direct scheduled bus services between selected places, 1929 and 1986	200
13.1	Farm size	209
13.2	Land under cereals, 1907–87	209
13.3	Hay and grassland, 1945–87	211
13.4	Stock	212
13.5	Landings of fish in tonnes—live weight and value	217
14.1	Industry in the Isle of Man, 1900	230
15.1	Isle of Man deposit base	240
15.2	Estimated number of persons employed: insurance, banking and finance	241
15.3	Income generated in insurance, banking and finance, 1969–88	243
15.4	Number of companies registered and removed from register	244

16.1	Numbers of visitors landed at Douglas in the late Victorian era	248
16.2	Tourism and seasonal male employment	251
16.3	Where visitors to the Isle of Man stay	251
16.4	Passenger arrivals on the Isle of Man	253
16.5	Scheduled sailings to Douglas from the United Kingdom	258
16.6	Numbers of premises and average size of accommodation by area of the Island, 1983	261
16.7	Grading of accommodation enterprises advertised in main brochure by main types of accommodation, 1987	263
16.8	Visitors' satisfaction with their accommodation	264
A.1	Isle of Man balance of payments estimates, 1983	269
A.2	National Income	270
A.3	Income generated in basic sectors from Manx sources	270

CONTRIBUTORS

VAUGHAN ROBINSON	Lecturer in Human Geography, University College Swansea
DANNY MCCARROLL	Research Geologist, University of Wales College of Cardiff
ROGER DACKOMBE	Lecturer in Environmental Science, Wolverhampton Polytechnic
GWYNETH DAVIES	Part-time Tutor in Geography, University College Swansea
W. DAWSON	Chief Financial Officer, Treasury, Isle of Man Government, Douglas
DAVID FREKE	Lecturer, Department of Classics and Archaeology, University of Liverpool
LARCH GARRAD	Assistant Keeper, Manx Museum, Douglas
RICHARD PRENTICE	Lecturer in Human Geography, University College Swansea
ROBERT QUAYLE	Formerly Clerk to Tynwald, Douglas
COLIN ROUSE	Lecturer in Physical Geography, University College Swansea

PREFACE

This book has grown out of a series of field trips organised by the Department of Geography at the University College of Swansea. Fieldwork has always been central to the discipline of geography at Swansea and we attempt to broaden progressively our students' horizons. The present series of field trips to the Isle of Man began in the early 1980s and has continued annually since then. The staff who participated in these trips repeatedly remarked about three things: first, the dialectical relationship in the Isle of Man between general and specifically local forces of landscape formation, and the way in which this relationship often produced unique local variants and trends; second, the way in which these local variants and trends produced a distinctive and very attractive sense of place; and thirdly, how there was no single text which did justice to this local personality. As a result, the organisers of our annual visit were continually faced with the problem of satisfying the curiosity of fifty enquiring students. Our response to this was to produce ever-larger field handbooks, and to involve local experts either as guest speakers or field-guides. Not surprisingly, after several years we proposed to Liverpool University Press that this same team of people might make ideal contributors to a book, a suggestion that Robin Bloxsidge warmly encouraged.

In producing the book we had three aims. The first was to provide a framework for the teaching of geography in the field. The varied landscape and unique history and culture make the Isle of Man an excellent 'field laboratory' for the integration of the disparate strands of modern geography. In our experience, students leave not just with a greater knowledge of the geography of the Island but with great affection for the Manx landscape and people. Second, we wished to provide an up to date account of the state of 'geographical knowledge' about the Island. Given the wide bounds of our subject such an aim is never realistic, but we hope that this eclectic collection is at least useful. Finally, we wished to produce a book which would also be of some interest to the general public. It is a sad reflection on academic geography that we have attracted so few amateur enthusiasts. We feel that by focusing on particular landscapes and places this can be remedied. By increasing understanding of the Manx landscape and people, one can only enhance the enjoyment of this unique and beautiful Island.

By way of justifying our decision to edit a book about a specific locality, when such works are profoundly unfashionable within the discipline, we have taken the liberty of preceding the substantive chapters by a short essay outlining some personal views on the current state of academic geography and how we feel it might best progress. It is intended to provoke thought and discussion amongst those who are interested. Those who feel it is out of place are more than welcome to ignore it.

In commissioning chapters we have deliberately not restricted ourselves to academic geographers. Rather we have sought views from whatever discipline or individual seemed appropriate. We have therefore assembled a panel of writers who have varied perspectives and experiences. None are 'mainstream' geographers. Rather we have material from an engineering geomorphologist, a Quaternary scientist, a geologist, The Assistant Keeper of the Manx Museum, an archaeologist, a former Clerk to Tynwald, the Isle of Man's Chief Financial Officer, a specialist in social policy and leisure studies, who also happens to be a bus and train enthusiast, a rural conservationist and a social demographer. As editors we have allowed each considerable freedom in choice of subject matter and approach. This is clearly apparent in the finished chapters, which vary a good deal in style and depth. We hope that they combine to provide some insight into the distinctive character of the Isle of Man.

Vaughan Robinson, Glais.
Danny McCarroll, Cardiff.
December 1989.

ACKNOWLEDGEMENTS

We have incurred many debts over the last seven years and it is almost impossible for us to acknowledge each of these individually. However invidious it is to select only some of these, we would particularly like to thank the staffs of the Reference Libraries at the Manx Museum and Douglas Public Library. We are also indebted to Merle Prentice who did a stalwart job of preparing the manuscript, Nicola Jones and Guy Lewis who drew all the excellent maps contained in this volume, and Alan Cutliffe who undertook all the photographic reproduction work. We also acknowledge grants from the Planning Studies Board at University College, Swansea and the Geology Department at University College, Cardiff both of which contributed to the cost of word-processing the text.

The following have given us permission to reproduce copyright material: The Geological Society of America and the authors for Fig. 4.2 from Shackleton and Opdyke (1976) and Fig. 4.8 from Eyles and Eyles (1984); The Geological Society of London for Fig. 3.3 from Simpson (1963) and Fig. 5.5 from Mitchell (1965); Yorkshire Geological Society for Fig. 3.5 from Dickson et al. (1987); G. S. P. Thomas for Fig. 4.6 from Thomas (1977); The Royal Society for Fig. 5.4 from Dickson et al. (1970); The Editor, *Boreas* for Figs. 5.5 and 5.6 from Coope and Brophy (1972); the Director, British Geological Survey for Fig. 3.2 redrawn from Wright et al. (1971) but with Crown Copyright reserved; and Fenton Photographic of Port Erin for the woodcuts and historic photographs.

Last, but by no means least, we could never have published this book without the cooperation and forebearance of the Manx people who annually put up with our temporary invasion of the island and our inane questions.

CHAPTER 1

A return to landscape and place: rationale for a new geography

Vaughan Robinson and Danny McCarroll

Introduction

Academic disciplines do not develop in isolation from the societies which nurture them and which also consume their output, either in the form of knowledge or of trained personnel. This is particularly true of geography. The academic origins of the subject can be traced directly to specific social and economic conditions of the nineteenth century and the ensuing cyclical changes in those conditions have been reflected by severe swings in emphasis within the subject. During this turbulent history there has been a consistent movement away from the original emphasis on places. The subject has lost its focus and has fragmented to the point where geography has been described as 'a collection of diverse and only loosely related groups of researchers travelling in no particular direction' (Taylor, 1985, p. 103). If geography is to survive as a rational framework for teaching and research, it must identify a new integrative core. We argue that a return to our traditional disciplinary concern with a sense of place and landscape would allow geography to move forward again, as a unified and valued discipline.

Geography in crisis; dereliction of duty at a time of need

The twentieth century has seen the dissolution of geography as a coherent discipline with a shared focus on the uniqueness of place. Human geography has divorced itself progressively from the lived-in world and has instead sought explanations in normative expectations, individual perception and the structure of society. Location and space have triumphed, place and character are frowned upon. Physical geography has moved gradually away from an emphasis on landscape towards an almost myopic preoccupation with process. Geomorphology, traditionally the mainstay of physical geography, has been described as degenerating into 'a minor branch of engineering' (Haines-Young and Petch, 1985, p. 200). Academic geography, with its unique origins and turbulent history is in crisis. There is little real overlap in the work of physical and human geographers and the sub-disciplines themselves have fragmented into discreet specialist groups. Even within the specialist groups of human geographers, factions fight to impose their particular perspective as a new orthodoxy (see Robinson, 1987, for an example of the diverse geographical perspectives that exist within one small part of the discipline). This has resulted not only in a lack of progress in the subject as a whole, but also in an increasing tendency towards nihilism that is often expressed in a very public manner. Whilst controversy and disagreement are inevitable parts of any intellectual endeavour, the guerilla warfare which seems to characterise some parts of the subject is being played out at a time when the discipline is under extraordinary external pressure.

2 THE ISLE OF MAN

Academic geography is experiencing external pressures both from below and from above. In schools, parents and students are increasingly assessing disciplines against a new set of criteria: these often centre upon the direct vocational relevance of a subject and therefore the skills, rather than knowledge, acquired. Central government has begun to intervene overtly in the definition of 'core' subjects and in the specification of curricula. And demographic trends, in Britain at least, are forecasting a 25 per cent reduction in the number of teenagers. Equally, higher education has had to respond to a new set of external pressures from above. Politically motivated cuts in public spending have led to underfunding of the entire University system. This has brought institutions and disciplines into conflict for scarce resources. Selectivity has favoured certain disciplines and certain universities, and has forced geographers to be far more pragmatic in their pursuit of recognition and resources. And many of these changes in political context occurred during a major world recession when, in line with tradition, geographers increasingly turned their attention to applied work or 'relevant' research. Paradoxically, internal turmoil and continually changing external pressures have emasculated geography at a time when society is in desperate need of the insights that geographers have traditionally offered and could still offer.

The twentieth century, and particularly the period since the Second World War, has undoubtedly seen a decline in individualism and a parallel loss of perceived identity. Fordist production methods, whilst making previously luxury goods available to the masses, have also led to product standardisation. This has been further exacerbated by the growth of multinationals plying homogenised products in previously distinctive markets. Supranational bodies have encouraged similar tendencies not only through their legislation, but also through their consolidation of fragmented markets. Even where national bodies have retained control of key aspects of everyday life, centrist tendencies have been to the fore, imposing, for example, minimum room sizes or development regulations. Advertising, the mass media and mass travel have made their own contributions, by either imposing tastes, fashions and expectations or by exposing more people to these. And finally, growing urbanism, or its influence, has helped to homogenise society.

The net effect of these forces has been to reduce the distinctiveness of place and help push man towards a state of anomie not unlike that expected of the profit maximising, economically efficient automaton so beloved of the spatial theorists. Relph (1976) even writes of 'placelessness' in the post-industrial landscape in which places are seen only for their functional and technical properties and potential. Thus at the very time that standardisation was being imposed on people and places, and the latter were losing their distinctive identities built up over centuries, the geographer was nowhere to be seen. In that sense we failed our society, for we neither acted as advocates for distinctiveness nor did we record the passing of places, landscapes and local cultures.

How then is geography to proceed as a coherent academic discipline which is also of some value to society? Some maintain that it cannot. Elliot-Hurst (1985), for example, has described geography as 'nothing more than a theoretical ideology based on technical practice alone' (p. 69) and as 'an untheorized point of entry to knowledge' (p. 85). To some extent, however, all academic boundaries are artificial and the wide bounds of geography allow us more freedom than most to manoeuvre into a more productive direction. We must begin, however, by dropping the pretensions that have been guarded for so long. The Kantian philosophical foundation, so beloved by Hettner and Hartshorne, which included geography with history as natural epistemological divisions has long been discredited. Geography has no divine right to exist. Nor is academic geography the heir to a centuries-old tradition. The discipline was

A RETURN TO LANDSCAPE AND PLACE 3

1.1 An evocative view of the steam locomotive 'Loch'. The preserved Victorian railway is just one of the many elements that helps to create a distinctive ambience and sense of place in the Isle of Man

imposed upon the university system (Taylor, 1985), with much opposition from the established sciences. The question, therefore, remains not *is* geography, but can geography *become* a coherent framework for teaching and research.

By focusing on real places rather than on abstract notions of space, or process for its own sake, geography can re-establish its original emphasis. However, whereas academic geography was created by states to further their nationalist and imperialist aims, the discipline is now in a position to turn that *cui bono* on its head. Johnston (1985a, p. 24) has stressed that 'geographers have disengaged themselves from studying and promoting the uniqueness of place, and consequently have contributed to a general ignorance of the world as a complex mosaic'. The universalist element has promoted 'not knowledge, not even information, but ignorance' (Johnston, 1985b, p. 331).

Geography must focus on places as they really are: attempting to understand how landscapes have formed and how they have changed and are changing in interaction with man; how cultures and economies have developed and responded. We must look beyond spatial patterns and appearance to understand what makes places special as well as how they reflect generalities. We must place geography in the vanguard of the struggle against placelessness and anomie and assert our role in breaking down nationalist preconceptions and stereotypes of other places and their people:

> There is a desperate need for a rekindling of geographical curiosity about all parts of the earth's surface, not for any simplistic, voyeuristic reason but because without a major educational and academic discipline which removes the blinkers on the societies that it serves, world understanding, and ultimately world peace, will become even more threatened than it currently is. (Johnston, 1985b, p. 326).

This is not a call for a return to the regionalism of the past, but rather to re-establish place and landscape rather than space and process. In terms of teaching, an emphasis on the study of landscapes and places will provide a focus for the wide range of physical and human aspects of the discipline. Such a broad education has advantages since:

> It can no longer be guaranteed that a set of specialised skills will remain useful for a working lifetime . . . In this situation we can expect the attractions of specialisation to diminish in the intellectual divisions of labour and more general training will become acceptable. Geography will be in a unique position to prosper from such a trend as an intellectually respectable alternative to inter-

4 THE ISLE OF MAN

disciplinary arrangements. (Taylor, 1985, p. 106).

Our graduates should depart, not with their heads full of ephemeral facts, of abstract models and processes, but with an enhanced understanding of what the world is *really* like. No one can ever understand all of the complexities that have interacted in shaping any landscape or place. If we can teach a little and inspire the desire to learn more, then—armed with their three great virtues of literacy, numeracy and graphicacy—geography graduates will depart with better qualifications than most.

As a coherent framework for research, a focus on landscape and place will serve to integrate and advance geography's many specialisms. In physical geography, Haines-Young and Petch (1986) stress that the lack of real progress reflects the absence of a scientific tradition. What it also reflects is a loss of focus on the real problems of understanding landscape. Most physical geographers have lost sight of the need to understand the history of landscape formation as well as the processes responsible. Whereas terrestrial geologists rarely look closer than the Late Cretaceous, few modern geographers look beyond the most recent glacial advance. A great gulf has developed in our understanding of landform development which can be rectified by applying our knowledge of process to particular places. Physical geographers must re-discover the desire to stand on hilltops and understand, or at least interpret, the view.

In human geography, both the 'spatial fet-

1.2 This view of Victoria Street in Douglas in 1890 illustrates well how the rapid growth of the tourist industry on the Island created an almost instant townscape characterised by a pleasing homogeneity of style

ishism' of the Quantitative Revolution and the subsequent divergent trends either towards the individualism of behaviouralist or humanist approaches, or towards the coarse economic reductionism of Marxian approaches, have left a massive void. Neither is explicitly concerned with the nature of real places. The former views places as 'concentrations of meaning and intention within the broader structure of perceptual space' (Relph, 1976). The latter sees place as a dependent and inevitable product of the broader processes of social change at work in our society. Conclusions are consequently either drawn at the gross societal level or at the minute individual level. While we clearly need such macro- and micro-level accounts these must be grounded in particular localities, with their differing characters. We need to study how similar socio-economic forces and human individuality are played out in different physical and human landscapes to different effect.

Attempting to understand particular landscapes and places should also serve to integrate geographical specialisms. The true character of a place cannot be understood by simple reference to its physical landscape nor to its basic human geography. Each part of the discipline has a contribution to make: explaining the past processes which helped shape the contemporary personality of a place; capturing all aspects of that physical and human personality; and helping to show how processes might reshape that personality in future.

It is perhaps appropriate that in arguing for a return to landscape and place in geography we have chosen to write about the Isle of Man, for during the very period when placelessness

1.3 The village of Cregneash near Port St Mary is now restored as an open-air museum of Manx folk-life. Similar thatched cottages can also be found in other parts of the island. Note the way that the thatch is tied down and how protruding slates on the chimneys act as flashing

was becoming ever more common on the mainland, the Island has progressively adopted a strategy which emphasises its own distinctiveness and its own unique solutions to general problems. Most of the chapters in the last three parts of the book discuss the way in which the Island's heritage and culture are being preserved and accentuated, not only in the Isle of Man but also in its diaspora. Chapter 9, on the constitution, goes further, since it describes how the Manx government has taken back powers and responsibilities from Westminster and used these to chart a distinctive and divergent trajectory for the Island and its people. By elucidating that distinctiveness and helping people experience, appreciate and value it, we have tried to make our own small contribution to stemming the tide of homogenisation, placelessness and anomie.

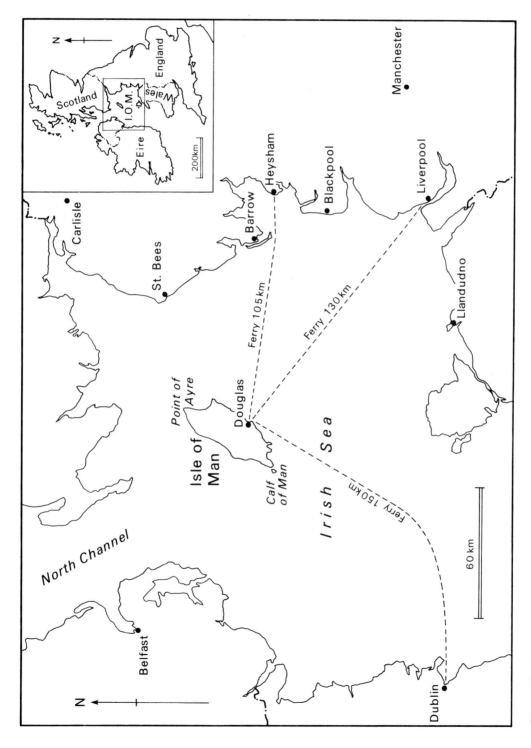

Fig. 1.1 The location of the Isle of Man

PART ONE

THE PHYSICAL ENVIRONMENT

Introduction

When viewed from afar, the Isle of Man resembles a displaced fragment of the Lakeland fells surrounded by the northern Irish Sea. Closer acquaintance reveals that, although there are similarities between Cumbria and Man, not least in geology, Man's insular status and position have bequeathed a unique geomorphological history which has produced a very diverse landscape. Deeply-incised slate uplands are flanked by drift-covered coastal plateaux and divided by a broad, glacially modified central valley. The fertile lowland limestone topography of the south-east contrasts with the exposed drift plain of the north. In Chapter Two the Manx landscape is described in terms of eight broad physiographic regions.

Although dominated by the Manx Slates which form the central hills, the Isle of Man boasts a range of lithologies and geological features which have not generally received the interest they deserve. South of the slate, red sandstone crops out around Peel whilst limestones form the south-east of the Island around Castletown. A range of igneous rocks occurs, including granites (with associated metamorphic aureoles), a diabase or gabbro and much younger (Carboniferous) intrusive and extrusive basalts, including pillow lavas. The granites and slate host a range of metalliferous minerals, including ores of lead, zinc, silver, copper, iron, nickel and antimony. North of the slate massif Permo-Triassic rocks, including saliferous marls, lie deeply buried beneath glacial drift. Although Ford (1987) has provided a short introduction to a few interesting localities, Chapter Three provides the only modern review of the solid geology of the Island.

Knowledge of the topography of the Isle of Man before the onset of the Pleistocene 'Ice Age' is sparse. Evidence of an early drainage system can still be seen and King (1976) has suggested some correlation of erosion levels with those recognised in Northern England. Unfortunately, the interpretation of long-term landscape evolution has been 'out of fashion' for some time (Chapter One) and until such work is recommenced it will be difficult to interpret the remnants of Man's pre-glacial landscape.

The legacy of glaciation is much clearer. Although probably not large enough to nurture its own cirque or valley glaciers, the Island lies directly in the path of the great ice sheets which have occupied the northern Irish Sea Basin at intervals during the last 2.4 million years. The northern plain consists entirely of material deposited during or since the Pleistocene, and at the northern extremity of the Island the very thick sequence proved in boreholes may comprise deposits representing three separate glacial episodes. The glacial deposits of the Island have been described in detail by Thomas (1976, 1977, 1985) and Dackombe and Thomas (1985). Rather than reproduce that work, Chapter Four seeks to view the landforms and deposits of Man from a historical perspective. Their interpretation has played an important part in the development of glacial theory and they remain central to the current debate concerning relative sea levels and the sedimentology of the complex sequences deposited during retreat of the last Irish Sea ice sheet.

Organic deposits preserved in depressions

in the Manx drift, particularly where they are incised by coastal erosion, provide some of the best sites in the British Isles for study of the flora, fauna and palaeoenvironment of the period immediately following deglaciation. The Manx deposits are particularly renowned for finds of *Megaloceras giganteus* (the Great Irish Deer) which are very rare outside Ireland. Holocene deposits in the Island have received less attention than they deserve, but have yielded some very interesting information, particularly regarding the environment of the early inhabitants. In Chapter Five, Lateglacial and Postglacial (Holocene) landforms and deposits are considered in turn. For consistency all dates are quoted in radiocarbon years BP, that is years before 1950 assuming a half-life of 5568±30 years (Lowe and Walker, 1984; Bradley, 1985). These dates are not comparable directly with the calendar dates often quoted in archaeological literature. Wherever possible, laboratory numbers are cited parenthetically, allowing dates to be traced and calibrated if required.

Where the coastline comprises unconsolidated Quaternary sediments erosion is relatively rapid and can threaten buildings, transport links and sites of scientific and archaeological interest as well as agricultural land. The extent of the problem is reviewed in Chapter Six together with some of the possible remedial measures.

The Isle of Man's varied geology and landscape, together with its insular status and position provide an extremely wide range of habitats for an area of only about 500 km^2. Conditions include the bleak summits of the slate massifs, wooded glens, the lichen heath of the Ayres, wetland Curraghs and steep coastal cliffs. The Island supports a wide range of wildlife and includes sites of National and International importance. Chapter Seven outlines the criteria by which important sites were defined in preparation for the Isle of Man's first Wildlife and the Countryside Bill and describes some of those sites which may be of particular interest to visitors. For the real enthusiast some species lists are appended.

CHAPTER 2

The Manx landscape

Roger Dackombe and Danny McCarroll

Introduction

Although the total area of the Isle of Man is only about 500 km^2, the diversity of scenery and landforms is such that eight broad physiographic regions may be defined (Figure 2.1). The Island is dominated by two upland massifs which are separated by the central valley, running from Douglas to Peel. The broadly triangular northern plain, composed entirely of glacial drift, lies beyond a steep northern escarpment, whilst fringing the uplands to the east and less prominently to the west are the rolling coastal plateaux. In the south of the Island, the eastern coastal plateau gives way gradually to the low, relatively flat plain of Malew which is overlooked in the south west by Mull Hill (alternatively known as Meayl Hill) and the Calf of Man.

The upland massifs

The northern and southern upland massifs are dominated by two high slate-cored ridges, guided by the dominant north-north-east to south-south-west Caledonide structural trend, which also accounts for the elongate shape of the Island. The western and lower of the two ridges runs from Mount Karrin in the north, through Slieau Dhoo, Slieau Freoghane and Beary Mountain. It is incised by the central valley at St Johns and continues through the southern massif via Slieau Whallion and Dalby Mountains to the steep slopes of the coast at Gob ny Ushtey. The eastern ridge can be traced from North Barrule through Clagh Ouyr to Snaefell, the highest point on the Island at 621m, and on to Greeba Mountain overlooking the central valley. It continues via South Barrule (483m), the highest peak in the southern massif, and abuts the coast at Cronk ny Arrey Laa.

In the northern massif, above about 200m and excluding the main ridges, the terrain forms a series of shallow upland basins and dissected high plateaux. Over most of the area drift, comprised exclusively of local lithologies, blankets the landscape resulting in a smoothly undulating topography. Although occasional bare rock surfaces can be found which display convincing striae, the northern uplands are conspicuously lacking in classical glacial landforms, and only one valley, Laxey Glen, shows any semblance of a typical 'U' shape. Hill peats are well developed in the centre of the northern uplands, particularly around Mullagh Ouyr and Beinn-y-Phott (Chapter 5).

The southern upland massif is generally similar to that in the north, though the closeness of the two main slate ridges reduces the amount of high plateau. Also, the western ridge is much more subdued than its northern counterpart, with only Slieau Whallion exceeding 300m in height.

Throughout the upland areas the rivers and streams are deeply incised into what appear to be preglacial valleys and are bordered by distinct terraces, the higher being the remnants of former solifluction features and the lower aggradational alluvial terraces.

The highest parts of the main ridges, par-

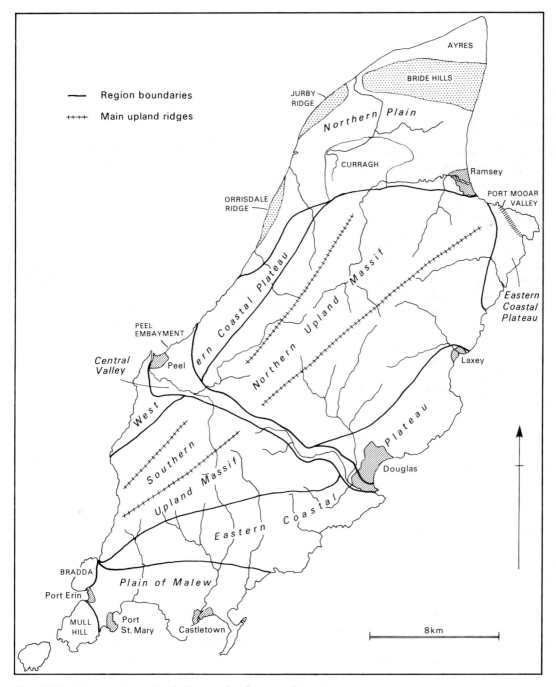

Fig. 2.1 The Manx landscape divided into eight physiographic regions

2.1 The general accordance of summit heights in the Manx uplands suggests the incision of a raised peneplain

ticularly on the flanks of Snaefell and along the North Barrule-Clagh Ouyr ridge are notable for the relatively high proportion of solid rock exposed. Altiplanation terraces, blockfields, boulder-runs and tor-like features all attest to the severity of conditions on these maritime highlands.

Temple (1956) mapped a series of terraces and associated slope breaks around the margins of the upland massifs and interpreted them as representing no less than nine Plio-Pleistocene strandlines ranging in height between 15 and 207m. It is difficult to establish to what extent these features reflect structural or lithological control, however, and this, taken together with their presence in areas of very restricted fetch, such as along the central valley, suggests that they might be more reasonably attributed to periglacial processes.

The coastal plateaux

The eastern coastal plateau is discontinuous and can be conveniently divided into three sections. In the north it comprises a roughly triangular area between Port e Vullen, just south of Ramsey, Maughold Head and the Dhoon, just north of Laxey. The topography is dominated by low rolling hills underlain by the flaggy formations of the Manx Slates. The streams which drain across the area are steep, but largely ineffective in altering their courses. The Dhoon Stream is particularly steep, having in its lower course several high waterfalls suggesting that the rock valley in which it flows has not yet adjusted to present sea level. In the north of the plateau the large Port Mooar Valley runs between the headland at Maughold and the lower slopes of North Barrule. The streams in its floor are markedly underfit, suggesting that it forms part of a truncated drainage system, probably of preglacial origin but undoubtedly much modified by glaciation.

The section between Laxey and the central valley at Douglas consists of a series of rolling foothills rising gently towards the upland massif. Most of the drainage is directed north to south, the one obvious exception being the southern arm of the stream which flows into Groudle Glen. Whether this alignment reflects the main structural trend or is a function of glacial modification is not clear. Temple (1956) has described a number of dry channel features in this area which cut across major spurs and often end in thin sand and gravel spreads. The highest is at 260m on the eastern spur of Slieau Meayll and some of the lower ones run parallel to the main ridge. Temple suggests that some of these may have discharged their waters into a lake in the area of the present Kerrowdhoo Reservoir.

The southernmost section of the eastern coastal plateau consists of a broad swathe of undulating topography that runs from the east coast between Douglas and Port Grenaugh towards Fleshwick Bay on the west coast. The eastern end of this tract is similar in character to the central section north of Douglas, but towards the west it changes to a sloping transition zone between the southern upland massif and the low plain of Malew. The south-west extremity of the Island forms a continuation of the coastal plateau topography, with Mull Hill and the Calf of Man, isolated hills of Manx slate, separated from each other by the Calf Sound.

The eastern section is again dominated by stream and valley courses which trend approximately north to south, but there are

2.2 The large channel which separates Bradda Hill from the southern upland massif is now occupied by a very small underfit stream. It may represent a remnant of a pre-glacial drainage system which has been altered by glacial and meltwater erosion

2.3 Glen Maye is typical of many steeply incised stream valleys bordering the upland massifs. Waterfalls suggest that the streams have not yet adjusted to present sea level

also a number of notable drainage diversions (Figure 2.2). At Oakhill, north of Port Soderick, a stream rising near Braaid and flowing south-east turns abruptly to the north-east to flow to join the Dhoo-Glass River on the outskirts of Douglas. From Oakhill the former course probably ran out to the coast in a southerly direction to Keristal. Similarly the Crogga River, just to the west of Port Soderick probably flowed through the narrow gorge at Crogga, now followed by the railway, to discharge at Port Grenaugh. If, as seems likely, these diversions occurred as more viable courses became available following withdrawal of the ice, it is interesting to speculate on whether the Santon Burn became ice-marginal in the Ballasalla area and was responsible for the extensive gravel spreads around the airport at Ronaldsway.

The western coastal plateau is far less extensive than its eastern counterpart and never exceeds three km in width. In the north it abuts the drift plain, from which it is clearly differentiated by the sudden appearance of solid bedrock in the coastal cliffs and low, rounded, rock-based hills inland. To the south it extends beyond the Peel embayment as far as Dalby, where the southern upland massif meets the coast. The area is drift covered throughout, and the deeper coastal glens are plugged with considerable thicknesses of slatey gravels and heads. In the northern section these are often seen interdigitating with foreign till and sands.

Drainage is generally by short steep streams running at right angles to the coast, but as on the east coast there is often evidence of drainage diversion associated with deglaciation. At Ballaleigh, just south of Kirk Michael, a narrow, sinous, steep-sided valley is cut into the solid across the main ridge crest. Over the valley at Cronk Urleigh a distinct flat-topped terrace suggests that it is part of a lake overflow system. Similar disconnected channels occur on the flanks of Slieau Curn and all eventually debauch into gravel spreads and probably represent an integrated ice-marginal drainage system.

The Peel embayment

This roughly square area of low, undulating topography extends from Peel inland towards St Johns. Although the Peel Sandstone formations are exposed on the coast, inland they are completely buried beneath thick drift. With the exception of the extreme coastal fringes, this is the only area in which the foreign deposits penetrate inland along the west coast.

Inland, the embayment passes into the western end of the central valley, from which it is separated at St Johns by coalescing spreads of gravel issuing from the tributary valleys of the Neb and Foxdale Rivers. To the

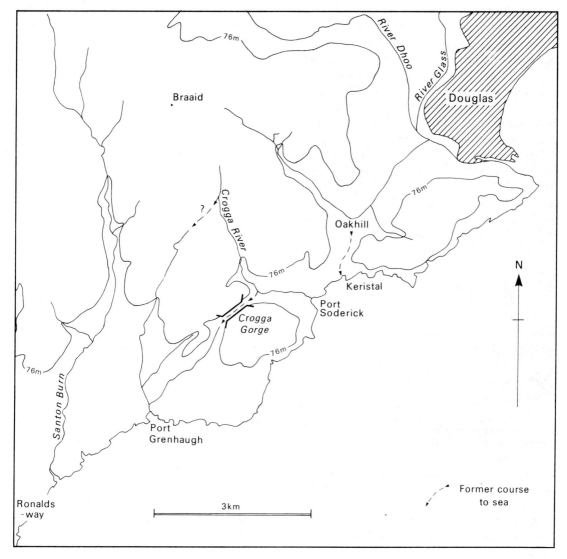

Fig. 2.2 Drainage diversions between Douglas and Ronaldsway probably occurred following withdrawal of the last ice sheet

west of this barrier the undulating platform of foreign drift comprises mainly low sandy and gravelly mounds trenched by drainage channels leading into the Neb.

The central valley

The central valley runs from the southern end of Douglas Bay through to St Johns, where it broadens and passes into the Peel embayment. Throughout, it has a broad, open 'V' section, the floor of which is infilled with mounds and terraces of drift. The valley is slightly sinuous, more so at the eastern end than in the centre where it appears to be fault guided.

The valley floor is widest at the western end where a broad tract of alluvium and peat lying at about 45m above sea level marks the position of Curraghglass. This former boggy lake was impounded by a dam of foreign drift which lies across the valley at St Johns. Fringing the basin are low mounds of foreign drift and subdued terraces of Postglacial flood gravels. Never wider than 0.5 km, the valley floor narrows progressively eastward to the watershed at Greeba, where alluvium abuts directly onto the steep valley slopes.

East of the watershed the valley floor falls very gently towards a further constriction at Union Mills, where 1–2m of silty alluvium and peat resting on slatey gravel may indicate the site of another former lake (Lamplugh, 1903). The course of the River Dhoo in this section is almost entirely artificial. Below Union Mills the valley falls more steeply and is more sinous, and the alluvial floor is again flanked by low terraces of Postglacial flood gravels.

The northern plain

The northern plain is a roughly triangular tract of land separated from the upland massif by a steep escarpment which probably represents the preglacial shoreline of the Island. The ground averages about 30 m above present sea level and everywhere drift buries the solid bedrock down to at least 30m below sea level. The northern plain can, itself, be conveniently divided into five separate regions.

To the north of a line running from Phurt in the east to Blue Point in the west lies the low gravelly foreland of the Ayres Postglacial raised beach. Mostly lying below 10m OD, the Ayres consists of a series of gravel ridges running sub-parallel to the present north coast and truncated by erosion along the eastern coast. On the northern coastal fringe a narrow, discontinuous belt of low sand dunes is poorly developed. Inland there is a thin cover of stabilised blown sand. The southern margin of the Ayres is marked by an abandoned cliff line, declining in prominence from west to east, the foot of which is obscured by low mounds of blown sand banked against it for much of its length. These cover the deposits of lagoons which were impounded by the early beach ridges.

Immediately to the south lies a crescent of morainic hills comprising three distinct sections. From Shellag Point in the east and running through the village of Bride to the valley of the Lhen are the Bride Hills. From the Lhen running south-west through Jurby Head towards the Killane River is the Jurby ridge which is cut obliquely by the coast and continues in a more southerly direction, towards Kirk Michael, as the Orrisdale ridge.

These hills are the most prominent feature of the northern plain and consist of a series of rounded mounds composed of tills, sands and gravels. They are highest and widest in the east, where they rise to 96m OD just to the west of Bride, and they decline to the west and south-west, where they seldom exceed 30m. There is little surface drainage, but the area is fretted by a profusion of dry valleys which form an integrated former drainage network. Only one significant stream still flows and that rises to the north-west of Bride and flows towards the north coast via Lough Cranstal, a boggy area of willow carr which contains at its base a series of marine/brackish water clays (Chapter 5).

Lying south and east of the ridges is an area dominated by platforms and subdued mounds of sand and gravel which descend towards the basin of the Curragh at the foot of the northern uplands. Lamplugh (1903) recognised three separate levels, most clearly seen in the east coast cliff-line, and interpreted these as the former shoreline terraces of the Lateglacial Lake Andreas. Once again there is little surface drainage. The few streams that drain the area are small, sluggish and follow a tortuous course towards the interior, where they join the Sulby River flowing along the foot of the northern uplands.

The Curragh proper is a roughly square basin of willow carr, underlain by peat, between the villages of Ballaugh and Sulby. Probably the site of a former lake basin (Ballaugh = Balla-Lough = Farm Lake) only occasional sandy islands provide dry sites in this otherwise water-logged area. Much of the low-lying peaty and alluvial land along the course of the Sulby River is similar in character and can be included in this division. Seldom more than 10m above sea level, the area is poorly drained and the present stream courses are largely artificial. Most of the Curragh basin drains to the west coast via the tiny Killane River. The more northern parts drain via the almost entirely canalised River Lhen, which is distinctly underfit in a broad flat-bottomed valley, probably cut originally by glacial melt-waters flowing towards the south.

The remaining division consists of a series of low spreads of slatey gravel, disposed as terraces and fans which fringe the foot of the upland scarp. The most obvious of these is the low flat fan at the mouth of Sulby Glen, the surface of which displays occasional abandoned channels. The Sulby fan serves to dam the eastern end of the Curragh basin and separate it from the other basins and the Sulby River, which are about three metres lower. The fan which issues from the mouth of Glen Dhoo sports a number of enclosed hollows, north-east of Ballaugh, some of which form small pools.

The plain of Malew

In the south of the Island, the eastern coastal plateau descends to a low plain which lies south of a line from Port Grenaugh in the east to Port Erin in the west. The gently undulating topography of this plain falls steadily from a maximum height of about 80m in the north to a flatter coastal strip in the south which is mostly below 30m OD.

Along the northern margin of the area the locally-derived drifts of the uplands give way to a broad gravel platform which descends both to the south and to the west. Most of the material in these gravels is locally-derived slate, but to the south and west it is progressively enriched with foreign erratics and locally-derived limestone.

Where the gravels are missing, the underlying limestone-rich till is exposed and in places upstanding knolls of limestone protrude through the drift cover. Around Ballasalla there are several low, rounded hills which have been interpreted as drumlins. Although mantled in drift, some of these hills display evidence of a rock core, such as that at Creggans opposite Ronaldsway airport, where Lamplugh (1903) observed a striated bedrock pavement beneath the drift.

As in the north of the Island, there is abundant evidence of temporary postglacial lakes, some of which persisted into historic times. To the south of Colby, for example, a low-lying area of ground is covered by a stony clay and alluvium. Similar materials were observed by Lamplugh in the Great Meadow area of Castletown which until historic times contained a shallow lake, and in borings around the airport as much as four metres of fine laminated silts are found overlying and infilling the dissected till surface.

Cumming (1853) reported finding sands with marine shells, which he interpreted as a raised beach, inland of Poyll Vaaish. Although there are a number of flat benches around these southern shores, the flat-lying attitude of the Carboniferous limestone makes it likely that some of them are controlled structurally. Others, notably on the Langness Peninsula where they are cut across the upturned edges of the Manx Slates, are more acceptable.

CHAPTER 3

Solid geology
Roger Dackombe

Introduction

Although a casual glance at the geological map of the Isle of Man (Figure 3.1) reveals that it is dominated by slates, there are a variety of geological features packed into a remarkably small area. The long coastline provides easy access to striking sections of rocks ranging from slates and greywackes of Cambro-Ordovician age, through Devonian fluviatile sandstones and conglomerates, Carboniferous limestones and volcanics to some of the finest Quaternary glacial sections in the British Isles (Table 3.1).

The Isle of Man is an inlier of Lower Palaeozoic greywackes, the Manx Group, in the centre of the northern basin of the Irish Sea (Figure 3.2). It forms part of the Caledonian orogenic belt and can be related to similar rocks in the English Lake District and S. E. Ireland. During the Caledonian orogeny the Manx Group suffered mild metamorphism, faulting and intense folding and was intruded by two late-orogenic granites. The Island forms an upstanding massif which protrudes through the younger cover of Carboniferous and Permo-Triassic sediments which extend eastwards into the Solway, Lancashire and Cheshire Lowlands and west into eastern Ireland around Dublin (Figure 3.2). The Carboniferous rocks extend on to the Isle of Man in the south-east around Castletown where a basal conglomerate succeeded by limestones and extrusive, basic volcanics lie unconformably upon the Manx Group.

On the west coast, around Peel, younger sediments consisting of red sandstones and conglomerates of Devonian age are down-faulted into the Manx Slates. The northern plain of the Island is thickly mantled with glacial deposits and the solid rocks are buried at least 40m below sea level. Despite this, the lure of an extension of the West Cumberland Coalfield on to the Island has led to a number of exploratory borings being sunk through the glacial drifts. These have proved a sequence of Lower Carboniferous limestones and shales overlain unconformably by Permo-Triassic sandstones shales and saliferous marls.

The geological evolution of the Isle of Man

The earliest geological events recorded in the Isle of Man are associated with the evolution of the proto-Atlantic Iapetus Ocean. This ocean basin, formed by the extension of oceanic crustal plates, began to open in late Precambrian times and has a long and complex history of sedimentation, closure and destruction. The Manx Group was deposited in this basin in late Cambrian and early Ordovician times around 500 million years ago at about the same time as similar sediments in Leinster and the English Lake District. These areas would probably have been part of the south-eastern continental slope in an ocean that extended from north-east to south-west through what is now Norway, eastern Greenland, north-west Britain, Ireland and into the eastern seaboard of North America. The deposits of the west side of

SOLID GEOLOGY 19

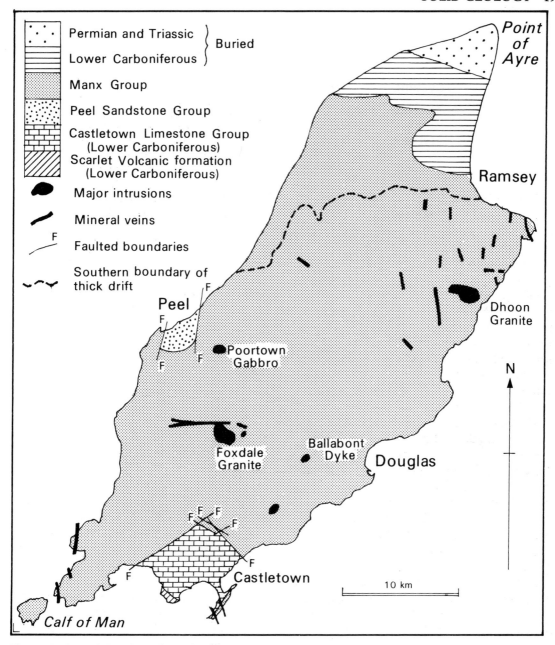

Fig. 3.1 Outline solid geology of the Isle of Man

Iapetus are now represented by the rocks of north-west Ireland, and the Southern Uplands and Highlands of Scotland.

As the Manx Group sediments were being deposited, Iapetus was becoming narrower due to thrusting of the oceanic crust beneath the continents, mainly on the western side of the ocean. Considerable deformation of sediments took place on the west side of the narrowing ocean to form the mountain chain of the north-west Highlands of Scotland and their extensions in the American continent.

20 THE ISLE OF MAN

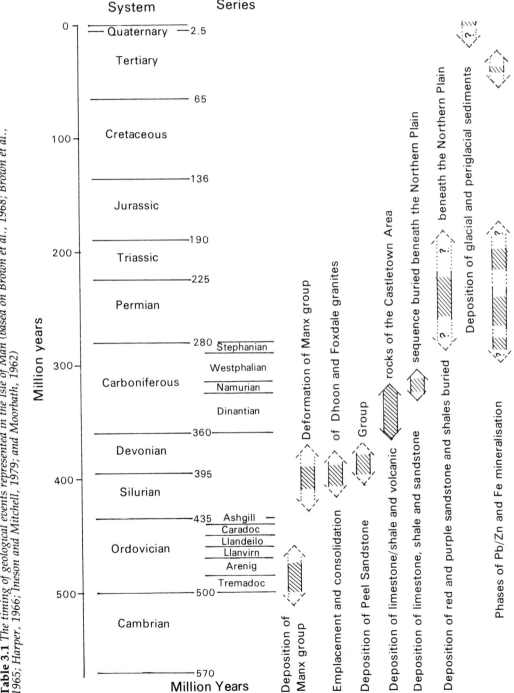

Table 3.1 The timing of geological events represented in the Isle of Man (based on Brown et al., 1968; Brown et al., 1965; Harper, 1966; Ineson and Mitchell, 1979; and Moorbath, 1962)

SOLID GEOLOGY 21

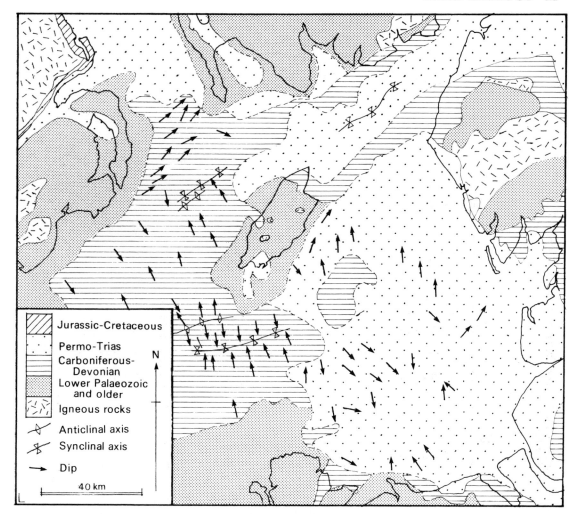

Fig. 3.2 The geology of the northern Irish Sea basin (based on Wright et al., 1971)

By about 500 million years ago, narrowing was occuring on both sides of the ocean at a rate of about 0.5 cm per year and by Wenlock times (435 million years) the ocean had closed completely.

With closure of the ocean, the sediments, including those of the Manx group, were intensely deformed and regionally metamorphosed during the Caledonian orogeny. In the later stages of deformation intrusion of hot molten magma which later solidified to form the granites of Foxdale and Dhoon caused localised thermal metamorphism. Similar events occurred in the Lake District with the deformation and metamorphism of the Skiddaw Group and intrusion of the Skiddaw granite and in north-west Ireland with the intrusion of the Donegal granites.

The regional metamorphism of the Manx Group has been dated radiometrically at between 360 and 414 million years ago and the solidification of Foxdale and Dhoon granites at between 370 and 315 million years. The latter date is rather young and probably reflects a mid-Carboniferous phase of mineralisation rather than the actual consolidation of the granite.

After the Caledonian orgeny, much of Bri-

22 THE ISLE OF MAN

3.1 The Carboniferous basal conglomerate lying unconformably on the Manx Group at The Arches, Langness

tain was subjected to rapid uplift and erosion under semi-arid climatic conditions. This is recorded in the Isle of Man by the Peel Sandstone which was probably deposited either as part of a major river system or in a mixed river plain and alluvial fan environment. There is some debate regarding the age of the Peel Sandstone, but there is a consensus that a Devonian (360–415 million years) age seems to fit best with the available evidence.

During Devonian times the mountains of the Caledonian orogenic belt were rapidly eroded and by early Carboniferous times (360–325 million years ago) a series of shallow marine gulfs with intervening land masses covered much of Britain and Ireland. In the Isle of Man the Castletown Limestones were deposited in shallow water conditions in a humid tropical climate. In later Carboniferous times, around 325–320 million years ago, major river systems spread into many of the marine basins and although rocks of this type are not seen in the south of the Island, the Carboniferous rocks beneath the northern plain are likely to have been deposited under these conditions.

In Upper Carboniferous times (315–290 million years ago) much of Northern Britain was covered by tropical swamps in which coal was deposited. These rocks are absent from the Isle of Man, probably due to erosion, and the Carboniferous river and delta sediments are succeeded by red sandstones, mudstones and saliferous marls. These rocks of Permian and Triassic age are known only from the deep boreholes at the Point of Ayre and form part of the extensive basin of Permo-Triassic rocks that occupy the Solway, Lancashire and Cheshire lowlands. Conditions during this time were probably of semi-arid desert and salt flats set amid arid mountains. Through-

SOLID GEOLOGY 23

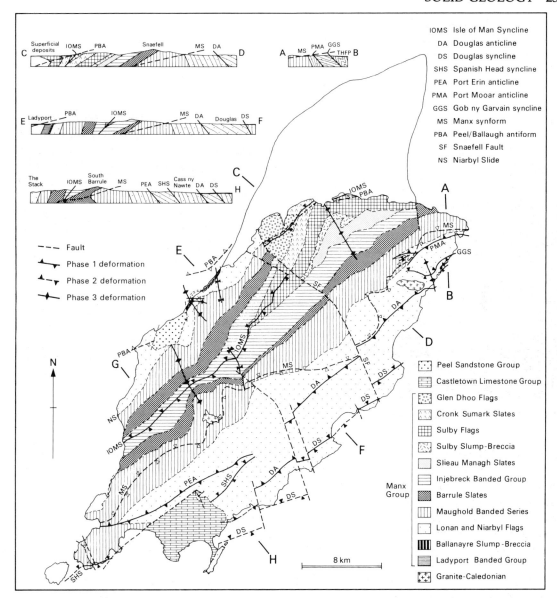

Fig. 3.3 The stratigraphy of the Manx Group (redrawn with permission from Simpson (1963), *Quarterly Journal of the Geological Society*, vol 119, pp. 367–400)

out the Carboniferous and into Permian times, volcanic and intrusive igneous activity was important in northern Britain. This phase is represented in the Island by the Scarlett Volcanic Formation, a suite of dolerite dykes, ashes and agglomerates and, less obviously, by the extensive base metal mineralisation of the Manx Group and the Dhoon and Foxdale granites. It is a common misconception that the granites were responsible for the formation of the rich lead and zinc ores of Foxdale and Laxey. In fact, they merely provided a favourable host for the deposition of these minerals from the circulating fluids associated with this mid-Carboniferous phase of volcanism.

The record of the last 200 million years is absent in the Isle of Man, except for that of the most recent events, those of the Quaternary glaciations, which are covered in detail in Chapters 4 and 5. The rest of this chapter deals with the rock formations of the Island in order of age.

The Manx Slate Group

The Manx Group is a thick sequence of monotonous slates and grits of Cambro-Ordovician age. Although descriptions of them figure in the earliest topographic accounts of the Island (e.g. Woods, 1811; Berger, 1914) the first detailed work was carried out by Harkness and Nicholson (1866) who compared them with the Skiddaw Slates of the Lake District. During the geological survey of the Island, Lamplugh (1903) recognised the complexity of the sequence and structure and defined four major formations contorted into a large complexly folded synclinorium. Parts of the north-east of the Island were mapped by Gillot (1955, 1956), but the most recent work is that of Simpson (1963, 1964a, 1965).

Stratigraphy

Simpson (1963) recognised twelve formations (Figure 3.3) with a total thickness of about 8000m. The lowermost formations are only exposed on the west coast north of Peel and consist of the thin Lady Port Banded Group succeeded by the Ballanyre Slump Breccia. Simpson follows Lamplugh in equating the two major flaggy formations, the Lonan Flags which crop out over much of the eastern half of the Island, and the less extensively exposed Niarbyl Flags. Although each formation is lithologically distinct the boundaries between them are gradational. The almost complete absence of macrofossil remains makes a detailed zoning of the sequence difficult and the latest micropalaeontological investigations (Molyneux, 1979) are at variance with Simpson's (1963) stratigraphy.

Sedimentology

The Manx Slates are a turbidite sequence and such variation as does occur may well be due to variations in the position of the Island relative to the source of sediment. The twelve formations can be conveniently considered in four groups.

The flaggy formations, the Lonan, Niarbyl, Sulby and Glen Dhoo Flags, consist of rapid alternations of thinly bedded sandy and silty greywackes. Siltstones and fine sandstones are dominant with subsidiary amounts of interbedded finer mudstone and occasional thick bands of quartzite. Colours range from grey, green and fawn in the coarser units to black and dark blue in the muds. Small scale current bedding, irregular slump folding and graded bedding are the most common sedimentary structures.

The banded formations, the Lady Port, Maughold and Injebreck 'groups', consist of striped alternations of light grey siltstones and darker blue to black mudstones. Coarser sediments are rare, although there are occasional bands of sandstone and quartzite. The latter are seen best in the Cregneish Peninsula where they are particularly thick. Sedimentary structures are limited mainly to a fine parallel lamination; other structures are rare.

The pelitic formations are the Barrule, Slieau Managh and Cronk Sumark Slates which comprise thick, monotonous sequences of uniform dark grey to blue-black shales and slates. There is little variation and the paucity of sedimentary structures together with the obscurity of the original bedding makes these formations singularly uninteresting. The two main ridges of the uplands are founded on the Barrule Slates while the Slieau Managh and Cronk Sumark Slates underlie the subdued uplands of the Island.

The Sulby and Ballanyre Slump Breccias consist of blocks of slate, siltstone and sandstone set in a blue-black muddy 'paste'. The size of the blocks varies but most are in the range 3–5 cm with less common blocks up to 30 cm. Gillot (1954) and Simpson (1963) both attributed these unusual units to sub-aqueous sliding and brecciation during deposition on steep slopes in an actively subsiding basin.

The age of the Manx Group

The extreme rarity of macrofossils in the Manx Group has led to some debate about their age and place in Caledonide palaeogeography. Bolton (1899) reported dendroid graptolite fragments and a possible imperfect cast of an Asaphid trilobite from Cronk Sumark. The former were identified as *Dictyonema flabelliforme* (discussion of Simpson, 1963) and thus a Tremadocian age was accepted for the Cronk Sumark Slates. This would place the lower formations in the Cambrian and the upper ones in the Ordovician. Downie and Ford (1966) examined an assemblage of palynomorphs from the Lonan Flags and suggested a Tremadoc-Arenig or younger age for the higher formations. More recently Molyneux (1979), while confirming a Tremadoc to Arenig age, has thrown some doubt on the validity of Simpson's (1963) stratigraphy.

Environment of deposition and palaeogeography

If a late-Cambrian or early-Ordovician age is accepted then the Manx Basin, together with Leinster Basin and the lower Ordovician deposits of the Lake District, could be considered part of the continental slope of the southern margin of the closing Iapetus Ocean. The presence of turbidites, thinly-banded suspension deposits and intensely disturbed slump breccias all point to a medial rather than either a proximal or distal setting. Slump folds within the flaggy formations, together with the thickness of the sequence, both point to rapid subsidence and sedimentation.

During the late-Ordovician and early Silurian volcanic activity commenced, first in the Welsh Basin and later in the Leinster and Lake District basins and became a conspicuous feature of the unstable shelf areas. It was related to the development of a south-east dipping subduction zone associated with the trench away to the north-west. In contrast with the Lake District however, the Manx Group contains only very occasional evidence of contemporaneous volcanism. Both Simpson (1961) and Lamplugh (1903) recorded the presence of andesitic tuffs and fine agglomerates at two sites in the Niarbyl Flags. Molyneux (1979) terms these the Peel Volcanic Formation and considers them to be of Arenig age. There are two probable explanations for the relative absence of volcanics in the Manx Slates. Firstly, Ordovician-Silurian volcanics tended to be associated with the shelf areas some distance from the trench itself. The Isle of Man may have been rather too near the trench to be affected. Secondly, volcanism in the Welsh Basin did not begin in earnest until the Late Tremadoc and in the Lake District until the Lower Caradoc. Thus deposition of the Manx Slates was probably largely complete before steepening of the subduction zone shifted the focus of volcanism north-westwards during Llanvirn-Llandeilo times.

The structure of the Manx Group

Although Lamplugh (1903) recognised the complexity of deformation of the Manx Group it was not until the work of Gillot (1956) and Simpson (1963) that the details of the structure emerged. In common with adjacent areas of the Caledonides the Isle of Man suffered a number of phases of deformation which culminated with the intrusion of granitic plutons.

Simpson (1963) recognised three phases of folding and two of faulting, each with its own characteristic stress system and structural style. Although the larger scale folds are rarely seen in outcrop, the congruent minor folds and associated cleavages reflect the geometry of the larger features and allow their reconstruction.

The earliest phase of deformation (F_1), caused by lateral compression from the south-east and north-west (Figure 3.4), produced folds with acute hinges and long straight limbs. In the hinges the beds are usually thickened and sometimes crumpled. Although originally vertical, the axial planes of these folds are now steeply inclined between 35 degrees and vertical. Cleavage (S_1) is well developed parallel to the axial planes of the folds and is often the most noticeable parting plane, especially in the fold hinges. The type of cleavage varies depending upon the lithology, with a slaty cleavage developing in the more pelitic beds and slip cleavage in the sandy units.

The main large-scale structure produced during this phase is the Isle of Man Syncline and this is the fundamental structure of the Island (Figure 3.4). The Isle of Man Syncline was refolded by the second deformation phase so that the eastern limb is now inverted and the axial plane dips to the south-east (Figure 3.3). Smaller fold-pairs of the same style exist to the east of the synclinal axis and the repetition of beds in this sequence of folds is responsible for the broad spread of the Lonan Flags in the east of the Island (Figure 3.3).

During the second phase of deformation the principal stress operated vertically, causing crumpling of the previous F_1 folds (Figure 3.4). The main structures formed were the Manx Synform and the complementary Peel-Ballaugh and Mull Hill antiforms. These folds, and their associated congruent minor structures, are more open and have a more rounded profile than the F_1 features. A further contrast is that the axial planes are inclined at low angles, generally in a north-westerly direction. Associated with the second phase of deformation was the development of a second axial planar cleavage. This later cleavage (S_2) is a strain-slip cleavage that overprints and often dominates the earlier S_1 cleavage.

During the third and final phase of deformation the principal stress returned to the horizontal but, in contrast to the first phase, was directed from the north-east to south-west (Figure 3.4) producing a set of folds or undulations with a south-east to north-west orientation. Again congruent minor structures demonstrate the open style and a third cleavage (S_3), dipping to the north-east, is developed. The main effect of this phase is to produce a series of undulations on the Manx Synform.

Following the final phase of folding two distinct sets of faults were developed (Figure 3.4). One series of transcurrent faults crosses the Island from south-east to north-west. Of these sinistral faults only the Snaefell Fault can be traced right across the Island, but many of the others are clearly displayed in the coastal cliffs as steeply inclined zones of brecciation. This first set of faults is probably due to a renewal of compression along a west north-west to east south-east line towards the end of the Caledonian orogeny.

A further set consists of steep normal faults of relatively small displacement trending between north-west to south-east and north-east to south-west. These traverse both the Devonian and Lower Carboniferous rocks of the Island and are probably of Armorican age. Faults of this set are responsible for the downthrown blocks which preserve these younger rocks in the Island.

Metamorphism of the Manx Group

It is possible to recognise three distinct episodes of metamorphism in the Manx Slates. The first is a syn-tectonic episode accompanying the formation of the F_1 structures and this

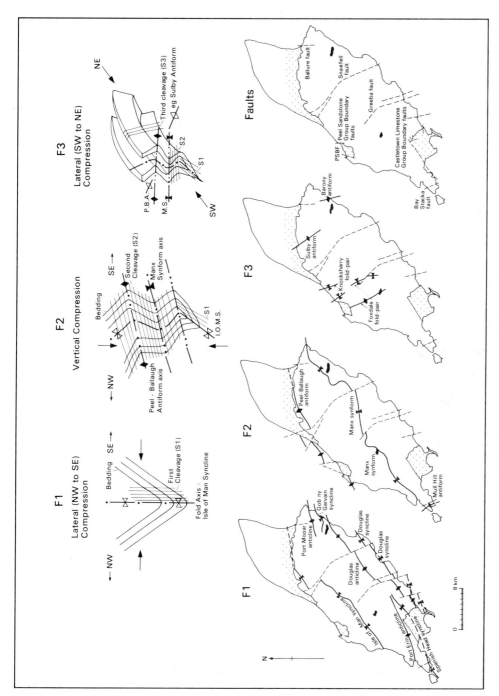

Fig. 3.4 The evolution of the structure of the Manx Group

produced a mild chlorite grade metamorphism throughout the Island (Simpson, 1964a). Its effects are greatest on the more pelitic units where the original shale has been converted into a dense felt of muscovite and chlorite. In the coarser, sandier units of greywacke, quartzite and flags the original quartz grains have become granulitised to form a matrix of smaller interlocking grains.

The second phase of metamorphism began during the F_2 refolding episode and extended into the static interval between the F_2 and F_3 fold episodes (Lamplugh, 1903; Gillot, 1955; Simpson, 1964a). More localised than the earlier chlorite grade regional metamorphism this phase is restricted to a belt, up to about six kilometres wide, extending from Maughold head in the north-east down the centre of the Island close to the hinge of the major F_2 Manx Synform to Bradda Head in the southwest. This metamorphism was responsible for the growth of small porphyroblasts of leocoxene, muscovite, biotite, cordierite, garnet, chloritoid and tourmaline. The rocks in this belt sparkle slightly as the light catches the new crystal faces.

The third phase of metamorphism is restricted to the thermal aureoles surrounding the intrusive igneous rocks and is considered with the intrusions.

The Caledonian igneous intrusions

The Manx Slates are host to a range of intrusive igneous rocks (Figure 3.1) including, as well as many dykes, two granites (the Foxdale and Dhoon), a compound acid/basic/ultrabasic intrusion (the Oatland Complex) and a diabase or gabbro (Poortown). Only the two granites have been investigated in detail and they are now known to be of Caledonian age. The Oatland intrusion is not considered here due to its poor exposure.

The Dhoon granite and associated granite elvans

The Dhoon granite is intruded into the Lonan Flags to the north of Laxey. The main intrusion, which is about two km by one km and elongated east-west, crops out in a number of quarries near the Laxey-Ramsey road and rises on to the high ground of Slieau Ruy to the west. Associated with it, and well exposed in the coastal sections are a number of 'acid sheets' (Simpson, 1964b). Lamplugh (1903) recognised these 'elvans' and commented on their distribution, restricted to the central part of the Island in a north-east to south-west elongated belt. The Dhoon granite is probably an elongated 'pod' which splits up at the extremities into the swarms of granitic 'elvans'. There is some support for this from the interpretation of the gravity pattern (Cornwell, 1972) which suggests a steeply inclined northern face to the intrusion and that the Dhoon granite is merely the surface tip of a much larger granite some six km by four km elongated east to west and extending to a depth of about five km.

Nockolds (1931) recognised two distinct types of rock in the intrusion. The main mass is a grey porphyritic microgranite consisting mainly of quartz, albite and biotite, with less common orthoclase, perthite and microcline. Muscovite is rare, a contrast with the Foxdale granite. Near to the margins the granite contains xenoliths of country rock, some partially assimilated, but most without any alteration.

After its emplacement the Dhoon granite was affected by the continuing deformation of the enclosing Manx Slates. The deformation is most severe in the western part of the outcrop but decreases in intensity to the east. The initial phase of deformation caused shearing of the granite along steep southward-dipping planes. This is well-displayed in the two main quarries where the shear jointing forms the steep inclined northern faces. The associated cleavage also penetrates the granite, giving gently inclined joints. Later folding developed in the western part of the granite and has a typical F_2 style which refolds the earlier shear planes. The third phase of deformation left only a mild imprint on the granite.

The 'granite' sheets or elvans associated with the main granite in the north are mainly porphyritic microgranites consisting of quartz, orthoclase, albite and occasional microcline. In the south the composition is similar, but the porphyritic character is less common. They vary in thickness between 30 cm and eight metres and although Lamplugh (1903) refers to them as dykes, Simpson (1964b) has shown that their emplacement was largely controlled by the pre-existing F_1 structures. Where cleavage associated with the first phase of deformation is not well developed the intrusive sheets tend to follow the original bedding, but wherever the cleavage is well developed they follow it and cut discordantly across the bedding. Almost all of these acid sheets are limited in lateral extent and can rarely be traced across country. Deformation after their intrusion has caused some of the acid sheets emplaced along the S_1 cleavage in the Barrule Slates to be flexed into typical open F_2 style folds. Throughout the Manx Slates the second phase of deformation was accompanied by the formation of quartz veins. The third phase left only a very mild imprint on the Dhoon granite and the acid sheets.

The Dhoon granite is surrounded by a contact thermal metamorphic aureole about half a kilometre wide in which the pelitic country rocks take on a silvery aspect due to the granulitisation of quartz grains and the growth of muscovite and chlorite flakes. These latter flakes are larger than those due to the earlier mild regional metamorphism. A small proportion of the thicker flags have recrystallised to a tough quartz-muscovite-chlorite hornfels. The main part of the aureole is impressed with a planar fabric emphasising the early F_1 fabric and turning the pelites and semi-pelites into contact phyllites.

The Foxdale granite

The Foxdale granite crops out as two exposures near the village of Foxdale. The larger of the two forms the low rounded hill of Stoney Mountain and to the east is a smaller outcrop separated from the main one by an area of low ground floored with Manx Slates. The broad extent of the metamorphic aureole (Lamplugh, 1903; Simpson, 1965), and the presence of the granite at shallow depths in many of the old mine shafts (Lamplugh, 1903), points to these two outcrops being continuous at depth and forming the apophyses of a much larger igneous mass underlying much of the south of the Island. Cornwell (1972) considers that the Foxdale granite is the tip of the South Barrule granite which is elongated north-east to south-west and suggests that this pluton is 10 km long by about four km wide and extends to a depth of about 10 km.

The granite is grey muscovite-granite consisting of quartz, orthoclase, microcline, and acid plagioclase and muscovite. It contrasts with the Dhoon granite in that the latter contains biotite which is uncommon in the Foxdale granite. Over most of its extent the intrusion has an even granitic texture with an average grain-size of 2–3 mm. Near to the margin the rock becomes a finer-grained microgranite which is sometimes porphyritic. Near the summit of Stoney Mountain massive quartz veins cut through quartz-feldspar and quartz-muscovite pegmatites which represent the apex of this cupola.

Simpson (1965b) investigated the Foxdale aureole and has shown that deformation of the envelope took place while thermal metamorphism continued during the consolidation of the granite. In contrast to the areas outside, the metamorphosed country rocks within the aureole show evidence of recrystallisation accompanying the development of the S_2 cleavage. Following the F_2 movements there was a static interval during which recrystallisation and the growth of new crystals continued under the influence of the enhanced temperatures around the granite. This phase is marked by the growth of porphyroblasts of almandine garnet, staurolite, cordierite, biotite, chlorite and muscovite. In the north-east further deformation follows the F_2–F_3 static interval due to renewed compression in the hinge of the Manx Synform producing a new flat-lying S_2 schistosity.

The northern fringe of the aureole is affected by the third phase of deformation with the Manx Slate being flexed into the open Foxdale fold-pair and a third axial planar cleavage (S_3) being sporadically developed. The heat from the granite had now largely dissipated and no further recrystallisation accompanied this phase of deformation.

The acid sheets associated with the Dhoon granite are metamorphised in the aureole of the Foxdale granite in exactly the same manner as the rest of the country rocks. Simpson (1965) has suggested, therefore, that the Dhoon granite and its 'elvans' were emplaced slightly earlier in the second phase of deformation than was the Foxdale granite.

The Poortown Augite 'Gabbro'

Despite being extensively exposed in the main roadstone quarry for the Island a little east of Peel (Figure 3.1), this intrusion seems to have escaped any detailed examinations. Hobson (1891) described it as part of a general survey of the igneous rocks of the Island. Lamplugh (1903) provides a description of a single thin section in which the rock is described as being made up 'largely of phenocrysts of augite' altered to chlorite around the edges, set into a groundmass of plagioclase, chlorite, ilmenite, sphene and a little actinolite. Watts (in Lamplugh, 1903, p. 304) suggests that the rock be termed an Augitic Lamprophyre.

The Peel Sandstone Group

The Peel Sandstone, consisting of red mudstones, sandstones and conglomerates arranged in a variety of associations, crops out in fine cliff sections to the north of Peel (Figure 3.1). Although it extends inland, exposures are rare and the contact with the Manx Slate Group is obscured by drift. The northern boundary is exposed in the cliff sections and is a fault of unknown throw. The stratigraphy, age, structure and origin of the Peel Sandstone have all been the subject of debate (Boyd-Dawkins, 1902; Lamplugh, 1903; Ford, 1971; Crowley, 1985), partly due to the poor inland exposure, the repetition of depositional faces, an abundance of minor faults and the lack of in-situ fossils.

Sedimentology

Although dominated by sandstones, mudstones and conglomerates, a range of sedimentary structures gives considerable variety to the sections. Within the sandtones planar and trough cross stratification, horizontal lamination and internally massive beds are common. More locally, especially near the northern boundary fault, disrupted horizons and slump over-folds occur (Boyd-Dawkins, 1902; Lamplugh, 1903; Ford, 1971). In the finer sandstones and siltstones water escape structures are common. Channel-fill structures can also be seen occasionally. Many of the sandstone beds contain occasional mudflakes and small pebbles.

One type of conglomerate consists of large, poorly sorted, angular or edge-rounded clasts of locally derived sandstone set in a sandy or muddy matrix. These units are of laterally restricted extent and often have erosional contacts with the adjacent beds. Fine examples of this type are seen in the headland to the north of Peel Bay. The second type of conglomerate is better sorted and consists mainly of rounded pebbles of exotic origin. Both types contain small mud-flakes and locally derived sandstone clasts.

Mudstones are generally subordinate to

sandstones and conglomerates and are usually massive, although laminated units do occur. Small well-formed dish and pillar dewatering structures are well displayed in this latter type. Polygonal nets of dessication cracks are seen where alterations of mudstones and sandstones are present.

Throughout the finer sandstones and mudstones there is sporadic development of calcareous concretions, usually of a nodular form but occasionally as vertically elongated masses. In sections dominated by thick mudstones, as for example at the northern end of Traie Fogog bay, the calcretion is sufficiently well developed to dominate the lithology.

Structure of the Peel Group

The Peel Group was unaffected by the intense deformation caused by the Caledonian orogeny and has a relatively simple structure. Throughout the sections the rocks dip towards the northwest at between 30 degrees and 60 degrees and are traversed by numerous faults. The absence of distinctive marker horizons and the frequent repetition of rock types makes detailed reconstruction of the structure difficult. It is likely that the majority of the faults have a limited throw with major movements being confined to the boundary faults.

Age of the Peel Group

The age of the Peel Sandstone has been a topic of much debate. Boyd-Dawkins (1902) considered it to be Permian, due to its general similarity to the Brockrams of the Cumbrian coast. Lamplugh (1903) equated the Peel Group with the Lower Carboniferous basal conglomerate exposed at Langness. This is equally unsatisfactory as the heavy mineral suites are different and the Peel Sandstone lacks the pebbles of Manx Slate that are so plentiful in the basal conglomerate. Gill (1903) and Lewis (1934) both noted fossiliferous pebbles derived from Ordovician and Silurian limestones in the Peel conglomerates while Ford (1971) considered that the general similarity of the Peel Group to the Old Red Sandstones of Ayrshire, Arran and Anglesey, taken together with the limited palaeontological evidence, favoured a Devonian age. More recently Crowley (1985) has drawn attention to the greater degree of deformation suffered by the Peel Group (dips of 40–50 degrees) compared with the Carboniferous Limestones of the south of the Island (10–20 degrees). The lack of any Caledonian cleavage in the Peel Group suggests that deposition must have occurred after the deformation of the Manx Group and the associated granites, but before deposition of the Carboniferous rocks. By comparison with the Lower Devonian rocks of the Lake District, Crowley has therefore suggested a Lower Devonian (Seigenian-Emsian) age for the Peel Sandstone.

Environment of deposition and palaeogeography

The range of lithologies and sedimentary structures, the red colouration and the frequent occurrence of dessication features are typical of alluvial sequences and an alluvial origin, probably by deposition in low sinuosity streams as part of a series of coalescing alluvial fans (Crowley, 1985), is supported by the limited cyclicity and the presence of channel-fill sandstones and conglomerates. The sandstone- and mudstone-clast conglomerates could be interpreted as the result of reworking of previously deposited and partially indurated sediment, the result of bank collapse or as locally developed mudflows. The calcreted horizons are due to pedogenesis in the over-bank environments under a seasonally arid climate with carbonate being deposited as moisture is evaporated at the ground surface.

The absence of Manx Slate debris and the west to east directed cross-bedding suggests that the Peel Group is composed of debris derived from a generally west or north-west direction. This is confirmed by Ford's (1971) analysis of palaeoslope direction based on the slump structure near the northern boundary fault. The angularity of much of the debris suggests that the distance to the source was small (Crowley, 1985) and together with the petrology and the interpretation of the sequence as representing a mixed alluvial plain-alluvial fan environment suggests that the Peel Sandstone is a small remnant of substantial package of sediments deposited in Lower Devonian times. Crowley (1985) suggests that this was derived from a sequence of Lower Palaeozoic shallow marine carbonates, volcanics and coarse clastics lying to the north-west of the Island, although Wright et al. (1971) have shown a deep Carboniferous basin in this area. An alternative view (Allen, 1974; Allen and Crowley, 1983) is that the Peel Sandstone is a locally uplifted segment of a major fluvial system which extended from the present day north of Ireland though to South Wales.

The Carboniferous Limestone and volcanic rocks

The Carboniferous outcrops of the south-east of the Island with their easily accessible coastal exposures and varied fossil content have attracted considerable attention. The Reverend Dr J. G. Cumming's description of the area (Cumming, 1846a) has been supplemented but largely confirmed by later workers. Lamplugh (1903) drew heavily on Cumming's work and Lewis (1930) described the palaeontology and correlated it with the mainland. Dickson (1967) was the first to attempt a detailed definition of depositional environents of the limestones and the volcanic sequence. A synopsis of Dickson's work has recently become more widely available (Dickson et al., 1987) and their terminology is used here (Figure 3.5).

Stratigraphy and sedimentology

The Langness Conglomerate Formation, the oldest unit, consists of 30m of coarse, red bouldery gravels with sandstone lenses. It is best displayed in Dreswick Harbour on the Langness Peninsula where it rests with marked angular unconformity on an eroded surface of red-purple stained Manx Slates. The conglomerate immediately above the slates contains large boulders and grain size decreases upwards. The angular pebbles and boulders of slate or quartz suggest a derivation locally from the Manx Slates. Both Lamplugh (1903) and Lewis (1930) suggested that the conglomerate is a beach deposit developed as the Carboniferous seas transgressed over the eroded Manx Slate landscape. However, the presence of scoured channels, plant bearing shales at the top of the conglomerate and the angular nature of the pebbles together with the red colouration suggest a terrestrial origin. Dickson et al. (1987) suggest that the unit was deposited as an alluvial fan issuing from the Manx Slate terrain to the north.

The Derbyhaven Formation consists of 85m of bedded limestones and shales exposed on the shores of Sandwick Bay and Derbyhaven north towards Cass-ny-Hawin and in the quarries around Turkeyland. The generally westward dipping sequence is extensively faulted giving a complicated outcrop pattern. Pale grey oolitic limestones with large wave-generated ripples and lacking shale partings or muddy matrix suggest deposition in shallow water with high wave-energy conditions. Away from the formation base thin crinoidal limestones and shales with corals and productid and chonetid molluscs suggest quieter, possibly deeper water. Shallow water features are present throughout and, although the varying quantity of land-derived detritus, reaching a maximum near the top of the formation, may be due to local deepening and shallowing, it is more likely to be due to the varying position of the river mouths and the varying conditions on land.

SOLID GEOLOGY 33

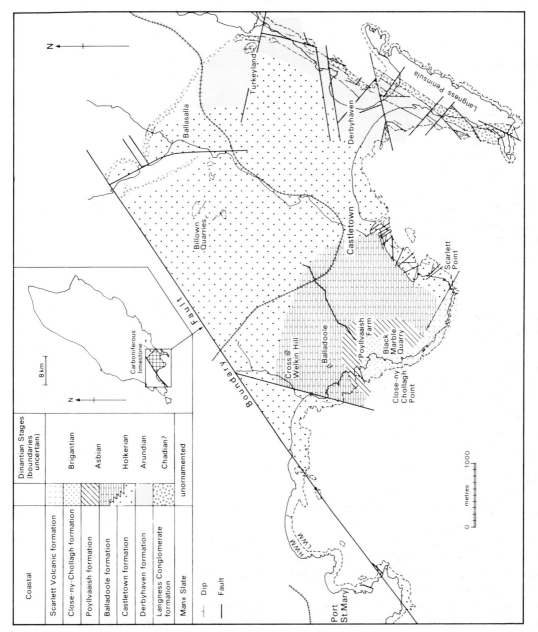

Fig. 3.5 The stratigraphy of the carboniferous limestones and volcanic rocks of the Isle of Man (from Dickson et al., 1987)

The Ballasalla Formation is only exposed in the disused quarries around Ballasalla and at Kallow Point, Port St Mary. While it is generally similar to the rocks of the Derbyhaven Formation, it is considerably thinner, of the order of 10m compared with the 85m of the coastal exposures.

The Castletown Formation is exposed on the shores between Castletown and Scarlett Point and Dickson (1967) recognised three divisions, the oldest of which consists of about 20m of nodular and irregularly bedded dark-grey, muddy limestones interbedded with black shales. The middle part of the unit is well displayed on the foreshore below the Scarlett Point visitor centre and consists of six metres of massive, mottled lime mudstones with little shale. The upper part of the formation is seen below the lookout tower at Scarlett Point and consists of 14m of evenly bedded, pale-grey lime mudstones with intervening thin black shales. Complete fossils are rare but crinoid debris is common. Poor sorting, a lack of detrital sediment, the good preservation of delicate corals and an assemblage of trace fossils characteristic of environments below the level of wave activity all led Dickson (1967) to suggest that the Castletown Formation was deposited during a period of progressive deepening of the water.

The Balladoole Formation is exposed on the foreshore immediately south of Chapel Hill, in the quarry at Chapel Hill and in the disused quarry at Cross Welkin Hill. The formation consists mainly of lenticularly bedded calcite mudstones but at Cross Welkin Hill quarry large biohermal mounds may be seen overlapped by thinner bedded limestones. These 'reef' limestones show no preserved coral framework but the diverse fossil assemblage and local eroded surfaces all suggest a shallow water environment.

The massive, parallel bedded pale-coloured limestones of the Poyllvaaish Formation are exposed between Poyllvaaish Farm and Salt Spring. They contrast strongly with the upper parts of the Balladoole Formation in that mound structures are poorly developed. The formation is famous for its rich, well preserved fossil fauna which includes brachiopods, bivalves, goniatites and gastropods. Slump structures suggest deposition on a significant submarine slope but still in a shallow water environment.

The Close-ny-Chollagh Formation is only seen in small outcrops around the black marble quarry at Close-ny-Chollagh Point and Poyllvaaish Farm where it consists of three main types of rock. In contrast to the other Carboniferous formations the dominant rocks are black shales and shaley limestones. These are finely laminated and contain an unusual mixed biota in which bivalves and goniatites represent marine conditions but terrestrial plant remains suggest proximity to land. The second type of rock consists of coarse grained, muddy limestones which contain bored and abraded grains and algal encrustations. These are features which are characteristic of shallow, turbulent water deposits, but are found here in a deep water setting. Dickson et al. (1987) suggest that this sediment was deposited in shallow water but subsequently redeposited into deeper water by turbidity currents.

The third rock type consists of pale limestone similar to the Poyllvaaish Formation. These isolated masses vary from small boulders to masses more than ten metres across and they were interpreted as 'patch-reefs' by Lamplugh (1903) and Lewis (1930). As they are older than the enclosing black shales Dickson (1967) suggests that they are blocks of Poyllvaaish material that have fallen from a steep, submarine escarpment into the younger, actively accumulating muddy sediments of Close-ny-Chollagh times.

The volcanic rocks

The Scarlett Volcanic Formation crops out in a narrow strip from Scarlett Point north-west towards Close-ny-Chollagh Point and out into Poyllvaaish Bay (Figure 3.5) and consists of a variety of basaltic lavas, agglomerates and ashes. Dickson et al. (1987) demonstrate that the whole sequence, while containing slumped horizons, is *in situ* and represents a phase of volcanic activity that took place partly concurrently with limestone deposition in a shallow marine environment.

At Close-ny-Chollagh Point the earliest volcanic materials slumped across poorly compacted black muds of the Close-ny-Chollagh volcanic breccias and ashes. The outcrop between Close-ny-Chollagh Point and Sclarlett Point is dominated by fine, green ashes, breccias with included limestone blocks and patches of basaltic pillow lavas. The pillow lavas, current bedding and thin, undisturbed lenses of limestone suggest that much of this material was deposited under water.

At Scarlett Point, columnar jointed amygdaloidal basalt forms The Stack and a rib of rock to the north-west around which are coarse volcanic breccias containing large boulders of baked limestone. The volcanics at Scarlett Point are separated from the Castletown Formation by a fault complex which brings down a small wedge of craggy dolomitised limestone from the Poyllvaaish Formation.

The structure of carboniferous rocks

The Carboniferous outcrop is structurally simple in comparison with the Manx Slates. The north-east boundary is complicated by faulting and the basal units are repeated north of Ballasalla (Figure 3.5). In the north-west a boundary fault runs from Kallow Point, Port St Mary, inland in a north-east direction towards Billown. Around the north-east and south-east edges of the outcrop the beds dip

3.2 Pillow lavas in the Scarlett Volcanic Formation. These features are formed by the rapid cooling of lavas erupted into water or wet sediment

3.3 Columnar jointing on the Stack

gently in towards Close-ny-Chollagh Point usually at angles of less than 15 degrees. The structure is a simple synclinal basin, plunging towards the south-west and faulted out on the north-west margin. Minor folding is rare except around Scarlett Point where good examples of monoclinal folds together with gentle domes and basins are seen plunging towards the south-west. Faulting is extensive with two major sets being identifiable, one trending north-west to south-east and another south-west to north-east. These complicate the outcrop pattern and are often the focus of dolomitisation.

Environment of deposition and palaeogeography

The limestone and volcanic sequence in the south of the Island is of Lower Carboniferous, Dinantian age (360–325m years, Figure 3.5) and represents the deposits of a marine transgression in a basin stretching through from Lancashire and Furness and into eastern Ireland. The conglomerates represent alluvial conditions and the limestones are deposited in shallow water with land never very far away. The 'reef' mounds of the Poyllvaaish and Balladoole Formations contain faunas indicative of shallow water, but the relative lack of terrigenous sediment suggests a decreasing input of clastic sediment to the basin or progressive isolation from the shoreline. The black sediments of the Close-ny-Chollagh Formation indicate some deepening in late Dinantian times and, although the shoreline cannot be placed with certainty, it probably lay to the north of the present outcrop.

The buried rocks of the Northern Plain

The northern upland terminates abruptly at a steep fault scarp which downthrows the Manx Group to the north where it is buried beneath thick glacial deposits (Figure 3.1). The bedrock surface beneath the drift lies at 60m below sea level and is remarkably even, which has led to the suggestion that it represents a marine platform of preglacial age.

Between 1891 and 1903 boreholes were sunk through the northern plain in a search for a continuation of the Whitehaven coalfield. They penetrated Carboniferous limestones, sandstones and shales and Permo-Triassic red sandstones and saliferous marls. The saliferous marls at the Point of Ayre were worked for brine but drilling did not penetrate to the Carboniferous strata below.

Starting in 1920, 20 more boreholes were drilled by the Coltness Iron Company around Andreas in search of haematite iron ore. From these Smith (1927) described a sequence of 600m of Lower Carboniferous shales, limestones, sandstones and conglomerates with occasional thin coal seams. Basement Beds of conglomerate and shale rest upon and fill hollows in the eroded surface of the Manx Group above which are thick limestones of the Lower and Middle Limestone Groups. The upper part of the sequence is dominated by shales with thin limestones, sandstones and coals of uncertain age but typical of Lower Carboniferous rocks.

The lack of success in finding commercially viable coal or iron deposits did little to encourage further exploration, but between 1985 and 1987 Riofinex North Limited (part of the RTZ group) drilled three boreholes in a new search for coal. The first two borings on the east coast penetrated sequences of shale and thin limestones which contained microfossils of early Namurian (Lower Carboniferous) age. A third borehole, at Phurt on the east coast, passed through Permo-Triassic purple shales and red sandstones to 430m below sea level, before entering dolomitic limestones and calcareous shales again of Lower Carboniferous Namurian age.

The failure of any of the borings to penetrate Upper Carboniferous beds suggests that beneath the northern plain of the Island, Lower Carboniferous limestones and shales rest uncomfortably on the Manx Group and are succeeded unconformably by Permo-Triassic sandstones and shales. The coal-bearing Upper Carboniferous Westphalian strata appear to be missing, either due to non-deposition or erosion.

Several of the borings passed through fault breccias or highly inclined beds which together with the juxtaposition of thick Permo-Triassic and Lower Carboniferous sequences suggests that the pattern of solid strata is complicated by faulting. Although the recent cores are still available, the absence of any detailed micropalaeontological data from the earlier borings makes it unlikely that any advance will be made on the 'conjectural' maps of Lamplugh (1903) and Smith (1927).

The Carboniferous sequence buried beneath the northern plain also includes Dinantian rocks, although they are very different from those of the southern outcrop. It seems likely that uplands composed of the Manx Group (the Manx-Cumbrian block) separated the early Dinantian northern and southern basins. The buried Carboniferous rocks under the northern plain are part of the Northumbrian-Solway gulf in which fluvio-marine conditions were succeeded by typical Yoredale-style deltaic-marine deposits. Although the borehole at the Point of Ayre penetrated over 800m of Permo-Triassic sediments little is known of their relationship to the mainland sequences. Gregory (1920) suggests that the saliferous beds are equivalent to those of Carrickfergus rather than the gypsiferous beds of west Cumbria and are therefore Lower Mercia Mudstones. He correlates the sandstones with the Triassic Kirklinton Sandstone and the Permo-Triassic St Bees Sandstone. The whole sequence is the product of a hot and mainly arid climate in which the sandstones represent the deposits of flash floods, the mudstones are the flood plain deposits and the salt beds the result of evaporation of ephemeral playa lakes set in a moderately rugged terrain.

Economic geology

Despite its small area the Isle of Man was well endowed with mineral wealth and in the nineteenth century two of its mines, Laxey and Foxdale, stood in the first rank in the British Isles for the production of lead and zinc as well as producing significant amounts of silver. Although these three metals were the most important, significant amounts of copper and iron ores and small amounts of nickel and antimony were mined (Figure 3.6).

Metalliferous ore deposits in the Island are in the form of discontinuous, but high grade, tabular fissure fills. The most productive veins are oriented north to south with a dip to the east, although the Foxdale lode strikes east to west and dips south. The host rocks are the Manx Group and the granites, with the latter being particularly favoured. There is a distinct clustering of productive sites along the axis of the Manx synform, particularly where it coincides with the Maughold Banded Group (Figure 3.6).

Lamplugh's (1903) proposal that all the metalliferous mineral deposits in the Island are younger than the pre-Carboniferous earth movements has been largely confirmed by Ineson and Mitchell (1979) who suggest that the Foxdale lode was emplaced at a time close to the Namurian-Westphalian boundary. The hydrothermal circulations involved also affected the northern half of the Island and were responsible for the mineralisation in the Laxey area. Ineson and Mitchell (1979) also confirm Lamplugh's (1903) view that there was no direct genetic link between the emplacement of the granites and the ores. Instead it seems that the Caledonian granites are the top of a larger deeply buried mass (Cornwell, 1972) which acted, possibly as a heat source, to focus hydrothermal activity over a considerable period of time.

Although most of the Manx lodes were emplaced during this Hercynian episode (290–279m years), in common with other primarily lead-zinc mineralisation in Derbyshire, the Lake District, the Northern Pennines and Southern Scotland, there is also evidence of Permian (270–240 Ma), Triassic (220–200 Ma) and Tertiary (40 Ma) metasomatic activity. These latter phases were restricted to the south of the Island.

The Manx Group, although often termed slates, are of little use for roofing slate due to the complex cleavages. It has been used extensively as a rough, and rather indifferent building stone, mainly derived from local quarries opened near the site.

The Peel Sandstone Group provides the only freestone found in the Island and was used extensively for detail work in the eighteenth and nineteenth century. Its poor durability means that it is now little used (Bawden et al., 1972) and much of the red sandstone seen in buildings today has been imported from the English mainland.

The Carboniferous limestones around Castletown have also been used extensively in buildings and in harbour works, mainly in the south of the Island. They have also been the main source of agricultural lime with extensive quarries at Turkeyland, Billown, Scarlett and Balladoole. Recently the limestones have seen increased use as an aggregate for foundations and road bases.

The main granites have all been used as sources of road aggregate and for kerbs and setts, although none are now so used. The extension of the outer breakwater of Douglas harbour used large blocks of granite from Foxdale as rip-rap and as infill in addition to the specially cast concrete blocks. The Poortown 'gabbro' is quarried as the Island's main source of aggregate for road wearing courses.

The saliferous beds encountered in the deep borings at the Point of Ayre were

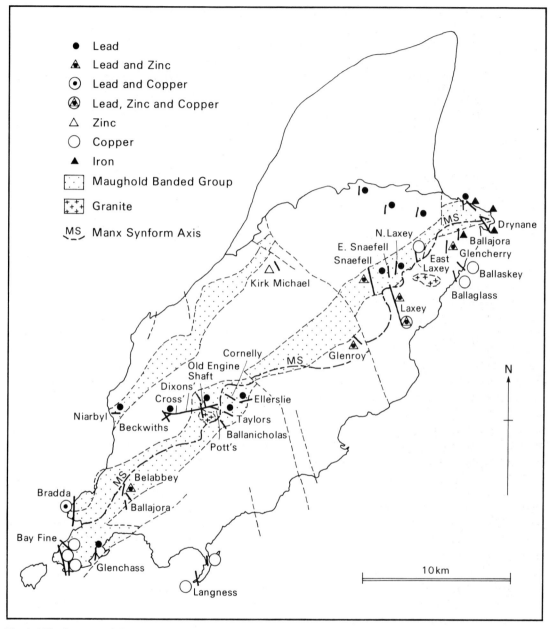

Fig. 3.6 The location of the Manx mines and their relationship to the structure and lithology of the Manx Group

exploited for brine which was pumped through the pipeline to the Ramsey works of the Manx Salt and Alkali Company where it was evaporated to give salt of exceptional purity. This undertaking ceased in the 1950s.

The only other geological products to be worked extensively are the superficial deposits of the north of the Island which are quarried mainly for sand and gravel for the building trade. Clay has been dug for brickmaking but yielded poor quality bricks due to the high calcium carbonate content.

CHAPTER 4

The Quaternary Ice Age in the Isle of Man: an historical perspective

Danny McCarroll

Introduction

The position of the Isle of Man, near the centre of the northern Irish Sea, places it directly within the path of the great ice sheets which have, periodically over the last 2.4 million years, advanced down that relatively shallow channel (Figure 4.1). The legacies of these ice sheets, both erosional and depositional, are manifest upon the Manx landscape and the Quaternary deposits, of the northern plain in particular, rank amongst the most complex and interesting in Europe. What is presented here is not, however, a comprehensive account of the current consensus of opinion upon the Quaternary history of the Island. Indeed such a consensus does not exist. Rather it is an attempt to emphasise the surprising importance of the Manx landscape and deposits in the continuing evolution of ideas concerning the relatively recent geological history of the British Isles.

The Quaternary Period

The Quaternary is the most recent period of earth history, beginning at the Pliocene/Pleistocene boundary, about 2.2 million years ago and extending to the present day. Although world climate is thought to have cooled progressively throughout the Tertiary, the Quaternary period is marked by pronounced climatic oscillations of high magnitude and frequency (Figure 4.2). Deep-sea cores, which provide a continuous record, suggest that these oscillations are cyclical and relate to changes in the earth's orbit around the sun. It has been shown that present climatic conditions represent the latest in a series of relatively warm 'interglacials', preceded and to be followed by cold 'glacials'. Although deep-sea cores retain evidence of at least twenty glacial/interglacial cycles, terrestrial deposits in Britain and Ireland provide clear evidence for only two glacial advances of greater extent than the last, Devensian glaciation (Figure 4.1). The latter is thought to have reached its peak about 18,000 BP, at which time an ice sheet extended down the Irish Sea basin as far as the coasts of Pembrokeshire and County Wexford.

By 14,000 BP the Devensian ice sheet had melted away in most of the Irish Sea Basin and peats were developing on the Lleyn Peninsula of North Wales (Coope and Brophy, 1972). Glacier ice may have disappeared completely from the British Isles before a short relapse to cold conditions about 11,000 BP (the Loch Lomond Stadial) when small mountain glaciers reformed. The time between the melting of the Devensian ice sheet and the return to warmer conditions at the end of the Loch Lomond readvance is known as the Lateglacial, and the last 10,000 years, through to the present day as the Flandrian or Holocene.

Glacial legacy

The thickest Quaternary deposits in the Isle of Man (Figure 4.3) lie beneath Point of Ayre, where boreholes reveal bedrock at −126m OD (Lamplugh, 1903; Smith, 1930). Between 65m and 35m below present sea level a series of fossiliferous silts and sands have been interpreted as marine in origin. This marine horizon is underlain and overlain by glacigenic deposits. To the south of Point of Ayre, drift thickness declines and glacigenic deposits rest directly upon a rock platform at between 41m and 53m below sea level. Further south,

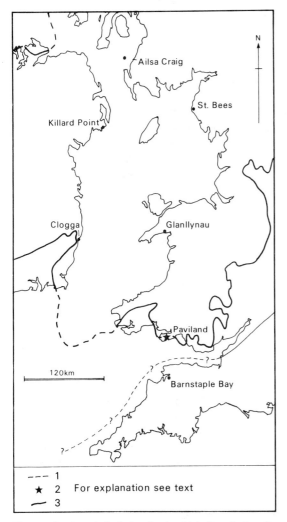

Fig. 4.1 Limits of glaciation in the Irish Sea during the Devensian (3) and a Post-Anglian (1) advance. A moraine at Paviland, on Gower, is thought to mark an intermediate advance

where the drift of the northern plain meets the slate massif at Ballure, south of Ramsey, a higher rock platform rises gently above the present beach. The deposits which stratigraphically overlie this platform, and which are exposed above sea level have traditionally been divided into two quite distinct groups; a high level suite comprising entirely local material and a very diverse 'foreign' suite comprising material of dominantly northern derivation (Figure 4.4).

The high level suite, or 'insular drift', is largely restricted to the uplands, though it does occasionally descend to the coast. On summits and interfluves it consists of a thin, angular rubble whereas on plateaux and in valleys thicker sequences often display a crude stratigraphy marked by variations in the proportion of silty matrix and by occasional beds of silt and of gravel. In many sections the junction between bedrock and drift is gradational and is often marked by a distinct layer of rock rubble (Dackombe and Thomas, 1985).

Although no erratics have ever been found within the high level suite, there is abundant evidence of the transport of local igneous lithologies from north to south. The clearest example is the Foxdale granite, which crops out at 203m on the northern side of South Barrule. Boulders of this distinctive lithology occur up to 280m above its present outcrop and are even found on the opposite side of the South Barrule ridge. Slate clasts in the insular drift are often clearly striated, and many bedrock surfaces display striations trending generally north-north-west to south-south-east (Figure 4.4).

The 'foreign' drift is restricted to the northern plain and lowland margins. On the east coast it rarely extends above 20m OD, though occasional erratics are found up to 75m. On the west coast the foreign drift is much more extensive and reaches a maximum altitude of 180m (Dackombe and Thomas, 1985). These deposits are almost entirely of northern origin and incorporate several distinct and traceable lithologies, including the riebeckite-eurite

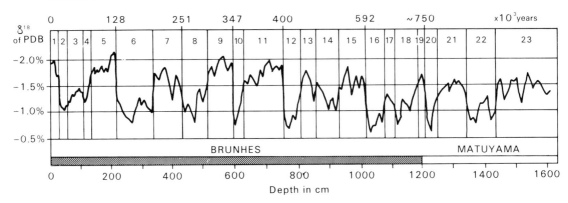

Fig. 4.2 Oxygen isotope and palaeomagnetic record of the last 1–6 million years in core V28–238 from the equatorial pacific (−1 degree north, 160 degrees east). Isotope stages are shown in the upper part of the diagram. Even stages are cold, odd stages are warm (after Shackleton and Opdyke, 1976)

which is unique to Ailsa Craig, off the southwest coast of Scotland (Figure 4.1).

The stratigraphy of the foreign drift (Figure 4.3) is complex. The lowest lithofacies exposed above sea level is often a dense silt with few stones (the 'Shellag and Wyllin tills' of Thomas, 1976, 1977). This is overlain by associated sands and gravels and in the north

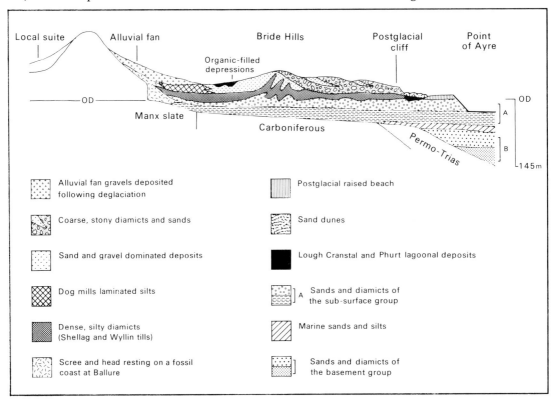

Fig. 4.3 Generalised Quaternary stratigraphy of the Isle of Man

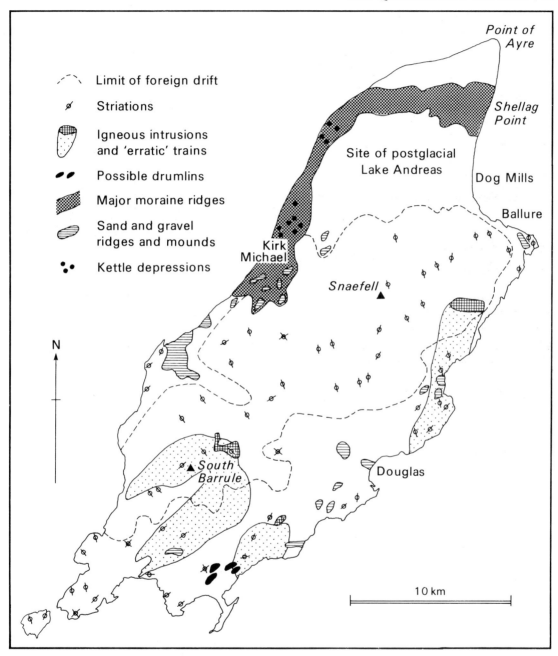

Fig. 4.4 Glacial landforms and deposits of the Isle of Man

of the Island the whole sequence has been deformed to produce the Bride Hills. This deformation is very clearly displayed in the coastal section of Shellag Point. To the north of the Bride Hills the Shellag deposits are overlain by more tills whilst cliff sections to the south are dominated by thick sequences of sand and gravel. On the east coast, around Dog Mills a sequence of laminated silts, clays and sands rests conformably upon the Shellag

4.1 Of the valleys draining the Manx uplands only the upper part of Laxey Glen displays some semblance of the U-shape typically produced by glacial erosion. The incised terrace in the valley bottom was produced by solifluction under periglacial conditions

till. Further south at Ballure, and on the west coast at Glen Mooar the foreign deposits interdigitate with beds of local material at the boundary with the slate massif. Inland, this junction is obscured by gravel fans issuing from the mouths of upland valleys.

The surface of the foreign deposits is often marked by ridges and small enclosed depressions. The latter are particularly common on the west coast around Kirk Michael, and where they appear in coastal sections they display a fill of calcareous mud and peat. Enclosed depressions also occur on the gravel fan issuing from the mouth of Glen Dhoo, north of Ballaugh.

Biblical floods and floating ice

The first detailed work on the superficial deposits of the Isle of Man was conducted in the mid-nineteenth century by the then Vice-Principal of King William's College in Castletown, the Reverend Dr J. G. Cumming. Although in the historic 'discourse of Neuchatel' of 1837 Louis Aggasiz had already presented the glacial theory to the scientific world, the concept of great ice sheets had not yet been widely accepted. Cumming, therefore, followed other British workers in interpreting the extensive superficial deposits of Man in terms of submergence, drifting icebergs and catastrophic floods.

In his earliest work, Cumming (1846) divided the deposits into four categories: boulder clay and erratic blocks, diluvium, drift gravel and alluvium. In the category of 'boulder clay' he originally included only the unstratified or poorly stratified deposits at the base of many of the coastal sections (the

4.2 The Bride Hills of the northern plain comprise mounds and ridges of till, sand and gravel incised by numerous dry valleys. They have been interpreted as a moraine produced by a minor re-advance of the last ice sheet and as a submarine ice-contact ridge. Much of the flat ground between the slate massif (in the distance) and the Bride Hills may have been occupied by pro-glacial Lake Andreas

Shellag and Wyllin tills), but in later papers (1847, 1853) he extended the category to include most of the lowland deposits. He recognised that some of the rocks within these deposits probably originated in Scotland and that many were scratched and striated. He also noted that the deposits often contain shell fragments and that some are 'so rich in lime to render them for the most part unsuitable for brick making' (p. 337). Similar shelly, calcareous deposits occur on both sides of the Irish Sea Basin and their origin had long been (and still is) the subject of debate.

Originally interpreted as clear evidence of the Biblical Flood, it was the English geologist Charles Lyell who first proposed a non-catastrophic mechanism which would account for the shelly fauna and northern, striated erratics of the superficial deposits of the Irish Sea coasts. He proposed that the material was deposited subaqueously by icebergs issuing from landborne glaciers and forming as 'ice foot' on arctic shores. Contemporary polar explorers had sighted huge icebergs as far as 1,400 miles from the nearest land and reported that they contained boulders and frozen mud. The 'drift' hypothesis conformed, therefore, to the principles of uniformitarianism, requiring only the long continued operation of a process known to be active today, rather than the catastrophic flood of the diluvialists.

Cumming (1846, 1847), therefore, considered most of the lowland suite to have been deposited subaqueously by the action of drifting ice. This would require a sea level about 120m higher than present, sufficient to divide Man into four islands surrounded by channels through which would pass the iceberg-laden waters. He envisaged a general drifting current from north or north-east and the thrusting of frozen gravel along and upon the coasts to explain the striation of bedrock surfaces.

The lowland or 'foreign' deposits of the Island could, therefore, be readily explained by the drift hypothesis. One feature of the high level suite, however, defied explanation: the transport of boulders of the Foxdale granite to altitudes considerably greater than the highest outcrop. Cumming could conceive of no extant process capable of such a feat and, therefore, invoked the action of enormous waves from the north and north-east. To these 'waves of translation' he also assigned the superficial deposits cloaking the uplands and partly filling the valleys. The gravel fans emanating from many of the glens he attributed to the subsequent redeposition, on the sea bed, of the diluvial deposits from the uplands.

It should not be assumed that the Reverend Cumming's acceptance of the diluvial theory to account for some of the Manx superficial deposits was in any way religiously inspired. Page (1969) points out that the influence of the Biblical Flood on geological thinking in the nineteenth century is often overstressed and Charlesworth (1957, p. 616) notes that the concept of the deluge was postulated 'not as part of a cosmogony but as a legitimate attempt to solve a specific problem'. Cumming, in fact, was clearly unhappy with this break from uniformitarianism, and in his search for a solution contacted Charles Darwin.

Darwin (1842) had already accepted the glacial theory as explaining many of the landforms and deposits of the British uplands, but advocated the drift-ice hypothesis to explain many of the superficial deposits found elsewhere. Whilst rejecting completely the catastrophic concept of the flood, he recognised that the most valid objection to the drift-ice hypothesis was the abundant evidence of the transport of erratics from a lower to a higher level. Citing examples of this phenomenon from Scotland, North Wales, the English Lake District and from North America, as well as the evidence supplied by the Reverend Cumming from the Isle of Man, he proposed an elegant mechanism which accorded both with the field evidence and the principle of uniformitarianism. Darwin (1848) suggested that such erratics were transported not by icebergs, but by the 'ice foot' which he had reported forming on polar coasts. He suggested that when shore waters freeze each winter, incorporated boulders may be transported onshore during storms and that the presence of erratics at elevations far above their parent outcrop suggests gradual submergence, with the boulders being pushed a little higher each year. Darwin (1848, p. 322) concluded:

> It is marvellous that Nature should have thus marked by buoys made of stone, the former sinking of the earths crust, and likewise, I may add, its subsequent elevation; and that on these blocks of stone the temperature, during the long period of their transportal, may be said to be plainly engraved.

This explanation, although quickly accepted, was to be succeeded by the concept of the great ice sheet. It remains as a sentinel that theories, even those produced by the greatest of minds, and apparently in accord with the available evidence, may nevertheless be quite wrong.

A great Ice Age, 'one and indivisible'

Whilst preparing the Geological Survey map of north Lancashire, R. H. Tiddeman (1872) was struck by the very strong evidence of glaciation. Striations and the distribution of boulders in the till suggested, however, that the ice did not flow from the watershed to the sea, as expected, but disregarding the local topography moved consistently towards the south-east. Tiddeman (1872) insisted that such a constant pattern of erosion and deposition could not possibly be assigned to the action of drifting ice and that it must be attributed to a landborne ice sheet. The direction of movement, however, required that there must have been some great barrier to the west which diverted the ice drainage of the Lune and Ribble catchments.

Tiddeman (1872) demonstrated that the evidence collected by the Reverend Cumming from the Isle of Man accorded very well with that from north Lancashire. The constant direction of striations and the transport of local and erratic lithologies all pointed towards a land-based ice sheet rather than the drift-ice or diluvial hypotheses proposed previously and, together with similar evidence from south Lancashire, Cheshire and Anglesey combined to suggest a great ice sheet flowing down the Irish Sea Basin.

The concept of a great ice sheet was soon accepted as explaining the unstratified calcareous shelly muds or tills of the Irish Sea coasts. The overlying stratified sand and gravel deposits and the upland sequence of the Isle of Man were still considered by many, however, to record a period of submergence. Horne (1874) and Ward (1880) both argued in favour of submergence following glaciation, regarding the absence of erratics in the uplands, the presence of shells in the sands and the generally flat topography of the northern plain as evidence in favour of marine action. Ward (1880) accepted Darwin's (1848) submergence model to account for the transport of the Foxdale granite boulders and the striations and, interestingly in terms of more recent work, assigned contortions and inversions of the lowland suite to the grounding of icebergs.

Percy F. Kendall (1894) was the first to reinterpret the Quaternary history of the Isle of Man entirely in terms of terrestrial glaciation. He stressed the regularity of the striations and the constant south-eastward transport of the insular boulders. He also traced some of the erratic lithologies of the lowland suite, including the distinctive riebeckite-eurite of Ailsa Craig. In addition, he proposed that many of the lowland landforms could be interpreted as of glacial origin, identifying the small hills north and north-west of Ballasalla as drumlins and the Bride Hills as either an esker or a moraine. The sands and gravels of the lowland suite he interpreted as proglacial in origin, recognising kame terraces and kettle holes near Kirk Michael.

Kendall (1894) denied any evidence of submergence either during or since glaciation 'except to the small extent indicated by the clearly-defined raised-beaches which are so generally recognizable round British shores' (p. 424). The shells occuring in the lowland suite he interpreted as derived from the floor of the Irish Sea and transported as clasts by the ice sheet, their generally comminuted condition and the mixture of species of diverse habitat mitigating against their interpretation as *in situ* fossils. It is interesting to note, however, that at Shellag Point Kendall did find several delicate yet complete bivalves embedded in stoneless clay, which he interpreted as a portion of the sea floor incorporated, transported and deposited by the ice sheet.

Perhaps Kendall's (1894) most important contribution was his proposal, despite the total absence of foreign erratics, of a glacial origin for the Manx upland suite. He suggested that as the ice sheet advanced upon the steep northern face of the slate massif, the basal debris-rich layers would have cleaved and passed around the flanks whilst the 'clean', erratic-free upper layers of the ice would pass over the uplands, incorporating the local rocks and thus producing the south-eastward trending features (Figure 4.5).

The Geological Survey map of the Island, together with a substantial memoir, was produced in 1903 by G. W. Lamplugh, the self-trained geologist who was later to become director of the Survey. Lamplugh (1903) added a great deal of detail concerning the

48   THE ISLE OF MAN

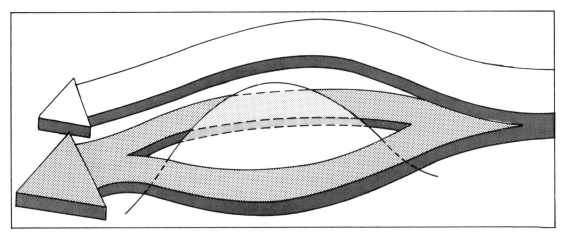

Fig. 4.5 Schematic diagram to represent cleavage of basal, debris-rich ice around the flanks of the Isle of Man whilst the uplands are overridden by clean ice devoid of erratics

Quaternary history of the Island, describing many of the drift sections, mapping striations and erratic trains and interpreting many of the depositional landforms. He interpreted the Bride Hills as a moraine ridge dissected by channels which carried meltwater from the ice-front southwards into a large lake (Lake Andreas). On retreat of the ice, the drainage of one of these channels was reversed, producing the Lhen Valley. The central valley of the Island he interpreted in terms of the breaching of a watershed by glacial meltwater.

It was during Lamplugh's survey that boreholes were sunk on the northern plain. One of these, at Point of Ayre, passed through 125m of drift, which included the very shelly mud and sand between 65 and 73m below sea level which might be interpreted as marine in origin. Because these sediments were underlain by further glacigenic sediments, however, Lamplugh (1903) regarded them as a displaced portion of sea floor incorporated and deposited *en masse* rather than as an *in situ* marine layer.

Lamplugh followed Kendall in interpreting the entire Manx sequence as the product of a glacial phase 'one and indivisible', with the cleaving of the basal debris-rich layers around the margins accounting for the absence of erratics from the Manx uplands, and minor fluctuations accounting for the complexity of the lowland drift sequence.

Subdivision and classification

Although Kendal and Lamplugh assigned the whole of the Manx succession to a single glacial episode, the concept of multiple glaciations had long been accepted elsewhere. By 1877 Geikie had recognised evidence for four distinct glaciations in East Anglia, and as early as 1864 the Scotsman James Croll had proposed an astronomical theory to explain climatic oscillations during the Pleistocene. Most of the evidence for repeated glaciations lay at the ice sheet margin, however, and the Isle of Man received little attention.

Smith (1930), presenting the results of a further 18 boreholes through the drift of the northern plain, was the first to propose a chronological subdivision of the Manx sequence. To the earliest advance he attributed the glacial deposits underlying the marine beds, which he considered to be interglacial and *in situ*. The deposits exposed above sea level he assigned to the most recent advance, and the underlying sequence he regarded as intermediate in age.

The tripartite subdivision was accepted by A. Marshall Cubbon (1957), for many years Director of the Manx Museum, who attempted to correlate the Manx sequence with that exposed on either side of the northern Irish Sea. This work was taken up by G. F. Mitchell, who looked in detail at the sequence exposed on the west coast of the Isle of Man near Kirk Michael (1956) and attempted a summary of the Pleistocene history of the Irish Sea (1960). Mitchell also accepted the general model of three glaciations, with the slate core of the Island exposed as a nunatak during the final advance, but argued in addition for a period of local glaciation to account for the presence of insular deposits in the coastal section at Glen Mooar. With some slight modifications he retained this subdivision in his later summary (Mitchell, 1972).

The inadequacies of the concept of subdividing Irish Sea glacial deposits on the basis of the number of till units, or 'count from the top' procedures was eventually demonstrated by D. Q. Bowen (1973), who presented a much simpler interpretation using interglacial raised beach deposits as stratigraphic markers. Bowen (1973) argued that the only glacial deposits which might pre-date the last advance anywhere north of County Wexford and Pembrokeshire are those underlying the marine sands beneath Point of Ayre. Like Kendall and Lamplugh, he attributed the complexity of many depositional sequences to minor fluctuations and other purely local factors.

The most detailed work to date on the Manx Quaternary deposits was that conducted in the early 1970s by G. S. P. Thomas. Thomas (1976) recognised that the sequence could be interpreted in terms of two or three glacial advances, based on evidence of interglacial marine conditions rather than lithology. Since there is no evidence of interglacial conditions stratigraphically higher than the fossil coast at Ballure, he interpreted the whole of the exposed foreign suite as deposited during the last glaciation. The underlying 'sub-surface' sequence of Smith (1930), which rests on a platform considerably lower than that exposed at Ballure, he interpreted as possibly dating from an earlier advance. The glacigenic sediments underlying the marine beds of Point of Ayre he assigned with more confidence a pre-last glaciation age.

In addition to this simple tripartite subdivision, Thomas (1976) produced a stratigraphic classification of the insular and foreign deposits exposed above sea level, defining eight formations and 51 members (Figure 4.6; Table 4.1). This is perhaps the most detailed drift classification ever attempted in the British Isles and, although the approach has since been criticised (Bowen, 1978; Eyles and Eyles, 1984), it was used by Thomas (1976, 1977) in interpreting the entire foreign suite and associated landforms in terms of minor fluctuations and readvances of the last ice sheet.

Several of the conclusions reached by Thomas (1976, 1977), however, contradicted the currently held views on the Quaternary history of the western British Isles. In particular, the evidence he presented raised questions regarding the geometry of the Irish Sea ice sheet, the timing of the last advance and retreat and the relative height of the glacial sea. Once again, the landforms and deposits of the Isle of Man became a focus of debate.

Thomas rejected Mitchell's suggestion that the Isle of Man might at some stage have supported a small local ice sheet, stressing that the area of accumulation is too restricted and the altitude insufficient to attract the necessary precipitation. Although accepting that the Island must at some time have been completely inundated by ice from the north (as attested by the transport of the insular granites), he concluded from a detailed study of the insular deposits and associated landforms that this was not the case during the last glaciation.

The suggestion that the slate massif stood as a nunatak above the last Irish Sea ice sheet had important implications, since it requires an extremely low ice-surface profile. In reconstructing the last Irish sea ice sheet Boulton et al. (1977) suggested an ice thickness in excess of 1600m over the Island (Figure 4.7). This reconstruction was based, however, on a rigid substrate model, and it has since been demonstrated that an ice sheet moving over deformable sediments, which would be the case in the Irish Sea Basin, would display a much shallower equilibrium profile (Boulton and Jones, 1979). Even applying this improved model, an ice thickness far in excess of the Manx uplands is required to prevent stagnation. Although Thomas (1976, 1977) proposed an alternative model, involving the thin edges of two largely independent ice sheets meeting along the flanks of the Isle of Man, this never gained support and the Manx upland deposits are generally regarded as tills produced by the last ice sheet and redistributed by solifluction following ice retreat.

The problem of the age of the Manx deposits arises from five anomalously old radiocarbon dates, in the range 18,400 BP to 18,900 BP, obtained from the base of kettle holes near Kirk Michael. A similar date (18,500 BP, Penny et al., 1969) has been obtained from beneath last glaciation deposits at Dimlington on the Yorkshire coast, and if both are correct this would imply that the Irish Sea ice sheet had retreated north of the Isle of Man before North Sea ice reached its maximum. Although Mitchell (1972) and Thomas (1976, 1977) argued that this might be the case, it has since been suggested that the Isle of Man dates may be erroneous. The moss from which the dates were obtained (*Drepanocladus revolvens*) is capable of subaquatic photosynthesis, which can lead to the incorporation of 'old' carbon from solution, giving rise to a 'hard water' error (Lowe and Walker, 1984). Support for a date of about 18,000 BP for the maximum of the last Irish Sea glaciation is provided by fossiliferous cave deposits sealed by Irish Sea till in the Vale of Clwyd (Rowlands, 1971).

The glaciomarine hypothesis

It is well established that as ice sheets build up, the concentration of water on land results in a lowering of the global ocean. Eustatic lowering during the last glaciation is generally thought to have been in excess of 100m (Lowe and Walker, 1984). Thomas (1976, 1977) sug-

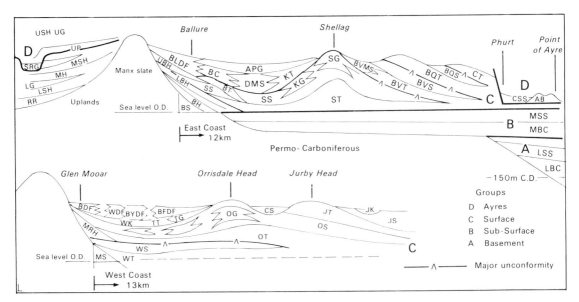

Fig. 4.6 Idealised cross-section of the Quaternary stratigraphy of the Isle of Man (after Thomas, 1976)

Group	Formation	Member	
Ayre	Point of Ayre	Cranstal Silts	(CSS)
		Ayre Beach	(AB)
	Curragh	Curragh Peat	(CP)
	Moorland	Upland Peat	(UP)
		Sulby Gravel	(SRG)
Surface	Ballaugh	Ballaleigh Debris Fan	(BDF)
		Wyllin Debris Fan	(WDF)
		Ballyre Debris Fan	(BYDF)
		Ballaugh Debris Fan	(BHDF)
		Ballure Debris Fan	(BLDF)
		Crawyn Sand	(CS)
		Jurby Kettles	(JK)
		Wyllin Kettles	(WK)
	Upland	Upper Stratified Head	(USH)
		Upper Gravel	(UG)
		Mid Stratified Head	(MSH)
		Massive Head	(MH)
		Lower Gravel	(LG)
		Lower Stratified Head	(LSH)
		Rock Rubble	(RR)
		Upper Blue Head	(UBH)
		Lower Blue Head	(LBH)
		Ballure Slope Wash	(BSW)
		Brown Head	(BH)
		Mooar Head	(MRH)
		Ballure Scree	(BS)
		Mooar Scree	(MS)
	Jurby	Andreas Platform Gravel	(APG)
		Trunk Till	(TT)
		Trunk Gravel	(TG)
		Cranstal Till	(CT)
		Ballaquark Till	(BQT)
		Ballaquark Sand	(BQS)
		Jurby Till	(JT)
		Jurby Sand	(JS)
	Orrisdale	Dog Mills Series	(DMS)
		Ballure Clays	(BC)
		Kionlough Till	(KT)
		Ballavarkish Sand	(BVS)
		Ballavarkish Till	(BVT)
		Ballavarkish Marginal Series	(BVMS)
		Orrisdale Gravel	(OG)
		Orrisdale Sand	(OS)
		Orrisdale Till	(OT)
	Shellag	Kionlough Gravel	(KG)
		Shellag Gravel	(SG)
		Shellag Sand	(SS)
		Shellag Till	(ST)
		Wyllin Sand	(WS)
		Wyllin Till	(WT)
		Ballure Till	(BT)
Sub-surface	Sub-surface	Middle Sands	(MMS)
		Middle Boulder Clay	(MBC)
Basement	Basement	Ayre Marine Silts	(AMS)
		Lower Sand	(LSS)
		Lower Boulder Clay	(LBC)

Table 4.1 *Formal stratigraphic classification of the Quaternary deposits of the Isle of Man (after Thomas, 1976)*

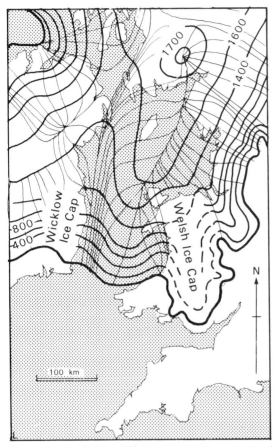

Fig. 4.7 The modelled surface topography (in metres) and flowlines of the Late Devensian ice sheet (after Boulton et al., 1977)

gested, however, that during retreat of the last ice sheet isostatic depression due to ice-loading may have been sufficient to result in a sea level little lower than that of today. The main evidence for this comes from the Dog Mills Series, which contain a rich micro-fauna (Wright, 1902; Wright and Reade, 1906) suggesting 'a cold, low-salinity, estuarine-intertidal environment, with an open water current-swept element' (Thomas, 1985, p. 154). More recently, Thomas (1985) has correlated the Dog Mills Series with lagoonal sediments beneath Holocene marine deposits in the Irish Sea south-east of the Island (Pantin, 1975, 1977, 1978). The suggestion that shallow marine conditions may have accompanied retreat of the Irish Sea ice complies with geomorphological evidence from coastal areas (Stephens et al., 1975; Stephens and McCabe, 1977; Synge, 1977, 1980) and has wide-ranging implications.

Although Thomas (1985) suggests that sea level during retreat may have been little lower than that of today, and that the Dog Mills Series, and perhaps the Shellag till and similar deposits elsewhere, may have been deposited in shallow water marine conditions, he interprets the rest of the exposed Manx sequence as terrestrial in origin. The concept of glaciomarine sedimentation in the Irish Sea basin has, however, been taken much further.

In 1984 Eyles and Eyles published a short paper in which they completely re-interpreted the exposed foreign suite of the northern plain as 'the largest subaerial exposure of last glaciation marine and glaciomarine sediments yet recognized on the British continental shelf' (Eyles and Eyles, 1984, p. 359). The Shellag till they interpret as marine mud with iceberg dropstones, and the overlying sands and gravels south of the Bride Hills as a generally coarsening-upwards marine sequence reflecting glacier advance and, therefore, increased proximity to sediment source. The deformation of this sequence to form the Bride Moraine is also considered to have occurred subaqueously, with high pore-water pressures in the fine-grained sediments, rather than deep permafrost (Thomas, 1984) accounting for the unusual structures displayed at Shellag Point. The sequence to the north of the Bride Hills, which Thomas (1976, 1977) interpreted in terms of repeated terrestrial re-advance are assigned by Eyles and Eyles (1984, p. 359) to a 'push-ridge-subaqueous-outwash depositional system deposited when processes of suspension deposition, density underflow, ice-rafting, highly variable traction-current activity, and sediment gravity flow were operative adjacent to a grounded marine ice margin' (Figure 4.8).

Although Thomas and Dackcombe (1985) published a comment and received a reply (Eyles, Eyles and McCabe, 1985), no agree-

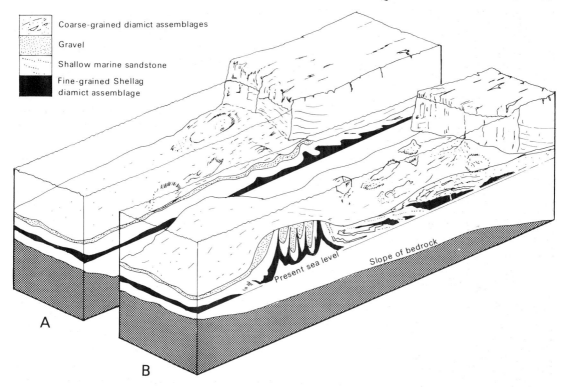

Fig. 4.8 Simplified schematic model for deposition of 'push-ridge-subaqueous-outwash' depositional system exposed on the east coast of the Isle of Man (after Eyles and Eyles, 1984)

ment was reached. One of the arguments used by Eyles et al. (1985) was that the stratigraphic classification erected by Thomas hindered realistic interpretation of depositional environments. They suggest that it is only on the basis of detailed description of lithofacies, contact relations, geometries and structures that environmental interpretation can begin. Countering this, Thomas et al. (1985) published a very detailed description of the sands, gravels and diamicts on the west coast, around Orrisdale Head, which Eyles and Eyles (1984) interpreted as distal glaciomarine. Using a facies analysis approach similar to that employed by Eyles and Eyles (1984), Thomas et al. (1985, p. 193) concluded that the sequence was deposited subaerially as 'a series of diachronous marginal sandur formed on an unstable, ice-cored supraglacial topography'.

The concept of high sea levels and glaciomarine sedimentation during deglaciation of the Irish Sea Basin has not been restricted to the Isle of Man. McCabe has been the main advocate, proposing that much of the Irish Sea drift of eastern Ireland can be interpreted as ice-proximal and ice-distal glaciomarine sequences (McCabe, 1987; McCabe et al., 1987). Eyles (1988, p. 34), has taken the argument further, suggesting that most of the complex drift sequences exposed around the British coastline can be interpreted as glaciomarine and that 'sea level, not climate, may have been the major control on ice sheet behaviour and sedimentation patterns'.

The great weakness of the glaciomarine concept is that almost all of the supporting evidence which has so far been presented is sedimentological. Although it is certainly possible to interpret many drift sequences, including those of the Isle of Man, in terms of glaciomarine models, this in no way implies that such models must be correct. Despite continuing opposition, however, the glaciomarine concept seems to have risen above the status of a theory and is rapidly gaining acceptance as a new paradigm in British Quaternary research.

In terms of the Manx landscape, there are a number of features which have yet to be fully explained in the context of a glaciomarine model. The topography around Kirk Michael and Orrisdale Head, for example, is characterised by conical mounds, flat-topped terraces, enclosed hollows, linear ridges and shallow channels. These features have been interpreted as kames, kettles, moraines and marginal channels, all indicative of a decaying terrestrial ice margin (Lamplugh, 1903; Mitchell, 1958; Dackombe and Thomas, 1985; Thomas et al., 1985). Such features, if they had been formed subaqueously, would be unlikely to survive passage of the littoral zone. On the coast north of Glen Wyllin, organic-filled depressions (Mitchell, 1958; Dickson et al., 1970) display very clear normal faulting of the surrounding and underlying glacigenic deposits, together with large water-escape features, suggesting the decay of buried ice. It is difficult to explain the burial of ice in distal glaciomarine sands.

Perhaps the most serious problem with the glaciomarine hypothesis is the amount of isotatic depression required and the dearth of evidence in support of such high sea levels. McCabe et al. (1986) have identified a glaciomarine delta at 80m OD in County Mayo, and it has been suggested that the deltaic deposits at 140m OD at Banc y Warren, near Cardigan, are also glaciomarine. Clear evidence of coastal features associated with isostatic readjustment are, however, largely restricted to the areas of thickest ice accumulation, and the stepped beach sequences which are so common in Scotland (Sissons, 1976) are not seen in the Isle of Man or farther south on the shores of the Irish Sea.

If the re-interpretation of the Manx landscape and deposits in terms of a glaciomarine environment is correct, then it may mark one of the most important contributions to British Quaternary studies since the concept of ice sheets was first proposed. Many problems remain to be solved, however, before we can consider abandoning the terrestrial model which has served for more than a century (Tiddeman, 1872) and return to the concept of a flood, albeit proglacial rather than biblical!

CHAPTER 5

Lateglacial and Postglacial environmental history

Danny McCarroll, Larch Garrad and Roger Dackombe

Introduction

A minimum age for retreat of the last ice sheet from the Isle of Man of about 15,000 BP is provided by a radiocarbon date of 15,150 ± 350 BP (Birm 754) from a kettle hole on the Jurby ridge (Tooley 1977, p. 33). The precise timing of the isolation of Man from the adjacent islands, however, is less clear. Advocates of the glaciomarine hypothesis (Chapter 4) would presumably argue for immediate isolation in a relatively high proglacial sea. The more conventional view is that land links were initiated upon retreat of a terrestrial ice sheet and later severed by a eustatically rising sea level.

Information on sea level following ice retreat in this area is sparse. To the north, Bishop and Coope (1977, p. 87) note that 'there is no evidence of the Lateglacial sea having inundated the present land areas adjacent to the Solway Firth', and they report a radiocarbon date of 9,640 ± 180 BP (Q–398) obtained from a peat layer about four metres below present high water mark at Brighouse Bay. To the south, estuarine and brackish water silts overlying till in Cardigan Bay have been assigned to the Lateglacial (Haynes et al., 1977) and, assuming minimal isostatic adjustment so close to the ice sheet limit, may reflect a sea level of about −60m (Lowe and Walker, 1984, p. 342). The recent recognition of periglacial features on the floor of the Irish sea north of Anglesey (Wingfield, 1987) adds further support to the concept of a Lateglacial low sea level. Since a deep trough (in excess of 100m) lies along the Irish coast (Figure 5.1), if a Lateglacial sea level of about −60m is accepted, then any land link with Ireland must have been broken soon after deglaciation. Water depths between the Isle of Man and North Wales, the Lake District and the Mull of Galloway, however, are relatively shallow (often less than 40m) and a land link may have persisted until much later. Allen

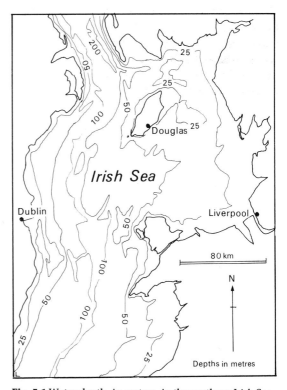

Fig. 5.1 Water depth, in metres, in the northern Irish Sea

(1976) suggests a severance date of about 9,000 BP.

The Late Devensian Lateglacial

The Late Devensian Lateglacial, usually abbreviated to 'Lateglacial', is the transitional period between arctic conditions which accompanied wastage of the last ice sheet and the much warmer conditions of the present (Flandrian or Holocene) interglacial. Throughout north-west Europe this transition involved rapid and acute fluctuations in climate associated with movements of the oceanic polar front.

Lowe and Gray (1980) have attempted a stratigraphic subdivision of the Lateglacial on the basis of palaeoclimate. They suggest a rapid climatic amelioration following ice retreat, culminating around 13,000 BP in a climate warmer than that of today. This was followed by a more gentle decline leading by around 11,000 BP to the re-establishment of cold conditions. This cold phase was short-lived, however, and by the beginning of the Holocene, at 10,000 BP, temperatures were rising steadily towards their peak during the climatic optimum, around 7–8,000 BP. The initial warm phase is known as the Lateglacial Interstadial or Windermere Interstadial (Coope and Pennington, 1977) and the final cold phase as the Younger Dryas or Loch Lomond Stadial. During the Loch Lomond phase a major ice readvance or recrudescence of glaciers occured in Scotland (Sissons, 1976) and small glaciers existed in Ireland (Colhoun and Synge, 1980; McCabe, 1987), the Lake District (Sissons, 1980), Snowdonia (Gray, 1982) and as far south as the Brecon Beacons (Walker, 1980).

The Manx record

Organic deposits dating from the Lateglacial have been identified from a number of sites in the Isle of Man, including the kettle holes inland and exposed along the coast north of Kirk Michael, and similar features near Jurby and in depressions on the gravel fan northeast of Ballaugh (Figure 5.2). Although many of the landforms and deposits of the Manx uplands probably originated in the Lateglacial, they can only be dated by association with the lowland organics.

The earliest organic deposits that have been studied in detail developed in a kettle hole exposed on the coast 110m north of Glen Ballyre (Figure 5.3). The base of this hollow was marked by a clay containing bands of the moss (*Drepanocladus*) which yielded the anomalously old radiocarbon dates (Chapter 4) of 18,900 BP (Birm–213) and 18,400 BP (Birm–270c) (Shotton and Williams, 1971, 1973). An organic silt 30–35 cm from the base has, however, yielded a date of 12,645 BP (Birm–214) and the upper part of a layer of mud, the base of which contained leaves of the arctic-alpine Mountain Avens (*Dryas octopetala*), yielded a date of 12,150 BP (GRO–1616) (Dickson et al., 1970; Dackcombe and Thomas, 1985, p. 107). This site lies to the north of the Ballyre alluvial fan and can be traced as a hollow on the surface. Kettles exposed along the coast further south (Figure 4.4) are covered by fan gravels and so have no surface expression. The youngest date obtained from organic deposits beneath these gravels (10,550 BP; Q–673), from a site 150m north of Glen Wyllin, places deposition of the fans towards the end of the Lateglacial.

Lateglacial landscape development

The Manx upland landscape is dominated by landforms and deposits fashioned and redistributed by periglacial processes. The periglacial legacy is so clear that Mitchell (1965) and Thomas (1976, 1977) considered that such conditions must have persisted throughout the Late Devensian cold stage. Evidence from elsewhere around the Irish Sea basin sug-

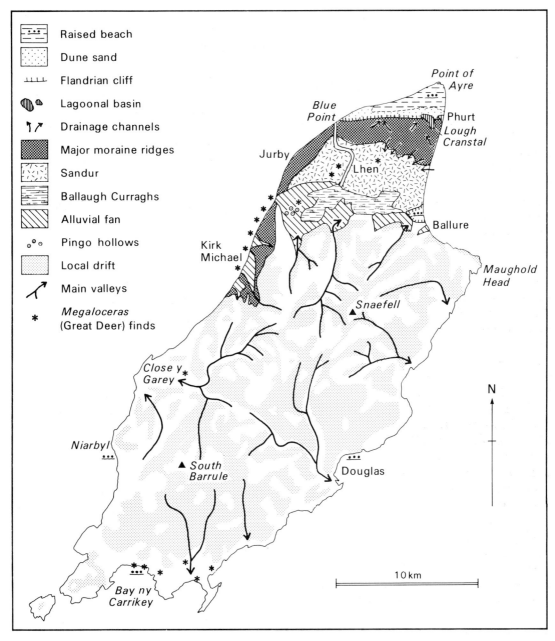

Fig. 5.2 Lateglacial and Postglacial landforms and deposits of the Isle of Man

gests, however, that the last ice sheet must have completely covered the Island, confining periglacial activity to the Late- and Postglacial (Chapter 4). Periglaciation would have followed immediately upon retreat of the last ice sheet, and must have begun in the Manx uplands whilst minor re-advances were still fashioning the drift topography of the northern plain. Cold conditions would have persisted until the abrupt warming around 13,000 BP and resumed during the Loch Lomond Stadial.

58   THE ISLE OF MAN

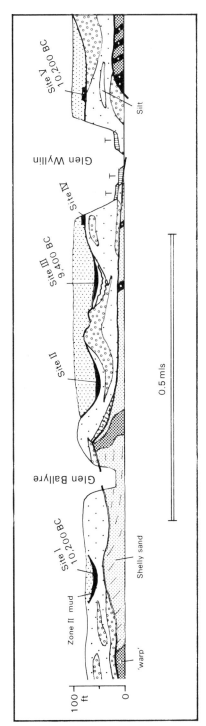

Fig. 5.3 Lateglacial organic deposits exposed on the west coast of the Isle of Man, north of Kirk Michael from which pollen, plant and beetle remains were obtained (With permission from Mitchell (1965), *Quarterly Journal of the Geological Society*, Vol 121, pp. 358–81)

The clearest evidence of periglacial activity lies in the thick local deposits which partly fill the upland valleys and form distinct terraces, resulting in a typically assymetric valley cross-profile. Thomas (1977) has demonstrated that these features are preferentially developed on slopes facing north-west to north-east and that the slope form and stone fabric are clearly indicative of periglacial action. Although the generally gradational boundary between bedrock and local drift is indicative of frost weathering, not all of the deposits need be derived in this way. Striated clasts and the transport of insular granites suggest that much of the material is probably redistributed till.

In most of the upland valleys the local drift masks a largely preglacial 'V'-shaped profile. The uncharacteristic 'U'-shaped cross-profile of the upper part of Laxey Glen can probably be attributed to enhanced glacial erosion. This valley would have formed a natural channel for basal ice armed with local clasts derived from passage over and around Snaefell.

Incision and erosion of the upland deposits must also have been initiated shortly after deglaciation, and the gravel fan issuing from the mouth of Glen Dhoo, and probably that from Sulby Glen, date from this period. The Sulby river, in particular, must have been considerably more powerful at this time, since the gravel fan now acts as a barrier, forcing the present river eastwards towards Ramsey. When the gravel fans were forming, the Sulby would probably have flowed into pro-glacial Lake Andreas, on the present site of the Curragh, which may have drained initially to the east along the channel between North Barrule and Maughold Head. Following the final retreat of the ice, the drainage of one of the proglacial meltwater channels incising the Bride Moraine was reversed and the lake drained via the Lhen Valley.

The organic-filled depressions on the surface of the Ballaugh Fan issuing from Glen Dhoo were originally interpreted as kettle holes produced by the melting of blocks of ice in the underlying glacial deposits, or brought down from the uplands with the gravel (Lamplugh, 1903). Organic deposits did not begin to form in these hollows until the Lateglacial Interstadial, however, and it is unlikely that buried ice would have remained for so long. It has since been suggested that the depressions may mark the sites of collapsed open-system pingos (Watson, 1971). These ice-cored hills develop in a permafrost environment where hydrostatic pressure forces water towards the surface and are common today in east Greenland, often occuring towards the base of gravel fans and in small groups. Due to their location on good arable land, the low bounding ramparts, which are typical features of most collapsed pingos, may have been removed by repeated ploughing.

The gravel fans of the west coast, around Glen Mooar and Glen Wyllin, must postdate those of the northern plain, since the former seal kettle holes containing a complete Lateglacial sequence. It is not clear why there should be such a marked disparity in the age of the gravel fans, but it is possible that those on the west coast bury older but less extensive features which correlate with the fans of the northern plain. In the hollows on the Ballaugh Fan, the Loch Lomond Stadial is marked only by alternating layers of sandy clay-mud, coarse sand and small pebbles suggesting, perhaps, that the rivers flowing towards the northern plain did not expand sufficiently to surmount the barriers built up during their deglacial maximum.

Although the fans mark two phases of enhanced activity, incision by the upland streams must have been active throughout the Late- and Postglacial. Erosion has worked its way steadily headward, with the lower parts of the major river valleys largely stripped of the fill of glacial and periglacial deposits while the valley heads are still deeply mantled. Most of the Manx river valleys contain sequences of terraces which record this progressive incision, but they have not been studied in detail.

5.1 Small organic-filled depressions, some forming small pools, on the surface of the gravel fan issuing from Glen Dhoo on to the northern plain have been interpreted as the remains of open system pingos formed during periglacial conditions following retreat of the last ice sheet

Lateglacial flora and fauna

The macro-fossil and pollen evidence of the Lateglacial flora of the Isle of Man has been studied in considerable detail (Mitchell, 1958, 1965; Dickson et al., 1970). Stratigraphic samples were collected from five of the kettle holes exposed along the coast to the north of Kirk Michael (Figure 5.3) and from one of the depressions north-east of Ballaugh. The Lateglacial flora recovered represents one of the richest yet obtained in the British Isles, including 114 taxa of flowering plants, one gymnosperm (Juniper), nine pteridophytes and 35 mosses. Of these, 25 speces were new additions to the Lateglacial flowering plant flora of Britain and 46 (28%) are no longer native to the Isle of Man. A very wide range of communities and habitats is represented, and the fossils obtained from the small samples that were examined must, moreover, represent only a fraction of the total Lateglacial flora, which may have included as many vascular plant species as the Island supports today (about 700, Allen, 1984).

Pollen diagrams were constructed from seven profiles and subdivided into three zones assumed to reflect vegetation changes in response to climatic fluctuations (Figure 5.4). Zone I was interpreted as reflecting a steady amelioration of climate following ice retreat. The flora included typically arctic-alpine species such as *Dryas octopetala* (Mountain Avens) and *Salix herbacea* (Least Willow), but in total 82 ecologically varied taxa are present, and it is possible that all of the Lateglacial flora may have arrived by the end of Zone I. The junction with Zone II is marked by a decline in the dominance of *Artemesia*, *Rumex*, *Salix* and *Empetrum*, and a marked rise in the pollen of *Graminae* (grasses) and *Betula* (birch). The pollen spectra of Zone II reflects a landscape dominated by grassland studded with flowering herbs and with scattered birch copses in sheltered places. The junction with Zone III is marked by a reduction in grass and the re-establishment of *Artemesia*, *Salix*, *Rumex* and *Cyperaceae*. The end of Zone III marks the junction between the Lateglacial and Holocene and is associated with an increase in grasses prior to the establishment of a forest cover.

The botanic evidence would seem to suggest, therefore, that the Lateglacial in the Isle of Man was characterised by two cold phases, with an arctic-alpine flora, separated by a warmer period of temperate grassland. This tripartite division is typical of the botanic record obtained from many Lateglacial sites in the British Isles, but the extent to which the vegetation changes reflect climatic conditions has been questioned, and it has been argued that a much more direct indication is provided by the analysis of beetle (*Coleoptera*) remains (Coope and Brophy, 1972; Coope, 1977).

Coleoptera provide excellent palaeoclimatic indicators because their remains are common in Quaternary organic and freshwater deposits, they appear to have maintained evolutionary and physiological stability throughout the Quaternary and many species are extremely sensitive to climate (particularly summer temperature) and are often associated with a restricted range of terrestrial habi-

5.2 The gravel fans of the west coast, seen here incised at the top of a coastal section, must date from the Loch Lomond Stadial (about 11,000 to 10,000 BP) since they seal kettle holes containing a complete Lateglacial sequence. They must, therefore, postdate the fans of the northern plain which support the remains of pingos formed immediately following retreat of the last ice sheet (before 15,000 BP)

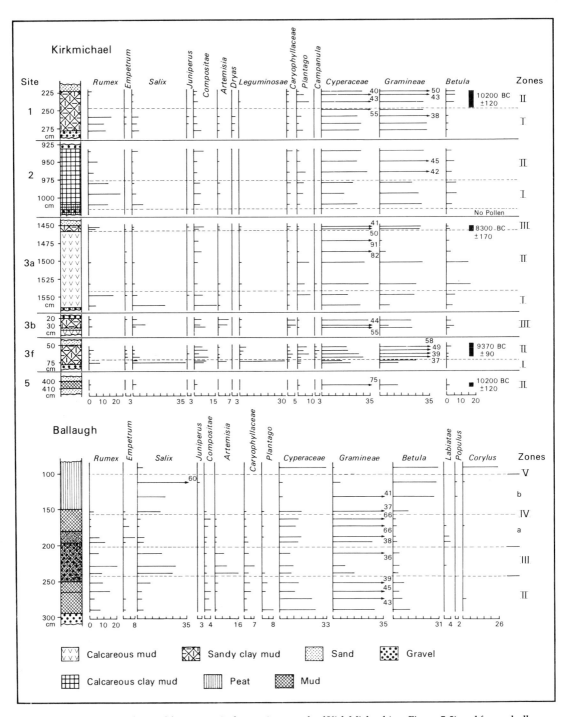

Fig. 5.4 Pollen diagrams obtained from organic depressions north of Kirk Michael (see Figure 5.3) and from a hollow on the gravel fan north of Ballaugh (after Dickson et al., 1970)

tats. Beetles are also much more mobile than plants and therefore provide a more precise indication of the timing of climatic change.

The beetle fauna preserved in the kettle holes on the west coast of the Isle of Man have been studied in detail by Joachim (1978), but this material has yet to be published in full. A short summary of the sequence preserved in the kettle exposed on the coast 110m north of Glen Ballyre is, however, provided by Dackombe and Thomas (1985, pp. 107–10).

The *coleoptera* obtained from this site included 194 species, of which 24 are not now native to Britain. The lower part of the sequence (0–35 cm) is dominated by species that today display a northern or highland distribution (e.g. *Bembidion hasti* Sahlberg and *Syncalypta cyclolepidia* Munster: Figure 5.5). This assemblage suggests an open pond surrounded by bare ground and a cold climate with an average July temperature of about 10 degrees centigrade. From 35 to 65 cm above the base there is a marked change, with northern species absent and the appearance of many which now occur only in the south of England or in southern Europe, including *Asaphidion cyanicorne* Pandelle (Figure 5.5). This assemblage suggests an average July temperature of at least 17 degrees centigrade. From 65 cm upwards, northern species reappear but the assemblages indicate a well-developed terrestrial and aquatic vegetation rather than bare ground. Gradual cooling is indicated from 65 to 85 cm. Above this there are few species present and the proportion with marked northern distributions increases. These include species such as *Boreaphilus henningianus* Sahlberg (Figure 5.5) which now occur north of the Arctic Circle or above the present treeline. The assemblage suggests very acid conditions and an averge July temperature of about 10 degrees centigrade.

When the faunal evidence is compared with the floral record obtained from the same site by Dickson et al. (1970), marked differences are apparent (Table 5.1). The coleopteran evidence suggests that the maximum temperature was achieved towards the end of Pollen Zone 1, and that when the more extensive vegetation of Pollen Zone II was becoming established the climate had already begun to deteriorate. This discrepancy between the faunal and floral record is typical of Lateglacial sites in the British Isles (Figure 5.6), and more detailed descriptions are available from Glanllynau in North Wales (Coope and Brophy, 1972) and St Bees in Cumbria (Coope and Joachim, 1980).

The Holocene

The Holocene or Flandrian is the period of time stretching from the end of the Loch Lomond Stadial, at 10,000 BP, to the present day. Like the Lateglacial, the Holocene can be conveniently subdivided on the basis of pollen stratigraphy. Zone IV, also known as the Preboreal, follows the Loch Lomond Stadial and is typically marked by the establishment of a birch woodland. In Zone V, pine forest becomes important in southern Britain but does not replace birch in Ireland until Zone VI; Zones V and VI together are known as the Boreal. In Zone VIIa, the Atlantic period, a more oceanic climate prevailed which is marked in the pollen record by a decline in *Pinus* (pine) and a corresponding increase in *Alnus* (alder). This is also generally accepted as the boundary between the Mesolithic and Neolithic periods. During the Atlantic, high mixed oak forests developed over much of Britain, though in the Scottish Highlands birch and pine remained dominant. The climatic optimum in Britain, with average temperatures about two degrees centigrade higher than today, is generally placed in the late Boreal or early Atlantic. The beginning of Zone VIIb, the sub-Boreal, is marked by a substantial elm decline. The junction with Zone VIII, which includes the present day, is indistinct, though it is generally associated with a climatic deterioration reflected by an increase in *Betula* (birch) pollen and a decline in that of *Tilia* (lime), *Hedera* (Ivy) and *Ilex* (Holly). Conditions also became wetter, initiating the fresh growth of *Sphagnum* in many bogs. Zones VIIb and VIII record the removal by man of much of the British 'wildwood' which had developed earlier in the Holocene.

It must be stressed that this subdivision of the Holocene on the basis of pollen zones is used here simply to provide a convenient framework against which to compare the results of work in the Isle of Man. It is now widely accepted that pollen zone boundaries are time transgressive (Smith and Pilcher, 1973; Huntley and Birks, 1983). Only the Lateglacial/Holocene boundary at about 10,000 BP and the Atlantic/Sub-Boreal (VIIa/VIIb) boundary, marked by the elm decline at about 5,000 BP, can be accepted as generally applicable throughout the British Isles. The Boreal Atlantic (VI/VIIa) transition, although time-transgressive (in relation to the spread of alder; Huntley and Birks, 1983; Smith, 1984) can probably be placed around 7,000 BP in the Isle of Man. Only when detailed pollen dia-

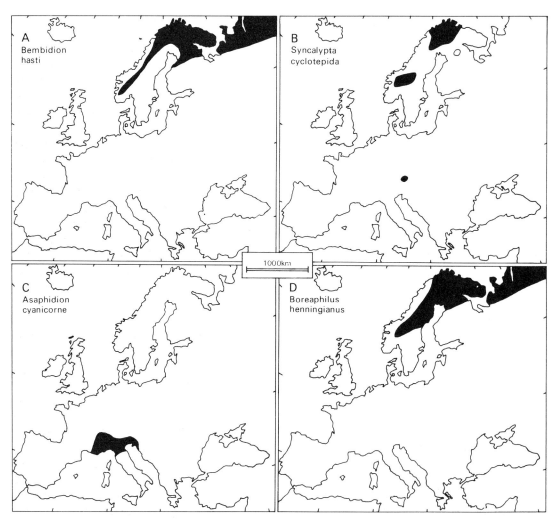

Fig. 5.5 Present-day distribution of beetle species found in Lateglacial deposits of the Isle of Man (based on Coope and Brophy, 1972).

LATEGLACIAL AND POSTGLACIAL ENVIRONMENT 65

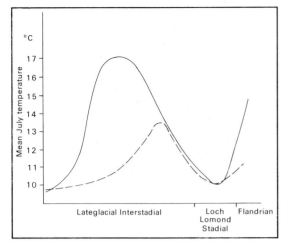

Fig. 5.6 Diagrammatic illustration of the variation in mean July temperature in lowland Britain during the Lateglacial, inferred from fossil coleoptera (solid line) and fossil pollen data (dashed line) (after Coope and Brophy, 1972)

grams supported by radiocarbon dates become available will it be possible to establish more realistic pollen assemblage zones for the Island.

The Manx record

Although Holocene deposits occur throughout the Isle of Man, they have received little attention and it is difficult to produce a coherent picture of environmental change. In this chapter, the deposits of three areas where some work has been done are described. First, the peats of Ballaugh and the Curragh, which began to accumulate early in the Holocene and, in the latter case, may have remained a site of organic deposition throughout the Postglacial. Second, the Ayres, extending beyond the raised 'Flandrian' cliff to the north of the Bride Hills (Figure 5.2), which has accrued entirely during the Holocene. Two sites of organic deposition, at Lough Cranstal and Phurt, have provided the most detailed palaeoenvironmental information to date. Third, the extensive upland peats which probably began to form relatively late in the Holocene. An attempt is then made to piece together the available information on the environment of the Isle of Man in prehistoric times, working through the Mesolithic and Neolithic before presenting the evidence for the diminution of the native Manx wildwood. Finally, the postglacial history of the Island's mammals is considered.

Ballaugh and the Curragh

The oldest Holocene organic deposits yet to be described from the Isle of Man accumulated in the 'pingo' hollows on the Ballaugh debris fan (Mitchell, 1958; Dickson et al., 1970). At 1.9m from the surface of one of these (Figure 5.4), the sandy clay mud and pebbly sand laid down during the Loch Lomond Stadial (Zone III) gives way to an increasingly organic-rich sequence of sandy muds and brown muds and, at 1.5m, wet brown peat with wood and leaf debris. The top metre of the deposits comprises brown amorphous peat. In the pollen record (Figure 5.4), the junction between Zones III and IV, the Lateglacial/Holocene boundary is placed at 2.05m, where *Artemesia* and *Salix* decline sharply and grasses (*Gramineae*) begin the increase towards their Postglacial maximum in Zone IVa. At this time there must have been very little tree growth, and most of the *Salix* pollen probably came from the tiny Least Willow (*Salix herbacea*). *Empetrum* (Crowberry) also reached its maximum during this stage. The change to Zone IVb is marked by a decline in grass pollen and an increase in that of *Betula* and *Salix*, probably derived from bushes and trees. *Ericaceae* also appears in Zone IVb. The opening of Zone V, at the Pre-Boreal/Boreal boundary, is marked at 1.05m by the appearance of *Corylus* (Hazel) pollen. Unfortunately, the rest of the sequence, which contained wood and plant remains, was not sampled.

The most complete Lateglacial and Holocene sequence in the Isle of Man is likely to have been preserved in the Ballaugh Cur-

FAUNA		FLORA (Dickson- et al)	
Interpretation	Faunal Unit	Pollen Zone	Interpretation
Very cold, acid bog	III	II (part)	Thermal maximum extensive vegetation
Cool, abundant vegetation			
Warm (17%C), meadow with occasional trees.	II		Gradual climatic amelioration, establishment of vegetation
Very cold (10%C), glaciers nearby, gradual establishment of vegetation	I	barren	Glacial conditions, little or no vegetation

Table 5.1 *Comparison of the Palaeoenvironment indicated by studies of the coleoptera and of pollen and plant remains from Lateglacial organic deposits in the Isle of Man (after Dackombe and Thomas, 1985)*

raghs. This extensive area of wetland and willow carr occupies the site of a proglacial lake and may have remained a site of organic deposition throughout the Postglacial. The thick wood peats, containing whole trunks of oak and Scots Pine, have been dug for fuel and have yielded Neolithic implements, including a corracle (Lamplugh, 1903), although this has not been preserved and its antiquity is questionable. Unfortunately, the biostratigraphy of this area has not been investigated in detail, the only published work being that of Erdtman (1925) who examined cores from the western part of the curraghs which only reached a depth of about two metres.

The Ayres

The Holocene deposits which have received most attention in recent years lie beyond the 'Flandrian' cliff line at the foot of the Bride Hills (Figure 5.2). At Lough Cranstal and Phurt, organic deposits record changes in Holocene sea level, climate and flora as well as yielding some evidence of the early human inhabitants of Man. The raised beach ridges of the Ayres are still accumulating, fed by erosion of the drift cliffs further south (Chapter 6).

Lough Cranstal at present contains a developing fen surface at about 9m OD (Chapter 7). Borings, however, reveal two distinct basins (Carter in Tooley, 1977; Tooley, 1978). The smaller southern basin contains terrestrial peats and detritus which suggest continuous fen conditions, whereas the larger northern basin is more complex. The lower parts of the sequence contain fresh water diatoms together with significant amounts of washed-in mineral material. Above this, a tenacious grey clay containing halophytic (salt-tolerant) diatoms and the remains of

LATEGLACIAL AND POSTGLACIAL ENVIRONMENT

Ruppia, a plant of brackish water ditches and salt marsh pools (now extinct in Man), suggest a marine connection, the commencement of which is dated at 7,825 BP (Hv–5226). A return to a freshwater diatom assemblage suggests that the basin probably developed into a lagoon separated from the sea by the earliest ridges of the Ayres raised beach; Marine-influenced deposition ceased at 7,370 BP (Hv–5225).

The brackish lagoonal sedimentation was replaced by deposition of lake muds containing pollen spectra indicative of a well-developed mixed oak woodland community. These are in turn overlain by terrestrial sediments in which cereals enter the flora in consistent frequencies and which may be linked with finds of Ronaldsway (Neolithic) artifacts on the lough edges, or earlier neolithic finds from the Phurt exposures.

On the coast east of Lough Cranstal the glacial sediments of the Bride Moraine pass beneath the beach and are replaced in the cliff sections by basins of lagoonal sediments and peats. Above beach level, three such basins are exposed. The most southerly, at Phurt (Figure 5.7), is similar in setting to the outer parts of Lough Cranstal, and the sediments within it may represent a similar or longer time-span. Below beach level, and only rarely exposed, is an earlier basin of similar extent.

The lowest basin, the top of which is at about present beach level, is cut into till and the banks display evidence of intense frost activity and of a former soil cover. The basin is filled with a sequence of sands and laminated clayey silts with an abundance of carbonised wood, including birch. Above the clayey silts and extending onto the banks is a thin but extensive peaty clay packed with the remains of vegetation, including well-preserved but compressed pieces of wood and frequent *in situ* stumps and large trunks of pine.

This 'forest bed' (Phillips, 1967) includes oak wood together with a pollen assemblage suggesting a discontinuous coastal woodland environment with a freshwater marsh or lake close by. Phillips (1967) suggests that the age of the bed is Pollen Zone VIa (Boreal). It is overlain by laminated muds and sands deposited in a freshwater but maritime environment, possibly spanning the Boreal/Atlantic transition.

The upper basins, exposed above the present beach, probably represent shifting lagoons amongst encroaching sand dunes. Phillips assigns the period of lagoonal sedimentation to the greater part of the Atlantic,

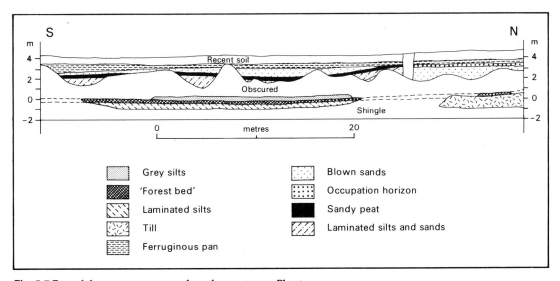

Fig. 5.7 Part of the sequence exposed on the coast near Phurt

after which a fall in sea level, and lower water-table, led to the deposition of freshwater organic muds. These grade laterally into silty detritus peat and, to the north, into a palaeosol (buried soil). Towards the basin margin this unit yields charcoal and a range of Mesolithic and Neolithic artifacts similar to those found around Lough Cranstal.

Above the northern bank of the basin and laterally continuous with the artifact-bearing layer of the southern basin is a layer of dark grey to black sand, of variable thickness, heavily cemented and containing abundant artifacts. This has been interpreted as an occupation horizon, delimited to the south by the peaty mud and to the north by a series of wedges interpreted as post holes. Pits extending beneath this horizon contain much charcoal. One such pit, holding Mull Hill Neolithic sherds and an abundance of carbonised pellets, had a clay 'lid', perhaps indicating grain storage. A saucer quern was found nearby. To the north this horizon appears as a dark brown sand with what may be the ghosts of tree roots, suggesting that the occupation site was a clearing in a woodland. Outside the occupation area the yield of artifacts is very much lower.

The evidence provided by these sites allows the reconstruction of the evolution of the Ayres in some detail, although there is still some uncertainty about the absolute dating of events and the precise interpretation of the pollen record at Phurt.

The first recorded event is the cutting of the 'Flandrian' raised cliff between Phurt and Blue Point, which must have preceded the initiation of shingle spit growth from Blue Point by a significant interval. The lower parts of the sequences at both Lough Cranstal and Phurt appear to have been deposited in lagoonal conditions in which there was some salt water influence. This suggests, in the case of Lough Cranstal, the existence of a shingle barrier through which seepage occured. Radiocarbon dates for the beginning and end of this phase (Tooley, 1978) are 7,825 BP (Hv–5226) and 7,370 BP (Hv–5225) respectively. At these times the sea would have stood at between −0.6m and +2.2m. Phillips (1967) ascribes the 'forest bed' at Phurt to a similar age, which coincides with the timing of high sea level in Wigtown Bay on the southern Scottish shore, which Jardine (1975) dates at between 6,600 and 7,200 BP.

The pollen data provided by Phillips (1967) require some reinterpretation, however, and it may be that too much weight has been given to the presence of salt-tolerant and salt-loving species in his reconstruction. The high level of the 'forest bed' at Phurt, the abundance of wood and tree pollen, together with an obvious association with Phragmites and Equisetum all suggest a freshwater lagoonal environment set amid lightly wooded sand dunes. The presence of pollen of halophytes probably reflects a coastal setting. If this is the case, then the 'forest bed' and the lagoonal sediments above it at Phurt are probably younger than the marine-influenced sediments at Lough Cranstal and equivalent to the phase of lacustrine sedimentation in the Lough. Throughout this period the Ayres would have been gradually extending to the north as successive shingle ridges were added.

The presence of early farmers in the area is marked by the appearance of cereal pollen in Lough Cranstal and of a number of artifact concentrations including the occupation horizon at Phurt. The earliest human traces include microlithic flints from just landward of the maintained hedge between Ballaghennie and Ballakinnag, and the overburden of the southerly gravel pit at the Point of Ayre. It seems that the boundary of the enclosed land and the heath of the Ayres coincides roughly with a line separating fresh, sharp flint implements from rolled, water-worn ones. This would suggest that in Mesolithic-Neolithic times the coastline of the Ayres lay about this division. The fossil soil at Phurt and the pollen data suggest that there may then have been a good tree cover on the slopes of the Bride Hills.

The upland peats

Although it appears that the Manx uplands, particularly the northern massif, are covered with peat, it is in fact fairly restricted and generally thin. Kear (1976) has shown that blanket peats exceeding half-a-metre in thickness are restricted to areas of gentle gradients above about 350m. Only around Beinn-y-Phott, Colden, the north-west slopes of Snaefell and the west slopes of South Barrule does the peat attain appreciable thickness, locally occuring as deep as three metres but an average is about 1.5m. Russell (1978) investigated sections in the peat on Mullagh Ouyr and showed that significant peat development here did not begin until about 2,850 BP.

In vertical sections the upland peats generally display two distinct zones. The lowest division is dark in colour and well humified, containing few plant macro-fossil remains but significant amounts of charcoal. Above this is a paler, less well humified unit, which is fibrous and contains plant debris but little charcoal. The lower unit rests on a palaeosol, and pollen from the soil-peat interface suggests that the vegetation at this time was dominated by grasses, heather and sedges with subordinate but important amounts of alder, birch and *Sphagnum* moss. Throughout the period of peat formation, the vegetation consisted of *Sphagnum*, heathers and cotton grass. Russell (1978) has suggested that the lower unit was formed during a period of deliberate and frequent hill burning.

Scattered wood fragments are frequently found at depth in the peat, and large tree trunks, thought to be mainly oak, have recently been found during ploughing for afforestation. Most of the concentrated deposits of fragments are of birch, which Russell (1978) suggests probably formed a thin, widely spaced scrub growing in sub-optimal conditions. Although alder, juniper, birch and hazel pollen are found in small quantities throughout the peat, the general lack of woody remains suggests that much of it was blown into the areas from the lowlands.

The cause of the initiation of peat accumulation in the Manx uplands is uncertain. The sub-peat soils show little sign of impeded drainage and there is little evidence for a significant climatic change, either in terms of increasing rainfall or decreasing temperature. Although climatic and anthropomorphic effects are difficult to dissentangle, it seems likely that peat accumulation was initiated due to the woodland and scrub clearance activities of Neolithic man, possibly to facilitate hunting or to simplify herding (Russell, 1978). Man was certainly active in the uplands at this time, as shown by the presence of hut circles and other remains. The reduction in charcoal in the upper levels of the peats and the evidence of a return to a scrub birch vegetation suggest a reduction in hill burning, possibly reflecting a wetter interval in the Bronze Age when grazing was dominant. The increasing abundance of grasses and weed pollen may point to an expansion of agriculture in later times. Arable cultivation was practised again in Medieval times to at least 1,000 feet, where slope permitted, as it apparently had been during the later Neolithic.

The environment of the early inhabitants of Man

Mesolithic

By the Atlantic period (Zone VIIa) Man was isolated as an island and the first human inhabitants had established themselves on its wooded shores. The coastal habitats may have supplied such a range of easily acquired foods over the whole year that the population became virtually sedentary. According to Woodman (1978), the inclusion among their microlithic blades (which were used to make composite tools) of locally-evolved needle and type-B hollow-base points, together with the absence of Cumbrian-type éclat écaillé and Irish-style axes and flint rods suggests

70 THE ISLE OF MAN

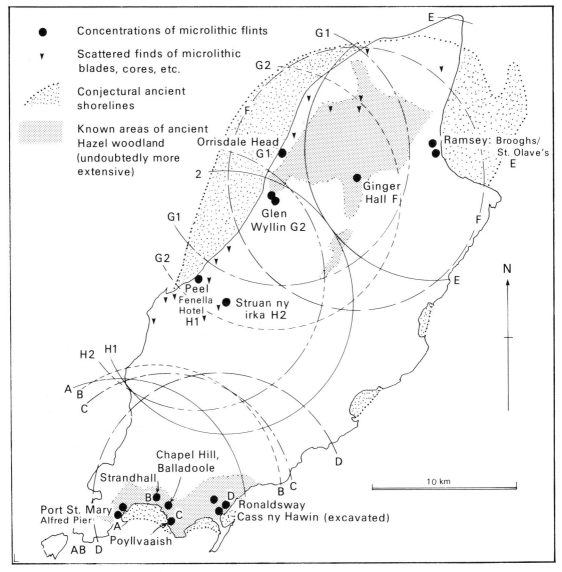

Fig. 5.8 Distribution of Manx Mesolithic sites. 'Base camps', marked by concentrations of microlithic flints, are surrounded by the notional 10 km territorial radius customarily assigned to hunter-gatherers

they lived in decent isolation from their neighbours on the adjacent islands and thus set the tone for their descendants.

The island on which they lived would have been larger than today. 'Forest Beds' containing Scots Pine and oak are occasionally exposed under the storm beaches of Bay ny Carrickey in the south and there has been very considerable loss on both sides of the northern plain (Chapter 6). The intertidal zone may also have been wider, which would have made more resources available to these strand-loopers.

The size of the Isle of Man is such that the notional territorial radius of 10 km customarily assigned to hunter-gatherers would virtually blanket it from the already known 'base camps' (Figure 5.8). If, as conjectured, the

Manx microlith-users lived mainly on shore foods and went into the wildwood only to gather hazelnuts (and, doubtless, other seasonal foods such as fungi), then their impact on the environment would have been minimal. Not so that of their successors who (on the evidence of finds) seem to have roamed over the whole island. No habitation site of Manx users of heavy-blade Mesolithic flint tools has ever been excavated, indeed none has been definitely located, despite the fact that the tools themselves are so abundant and widespread.

There is a general bias towards a western and northern distribution of mesolithic sites in Man which probably reflects that of foreign drift, which is the sole source of flint for toolmaking. Recent field-walking has revealed an almost unbelievably dense concentration of finds in Andreas and Bride, continuing down the west coast on the lighter soils through Jurby and Ballaugh. However, finds in Maughold, Lonan, Onchan, Braddan, Marown and Santon remain scanty. Whereas in Andreas or Bride a flint turns up on average every half acre, in German or Patrick it is every ten acres and in the under-populated parishes a whole season's research may yield only a few scraps.

The explanation of the distribution of sites is undoubtedly related to the fact that the woodland on the lighter soils is likely to have been thinner because these soils dry out so readily. Indeed, at the present time irrigation is required by market garden crops on the Jurby plateau and summer drought stress must have been even more intense before the onset of the increase in rainfall in the Atlantic period. Thus the use of fire to drive game would have been facilitated, particularly when drought conditions prevailed.

Neolithic

The Atlantic period ended about 5,000 BP to be followed by the cooler sub-Boreal (Zone VIIb). It has been suggested that the Strandhall submerged forest was drowned at this time and it may well be that the numerous darraghs (the Manx term for bog oak) of the Curragh basin were also killed then by waterlogging. The dead trees, which included both Scots Pine and oak, would have stood for a while and were then blown down by exceptional storms, so that the fallen trees found buried in the peat are aligned with each other. New ideas (and possibly a few new people) arrived and the first farming communities became established. They were equipped with stone axes, mainly from the Langdale (Group VI) factories, which served both to clear vegetation and break up the ground.

Unfortunately, the Mull Hill Neolithic farmers are best known for the elaborate megalithic burial and ritual centres, such as Cashtal yn Ard, King Orry's grave and the eponymous Mull Hill 'circle' (Chapter 8). Most of these were excavated some time ago which means that scant attention was paid to environmental evidence. It is hoped that more information will be obtained from the site at Phurt when analysis of the pollen and plant remains is completed. The distribution of the known sites probably indicates that they lived mainly in areas which were subsequently to be cultivated intensively. Their surroundings are likely to have been extensively cleared of woodland since they may have a quite sophisticated alignment relating to the heavens, which would have required a clear view.

The establishment of the first farms is usually marked in the pollen record by a decline in tree pollen and the apperance of that of ruderal weeds and cereals. Around 5,000 BP there is an almost universal decline in elm pollen in north-west Europe and the evidence from Lough Cranstal shows that in this the Isle of Man was no exception. For a while it was customary to attribute the decline to both large-scale human clearance and the feeding of elm foliage to stock. However, the current epidemic of Dutch Elm disease which has so drastically reduced elm numbers over most of north-west Europe has resulted in the pollen decline being attributed to a previous similar catastrophe. This is particularly interesting because the Isle of Man was still free of the disease in 1988, possibly because it is so seldom warm enough for the beetle which carries the destroying fungus to fly freely. This safeguard would not have existed during the warmer climate of the Neolithic.

Traces of man's activities, in the form of the pollen of dandelions and Ribwort Plantain, occur both before and after the elm decline, but cereal pollen is present at Lough Cranstal only afterwards. This may be a further indication that the Mesolithic users of heavy-blade tools had already altered the wildwood.

Sites of the later Manx farmers (of the 'Ronaldsway' Neolithic culture, Chapter 8) are proving to be considerably more numerous than had originally been supposed. Although those investigated in recent years have all been no more than rubbish dumps, with a maddening scarcity of recoverable environmental evidence, they are widespread. Site distribution (Figure 5.9) shows that a considerable part of the potential arable that could be utilised without elaborate drainage may already have been cultivated before the first metal tools arrived. Population pressure may have been such that clearings became permanent, as on the coastal plain of Cumberland. Hill woodland seems to have been thinned to improve the hunting and for (and by) grazing.

Diminution of the wildwood

A more pastoral economy came into favour after bronze came into general use. On the basis of Russell's (1978) date for birch twigs embedded a little above the base of the peat near the Bungalow (2,850 BP), it may be suggested that much of the higher hills had become open heath with scattered birch copses. Turbaries on Beinn y Phott, Mullagh Ouyr and Snaefell have also yielded birch. Sizeable oak has been found at 210m on Mount Karrin and at about 210m on Slieau Managh, yet a Bronze Age burial was found at about 240m in Druidale and other traces at 229–274m on Black Mountain. Bronze Age finds are otherwise evenly distributed over the Island.

Despite many rescue trials dug in recent years, virtually nothing is known of the timber needs of the Neolithic peoples, since the only known habitation site with structures is the eponymous one destroyed by the airport. Still less is known of Bronze Age timber use, apart from metal working, until the move to the summit hill fort on South Barrule. Its occupation apparently coincided with the Sub-Atlantic (Zone VIII) deterioration of climate. Its defences included a timber *Chevaux de Frise*, implying access to good-sized, probably coppice, poles. The continued survival of considerable areas of woodland with oak standards growing in reasonable shelter and on fairly fertile soil is also implicit in the existence of the great round houses excavated by Bersu (1946, 1977) at Ballacagen and Balla-

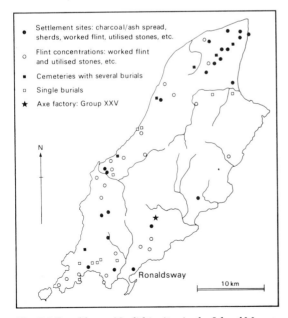

Fig. 5.9 Ronaldsway Neolithic sites in the Isle of Man

norris in Arbory. He stated that some two hundred trees were required to build one such structure and that the trees used had grown very slowly and were up to two hundred years old. The very few timbers now surviving (the excavations pre-dated radiocarbon dating and dendrochronology) show less than thirty years growth and it seems more reasonable to assume that the Medieval practice of using the smallest trees that would do the job was followed. On the basis of experiments at Butser experimental farm, these would probably be poles of seventy or so years growth. Rackam (1980) suggests that a maximum of 35–40 timber trees per acre, or a hundred or so to a hectare, is practicable with normal woodmanship, so the existence of even a few of these houses demonstrates that Man was far from being deforested. It may be no coincidence that a hill near Ballacagen and Ballanorris is called Cronk Darragh (Oak Hill in Manx). Promotion of fresh straddles in the coppice would have catered for repairs and it seems that Bersu's estimates of timber life were pessimistic, being based on posts in the open, rather than those deep within a structure.

The majority of other excavated Iron Age sites have been found to have smaller structures, often including a good deal of stone. However, even a small workshop, like that excavated at Braust, Andreas (on the stoneless northern plain) might have massive double post holes which would have required an acre of trees. Manx-style sod hedges and lankets (hobbles) on the stock would have served to keep them from the crops but not to keep the wolf from the fold. Many large coppice poles must have been required for close-set palisade fences and these were also incorporated in the outer defences of such sites as the promontory fort at Close ny Chollagh, Malew. They must also have been used to lace the stone ramparts of the small forts, all undated, on Cronk Sumark, Maughold Head and near Castleward, since all three have yielded vitrified slate resulting from the burning of the timber component of their defences. Experiments at Butser have also shown that a great deal of lighter coppice material, mainly birch and willow, would have been needed for woven fencing and this would have been true as soon as farming started. A reasonably large area of close-canopied woodland was needed to produce tall straight trunks but even scrub of birch and willow, such as might be found on the hills or wetlands, could be burnt into charcoal to fuel industrial processes. Reynolds (1987) has also suggested that old fencing was recycled into charcoal. A majority of excavated Manx Iron Age sites have produced scraps of crucibles, moulds, pattern stones and other evidence of small scale metal-working. Appropriate 'fuel ash residue' is known from Rhendhoo and Kerrowkneale in Jurby and also from the Bronze Age component of the Ronaldsway 'village' site partly obliterated by the airport's main runway.

A founder's hearth with a sword mould fragment from Crawyn Brooghs and a scrap metal hoard from Lonan prove bronze working in the Island. Knowledge of smelting of iron may have been another 'passenger' with Christianity, like the use of waterpower, warp-weighted looms and literacy. Peter Gelling excavated an iron-smelting and metal-working site, with 'E' ware at Kiondroghad. There have been recent surface finds further up the Lhen on Rhendhoo, while the incompletely excavated and unpublished Purt e Candas near St Johns is another contemporary manufacturing site. These would all require fuel, but good coppice management could have ensured continuity of supply. The iron ore (haematite) came from Maughold. There is no indication of any use of bog iron, altough iron ore erratics may have been used if found.

It has become almost an article of faith to suggest that the 'native' Celts had already denuded the Island of its timber by the time of the establishment of the Norse kingdom. It is, of course, true that the pattern of sheading, parish and farm boundaries (that was to be preserved as a fossil because it became the basis of the Derby Lords' rent collection) suggests that the last had substantially been carved out of the wildwood. However, much of the 'waste' must still have held woodland, however scrubby, since iron smelting and working continued. While there are traces of coal from Braddan and it was possible to use peat as a supplement, the evidence from the sites explored to date is that charcoal was the main fuel.

Unfortunately, Manx documentary sources are very scanty and tell us virtually nothing about the composition of the remnants of the wildwood that survived into Medieval times. A reference in the *Chronicles of the Kings of Man and the Isles* to the use of '. . . a wood on the sloping brow of the mountain which is called Sky Hill' to hide three hundred of Godren Crovan's men in 1079 suggests that it was the north-west of the hills where most trees survived, a supposition to which the distribution of less degraded soils and plants such as Wood Sanicle (*Sanicula europaea*), Three-nerved Sandwort (*Moehringia trinervia*) and a Speedwell (*Veronica montana*) lends credibility. Another reference from the *Chronicles* has often been quoted as proof of the absence of timber trees, but George Broderick's (1979) transcription and translation makes it clear that this contention is unjustified. King Magnus of Norway indeed 'constrained and compelled the men of Galloway' to cut timber and take it to the shore for the construction of fortresses and built fortresses in Man 'which to this day still bear his name' but the chronicler's stress is on the tribute exacted from Galloway and there is no hint that its timber was used in Man.

In the *Chronicles*' 'miracle about St. Mary' dating to about 1249 there was still a wood at Myerscough (i.e. the Curraghs). This place name appears again in the boundary between the lands of Rushen Abbey and the lands of the monks of Myerscough in which Myerscough is again called Boscum, as also another wood, or thicket, called Kor. Megaw has argued (Davey, 1978) that this particular version of the Abbey's bounds should be dated c.1280 AD. It also mentions a place called Hesca na appayze which R. L. Thompson suggests indicates land formerly covered with trees. He also suggests that Munenyrzana usually identified as the seemingly oddly named Mullenlowne, was yet another thicket from the Gaelic Muine.

The history of mammals

The Lateglacial deposits of Man have yielded definite evidence of only one mammal, *Megaloceras giganteus*, the 'Giant Irish Elk'. In fact, this remarkable creature did not belong to the Elk or Moose family (*Alces*) but was a deer of the sub-family *Cervinae*. Standing more than six feet tall at the shoulder, it was only slightly smaller than the American Moose (*Alces gigas*), but its most striking feature was its enormous palmate antlers. In large stags these commonly spread ten to twelve feet and weighed as much as ninety pounds.

The first scientific description of *Megaloceras giganteus* (formerly known as *Cervus giganteus*) was from Ireland (Molyneux, 1697) and the first description of similar finds from the Isle of Man was by Feltham (1798) (see Lamplugh [1903] for a review of the early Manx finds). The first almost complete skeleton was obtained in 1819 from Loughan Ruy, one of the 'pingo' hollows on the Ballaugh gravel fan. A local blacksmith reconstructed the skeleton, probably from several individuals, using a horse as his model and exhibited it for payment until, after a lawsuit, the Duke of Athol claimed it and presented it to the museum of the University of Edinburgh. It is this reconstructed skeleton that is figured by Cuvier in his *Ossemens Fossiles*. Unfortunately, it was later found that, as well as distorting the shape of the spine, the blacksmith had included some horse bones in the reconstruction, including the pelvis!

Irish Elk remains have since been recovered from a number of Lateglacial sites in the Isle of Man (Figure 5.4), but the best preserved and most complete is that recovered by a committee of the British Association, together with the

5.3 The virtually complete skeleton of a Great Deer excavated at Close y Garey, near St Johns, in 1887 which is now in the Manx Museum. This shows it before 1922 when it was housed in Castle Rushen. Reproduced by courtesy of the Manx Museum

Isle of Man Natural History and Antiquarian Society from a marl pit between Peel and St Johns. This near-perfect specimen was reconstructed and housed in Castle Rushen before being moved to the Manx Museum, Douglas, where it is now displayed to full effect.

The only pre-Lateglacial site in the British Isles at which the remains of *Megaloceras giganteus* have been recorded is Castlepook Cave, County Cork, which is believed to be Middle Devensian (Stuart, 1977). All other finds have come from deposits of Zone II (Windermere Interstadial) age. Their abundance in Ireland and the Isle of Man at this time, compared with their relative scarcity in England, probably reflects the difference in vegetation. The enormous antlers would have been a disadvantage in the birch woods of England whereas the herb and grassland of Ireland and Man presented ideal grazing.

The timing of the extinction of the 'Giant Irish Elk' has long been the subject of debate. In the Isle of Man, as in Ireland, no remains have been found in deposits younger than Zone II, and it is generally agreed that the animal was extinct in this area before the arrival of man. Elsewhere it may have survived into historic times, Sizer (1962) and Kurten (1968) citing finds from Syria and southern Russia indicating possible survival until about 3,000–2,500 BP (Gould, 1974).

The 'Giant Irish Elk' also has an important place in the history of evolutionary theory, since its extinction was one of the main weapons employed by evolutionists opposed to Darwin's idea of natural selection. They argued that *Megaloceras giganteus* was the product of a long evolutionary series marked by ever-increasing antler size, culminating in a beast with antlers so large, heavy and unwieldy that they must have been useless for fighting and, therefore, a disadvantage. Gould (1974) has demonstrated that there is no evidence of a long evolutionary trend, *Megaloceras giganteus* appearing quite suddenly in the fossil record, and that the enormous antlers, when compared with those of other deer, are actually in proportion for the animal's body size (those of the American Moose are smaller than expected). Most importantly, Gould (1974) suggested that the antlers of the 'Giant Irish Elk' were not used for fighting, but in ritual battles and sexual display. This is supported by the unique orientation of the palms, which are fully displayed when the animal looks straight ahead, and by the backward projection of the tines, useless for fighting but displaying their full length when the head is bowed. Larger antlers, in this case, lead to greater success in breeding. The extinction of *Megaloceras giganteus* from the British Isles, therefore, was probably due to the marked change in climate and vegetation at the end of the Windermere Interstadial rather than the necessary outcome of an evolutionary cul de sac.

The Lateglacial deposits of Man have not, so far, yielded any evidence of reindeer, which is commonly associated with *Megaloceras giganteus* in Ireland, or of horse, which was present in England (Stuart, 1977). There is one reference to the discovery of what may have been the antler of a true Elk from near Ballaugh (Hibbert, 1825), although this was dismissed by Lamplugh (1903) because of lack of remaining fossil evidence. The Elk is known from Lateglacial deposits in England but not Ireland (Stuart, 1977).

During the Holocene there is evidence for a much greater diversity of mammals (Garrad, 1978). Red Deer were certainly present. There are jaw bones from the Lhen at the Guilcagh and the (possibly sub-Boreal) 'forest beds' at Standhall (preserved in the Manx Museum) and Red Deer remains have been found on a number of archaeological sites. Their descendants survived into historic times. There have been several attempts to establish Fallow Deer, not native to the British Isles, while there is a single archaeological record of Roe, the other native species in Britain, from Chapel Hill, Balladoole.

Predators present certainly included the Wild Cat, bones of which have been found on several sites ranging from the Mesolithic/Neolithic midden at Port St Mary to the first century BC/AD midden in the cave at Perwick but not, as yet, later. The resemblance of the Manx Stoat (confusingly called weasel by the Manx) to the Irish version is such that they are likely to be closely related and descended from a common ancestor. If Sir Arthur Keith's tentative identification of a tooth from Castletown is accepted (Stenning, 1928) there were probably wolves, but the absence of any oral tradition would suggest that if so, they were wiped out early in historic times. The same is likely to have been true of foxes, but they survived long enough to give a name to a hill and shieling in Arbory (Cronk and Eary ny Shiannagh) and a leap on the cliffs above Bay Fine in Rushen (Lheim ny Chynnee). Ellan a Voddij (Dog's Island) in the Ballaugh curraghs may refer to otters (Moddey doin in Manx) although Carrick y Voddey (Dog's Rock) in Maughold may rather refer to wolves. As regards the past presence of badgers there is no evidence. Polecats could well have been present, although the present large population has been heavily contaminated by ferrets brought in to hunt the rabbits, themselves brought in by the Lords of Man to increase the number of nest sites for their precious Shearwaters (known as puffins) on the Calf.

T. Lloyd Praeger, the great Irish naturalist, was quite convinced that Man once had native hares but these must, he said, like its stoats, be related to those native in Ireland, i.e. they were a Blue or Mountain Hare which did not turn white in winter. These had then died out at an unknown date. Bone evidence has been lost but it seems quite probable that, since Man retained its landbridge to England until quite late, it also acquired the Brown Hare. These then were probably the ancestors of the present Brown hares in the Island, whereas the modern Mountain Hares are known to be the descendants of 1950s imports brought in for sport, Manx hares having been preserved at least as early as 1417. Although there is a well known folk tradition which suggests that Hedgehogs came on a ship wrecked about 1805, bones were found in the Close ny Chollagh promontory fort. Like most of the larger islands of Britain, Man has Pygmy Shrews which may have been denizens of the wildwood. Other rodents arrived later, but Long-eared Bats, which like to take moths from tree foliage and often roost in trees, are likely to have arrived early. Natterer's and Pipistrelle bats are also now present and Whiskered were recorded in the past from Braddan where a colony numbering some thousands was exterminated within the last twenty years. (Man, uniquely in Europe, afforded bats no protection as late as 1988!). There are no voles in Man. There is no evidence concerning Red Squirrels or Pine Martens, but the greater nut yield in antiquity would have offered a suitable home for the former. Moles are absent and probably never arrived.

CHAPTER 6

The Isle of Man's unstable coast
Colin Rouse

The subject of coastal erosion is of no consequence to the vast majority of people, but throughout the stretches of sparsely populated coastline the loss of valuable land is of vital concern to the individual owner. For centuries, therefore, where there exists a localised concentration of population or wealth, remedial works in the form of robust bulwarks or promenades have been constructed to protect against the inroads of the sea.

The greater part of the Island, about 80% in fact, is composed of competent rock which presents a robust barrier against the sea, so that during the span of a human life the effects of erosion are imperceptible. However, on the remaining 20% of the coastline, consisting of less competent Pleistocene and Holocene deposits, the recession of the land face is obvious. The erosion is not confined to the north of the Island (see Figure 6.1) since isolated pockets of recent deposits are also being eroded in the south, for example along the seaward boundary of the Gawing estate (SC225688) and on the coast between the foot of Fishers Hill and Poyll Vaaish (SC 245675). A considerable amount of money has been spent on coastal protection works in the south both by local authorities and private owners, this being consistent with a concentration of population and development of the area. The rocky coastline, together with such protective works that exist, exercise a measure of control against marine erosion over most of the Island's coastline. However, the northern drift lowland extending from Glen Mooar (Figure 6.1) around the Point of Ayre to the Vollan is left unprotected, the coastline is in retreat and the effects of erosion are obvious.

Over the years, numerous opinions and reports on coastal erosion in the northern drift lowland have appeared. Some have been made public whereas others have never emerged from committee rooms. These milestones are summarised by I. P. Jolliffe in his comprehensive report (Jolliffe, 1981) and in Table 6.1.

According to Renshaw (1975), records of coastal erosion in the Isle of Man exist for at

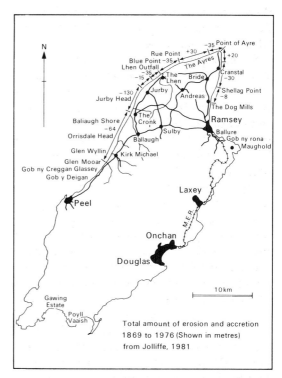

Fig. 6.1 Coastal erosion and accretion in the north of the Isle of Man

Table 6.1 *'Milestones' in the investigation of coastal erosion, IOM*

Year	Event
1892	A committee of Tynwald appointed.
1897	A committee of Tynwald appointed and issued a report (Comm. Tynwald, 1897) which said that 'cost of works is seldom, if ever, justified by the value of the land saved'.
1914	Experimental groyne built at Kirk Michael, 300 feet long and £300 at Government expense. Its life was short and unsuccessful.
1917	Report by Mr Blaker on cost of foreshore protection along NW coast was given as £40,000.
1921	A committee of Tynwald appointed.
1922	Report by Mr Blaker on costs of control.
1923	Report by Mr Cowin on costs of control.
1924	A committee of Tynwald appointed and issued a report (Comm. Tynwald, 1924) recommending construction of 154 groynes along NW coast.
1936	Report by Mr Brown on failure of Kirk Michael groyne.
1946	Memo on coastal erosion mostly concerning itself with control and costs in the Jurby area.
1949	Letters to the Government surveyor from Mr Brown.
1950	H. G. Nesbitt: report on extent of erosion (incorporated into 1951 report).
1951	J. C. Brown: report on extent of erosion submitted to I.O.M. Harbour Commissioners (Brown, 1951).
1952	Governor had decided not to proceed with any legislation on coastal erosion.
1965	Detailed survey by R. K. Gresswell of the University of Liverpool (Gresswell, 1965).
1975	J. C. Renshaw; unpublished report on history of erosion (Renshaw, 1975).
1979	I. P. Jolliffe: preliminary survey.
1980	W. Jones: unpublished report on littoral transport around Point of Ayre (Jones, 1975).
1980	P. Burrows: unpublished research on coastal erosion of boulder clay cliffs.
1981	I. P. Jolliffe: main report on coastal erosion around northern coastline of the Isle of Man (Jolliffe, 1981).
1982	Report of Tynwald on Jolliffe Report (Tynwald, 1982).

Source: Jolliffe, 1981.

least 250 years but none of the writers gave any details of the rate of erosion. It was not possible to hazard a reasonable guess on rates of erosion until the 1869 Ordnance Survey was published, although some idea of the extent of the erosion may be acquired from old records. In 1774 the Reverend Crellin recorded that the Chapel of St Bartholomew (near Kirk Michael) was on the brink of the cliff and graves and skeletons were protruding from the cliff face (Crellin, 1938). By the mid-nineteenth century the Keeill itself had disappeared into the sea but part of the burial ground remained. In 1855 a Dr Crellin found a lintel grave containing a skeleton at the cliff face, but by 1911 no trace of the burial ground remained (Manx Arch. Survey, 1911).

In 1892 a committee of Tynwald was established to consider the coastal erosion problem, but no report was issued. Then, in 1897, another committee was established and Lamplugh (who had just completed the geological survey) was asked to draw up some notes on coastal erosion. It was discovered that since the 1869 survey the area of most rapid retreat was on the west coast, in the parish of Jurby, where 45m had been eroded in 28 years (Comm. Tynwald, 1897). To the south, near Orrisdale and Kirk Michael, the loss over the same period had been from 18m to 27m. No figures were available for the erosion that was seen on the east coast.

Several local farmers, who had farmed along the Michael parish coast had memories going back more than 50 years and described the beach covered in boulders. Cumming (1848) makes mention of a lime kiln at Glen Trunk which used boulders from the beach and the cliffs with a thick cover of grass which was grazed by sheep during the 1830s. The cliffs became steeper and lost their grass cover as erosion caused the loss of 50m in 50 years. Tithe plans of 1785 and 1840 show very little change, confirming the farmers' recollections. Rapid erosion seems to have commenced between 1840 and 1897 and local opinion blamed the Highways Board for removing large boulders from the beach. Undoubtedly,

6.1 At the Chasms, southwest of Port St Mary, huge joint-bounded blocks of Manx slate are slowly rotating and sliding towards the sea

the removal of material by the farmers for lime burning and construction was also a contributory factor (Comm. Tynwald, 1897). The Committee produced very few recommendations for landowners as to how they should protect their land. One recommendation was to discourage the removal of beach material for construction; another the suggestion to build an experimental groyne at Kirk Michael (not to be attempted until 1914). Overall it was not felt that coastal protection work was cost-effective.

Another committee of Tynwald was established in 1921 and another in 1924. The latter produced a report which incorporated the surveys of Blaker and Cowin (Table 6.1). The report contained a definitive scheme for coastal defence of the north west coast, essentially Blaker's suggestion of a groyned foreshore along the whole stretch of coast. The report proposed that 80% of the cost was to be covered by the Manx Government. The scheme was, however, rejected and no legislative action was taken. In any case, the experimental groyne erected at Kirk Michael failed by 1936 (Table 6.1).

Erosion damage is undoubtedly due to sporadic events rather than continuous retreat. Brown (1951) showed that erosion north of Cranstal had removed a two-mile-long strip of land two metres wide in a single storm on 11 December 1946 when a Force 8 gale blew from SSE and coincided with high water. On 7 and 8 January 1947 there were again storm surges higher than predicted and a further loss of a metre of land. Brown (1951) suggested that there had been a six-metre recession over the previous 80 years at Cranstal but that this had taken place sporadically.

Brown's report of 1951 incorporated the

survey results of Nesbitt (Table 6.1) along with his own and research of the 1869 Ordnance Survey map. This allowed him to divide the coast into sections with different mean rates of erosion and accretion (Table 6.2). The rate of recession was found to be just over twice as great on the western as on the eastern coast (Brown, 1951, p. 16). This is explained by the greater frequency of gales from the north-westerly sector.

A Harbour Board survey carried out in March 1973 pointed out that in the Dog Mills area (Figure 6.1) the average rate of erosion from 1950 to 1973 was 0.6m/yr compared with 0.3m/yr between 1869 and 1950. This was also indicated by Brown's reappraisal of the erosion at Cranstal between 1950 and 1956, when it averaged 0.7m/yr compared with 0.07m/yr between 1869 and 1950 (Jolliffe, 1981, p. 78). The Bride parish commissioners were expressing serious concern in 1951 about the coastal road and cottages at Cranstal which in 1925 were 100m from the sea, but by 1951 had lost their pathway and were a mere five metres from the sea. The cottage nearest to the sea lost its seaward gable in October/November 1961 and only a portion of the back wall was left standing by March 1962 (Jolliffe, 1981, p. 82). By 1975 it had disappeared completely and the gable end was all that remained of the second cottage. This now lies on the beach.

When Thomas was researching his Doctorate on the glacial deposits of the northern drift lowland (Thomas, 1973), although primarily interested in the fresh exposures of drift, he remarked upon the areas of current maximum

6.2 The interbedding of clay-rich and more permeable sediments in the drift cliffs of the northern plain results in some unusual features. Here concentration of water above a relatively impermeable diamict (till) has caused saturated sands to flow onto the beach

6.3 Breached sea defence wall at Gansey, Port St Mary. Despite extensive protection, the sea occasionally illustrates its capacity to erode as it did here in Spring 1989

6.4 Property lost to the sea in Kirk Michael parish. This plate was taken in Spring 1989 and illustrates a localised dramatic loss of property. Is the notice for the benefit of the sea?

erosion as being between Kionlough Glen and Dog Mills and secondly north of Shellag Point. He, like others, pointed out that the longitudinal drift was in a northerly direction on both the east and west coasts. Also, this movement had been going on for some 5,500 years. It was stated that a reduction of beach mining north of Cranstal (as advocated by Brown [1951]) would have very little effect, since the beaches between Ramsey and Cran-

Table 6.2 *Rates of coastal accretion or erosion*

	Extent of erosion	Average width lost	Annual loss	Length of sector
Dog Mills to Point of Ayre	12.5 hectares	17.7m	0.23m	7.73 km
Point of Ayre to Rue Point	4.86 hectares	22.3m	0.25m	6.44 km
Rue Point to Glen Mooar	69.66 hectares	38.4m	0.48m	17.87 km
Rue Point to 3.70 km north	8.1 hectares accretion	26.2m gained	0.30m gained	3.70 km

Source: Brown, 1951.

stal were so thin (boulder clay being exposed for several miles in 1972–73) (Jolliffe, 1981, p. 83). Possibly the Ramsey breakwaters prevent the small quantities of beach material moving northwards from Ballure and the south. Thomas pointed out that prior to the turn of the century a substantial shingle beach afforded protection to the base of most of the east coast. The cliffs were of low angle and grassed over. The construction of the Point of Ayre to Ramsey brine pipeline and the removal of the shingle for aggregate led to an outflanking of the beach barrier and transportation of the remainder to the offshore zone.

The localised visual evidence of land loss to the sea is demoralising to the local residents, particularly since human action is often seen to be the precursor to the erosion. Sometimes the evidence is locally dramatic like the old corn mill at Glen Wyllin, which was in use as a fish hatchery until 1971; in 1951 it was six metres from the cliff edge, but by 1975 it stood on the edge. A small concrete wall built in the 1960s to protect it was 20m offshore by 1981 (Jolliffe, 1981, p. 84). Increasing awareness and public concern led to pressure for another report on the coastal erosion in the north of the Island. This was prepared by I. P. Jolliffe (1981) and formed the basis of the report of Tynwald in 1982 (Tynwald, 1982).

In this major report Jolliffe (1981) quotes Renshaw (1976) for the position of the coastline for 1869, 1950, 1957 and 1976 and the overall figures for gain or loss are listed in Table 6.3. These figures show that the rate of erosion was highest in the parishes of Jurby and Ballaugh (i.e. 1.2 m/yr) and that the rates vary around the coastal extent of the north. At the Point of Ayre there has been accretion of a mere 20m, although the high water mark has periodically fluctuated by as much as 60m (Jolliffe, 1981, p. 91), thus giving some observers the impression that the point is eroding and others that it is accreting more rapidly than the average figures. The original lighthouse was deliberately built inland in case the Point was eroding. Two minor lights have been built some distance from the main

Table 6.3 *Amount of coastal accretion (+) or erosion (−) 1869–1976*

Length of coast (km)	Av. width (m) of strip lost	Area of strip lost (sq.m)	Av. rate of location recession (m/yr)	
7	−64	−512,000	−0.7	Glen Mooar-Cronk
2	−130	−260,000	−1.2	Cronk-Jurby Head
1.5	−15	−22,500	−0.14	Jurby Hd.-Sartfield
2	−35	−70,000	−0.33	Sartfield northwards
2	−35	−70,000	−0.33	Lhen-Rue Point
2	−35	−70,000	−0.33	Point of Ayre w'ward
5.5	−30	−165,000	−0.28	Kionlough northwards
1	−8	−8,000	−0.07	Kionlough-Dog Mills
23	−51.2	1,777,500	−0.48	Average for all areas of erosion
1.5	+35	+52,500	+0.33	Llen southwards
3.5	+30	+105,000	+0.28	Rue Pt northwards
0.5	+28	+10,000	+0.19	Point of Ayre-south (east coastside)
5.5	+30.5	+167,500	+0.28	Average for all areas of accretion

Source: Jolliffe, 1981.

light, the latest of 1950 was built on a shingle ridge on the HWMT of 1869, although the sea has never come closer than nine metres to it since 1950.

The average rates do not, however, portray the true nature of the coastal erosion. The average rates are generally low compared with similar areas of soft cliff sites in Britain, even though the rate for the north-west coast is something like 2.2 times greater than the north-east coast. The evidence outlined above indicates that erosion is very sporadic and the rates of the last few decades are greater than the earlier period (Brown, 1951; Jolliffe, 1981).

The remedial strategies outlined by Jolliffe are fourfold (Jolliffe, 1981):

1. 'Do nothing'—which assumes that the coast is in a state of dynamic equilibrium and shore hazard zones will exist. This approach is the least expensive since compensation is cheaper than wholesale remedial action. Future development could be prohibited from the hazard zone and past average rates of erosion used to forecast areas at risk.
2. 'Maintain the status quo'—a traditional approach utilising conventional engineering hardware, notably walling and groyning, but more recently artificial beach nourishment. This would involve high capital and recurrent costs and can be said to be geomorphologically and visually intrusive.
3. 'Build-forward'—nourish the beach, either from offshore, or alongshore, or reclaim land from the sea. This approach is expensive, rarely a general solution but often attractive and effective at a local level.
4. 'Build-off'—here one must protect the coast with an offshore breakwater of some kind. Again, it would be very expensive and only effective at a local level.

The 'structured approach' to erosion abatement presents several problems. Groyning would have to be carried around the entire extent of the coast to avoid the problem of end scour and may still cause problems elsewhere. Also the groynes would intrude into what Jolliffe calls a unique structureless dynamic equilibrium system (Jolliffe, 1981, p. 9). Walling was also dismissed as being ineffective apart from low cost methods that might be used to protect sensitive points. Generally, walling would create many problems outside of the immediate local area.

The 'non-structured approach' also presents some difficulties. To reshape the cliffs would involve enormous quantities of material and be prohibitively expensive, although this cost could be reduced by providing toe protection to the cliffs and allowing them to degrade naturally. Beach nourishment could be practised with material brought from the Point of Ayre, although access points would have to be created. The problem, however, arises as to the paucity of material at the Point of Ayre due to the commercial removal.

In summary, Jolliffe (1981, p. 17) stated that it does not make economic sense to employ any structured approach; rather, systems should be left alone and buildings and roadways resited. However, he does make exception for the Ballure cliff near Ramsey which is threatening the Manx Electric Railway. He also recommends repairs or improvements to broken or badly sited field drains. The recommendations of the report were incorporated into the report to Tynwald in 1982.

Coastal erosion in the north of the Island can thus be seen to be a very long standing and emotive issue and the long catalogue of reports and apparent lack of activity must frustrate those directly affected by erosion. On the other hand, the lack of a cost-effective solution must equally frustrate the authorities who have long endeavoured to find some magical solution.

CHAPTER 7

Nature conservation

Larch Garrad

Conservation-orientated societies, within the Isle of Man, include the Society for the Preservation of the Manx Countryside (founded 1935), the Manx Ornithological Society (founded 1967) and the Manx Nature Conservation Trust (founded 1973). Partly as a result of the restricted population base they are ill-equipped to substitute for government agencies. Formalised nature conservancy work is in its infancy in the Isle of Man and the legislative framework present in the adjacent islands does not yet exist, although a first Wildlife and the Countryside Bill has reached the House of Keys. As initially envisaged, it will create no nature conservancy officers within the Isle of Man civil service, and there will be no additional funding for land holding or management, or for scientific research. Provision for habitat protection would be minimal.

As part of the background for the Isle of Man Development Plan, the Local Government Board (as it then was) and the Manx Museum commissioned an ecological survey from the UK's Nature Conservancy Council (NCC)/Institute of Terrestrial Ecology (ITE). The findings of the 1975 report, *Nature Conservation in the Isle of Man*, were duly incorporated in the *Development Plan*. Extensive areas were cited as being of scenic value, an important consideration in view of the tourist trade, while two areas, the central Ayres and the Ballaugh/Lezayre Curraghs, were designated as sites for potential Manx national nature reserves. Other, mainly smaller areas were called 'sites of special ecological importance'. It was envisaged that these last would be given a comparable statutory status to the UK's Sites of Special Scientific Importance (SSSIs), with the staff of the Manx Museum being increased to cover such matters as scientific evaluation, notification and the creation of management plans.

The Manx Museum has collected Manx biological records since it moved into its present Douglas headquarters in 1922 and is now officially *Biological Records Centre* for the Isle of Man (Vice County 71) under an arrangement with the main centre for the British Isles at Monks Wood (although NCC/ITE do not normally work here). As part of its record centre duties, it has acted as watchdog for the listed sites, but existing legislative protection is so inadequate that some have been destroyed or irreparably damaged. Some sections of the Isle of Man Government have, however, scrupulously observed the boundaries of the designated areas, most notably those concerned with planning, harbours and forestry. The last is now working to a very carefully thought out long-term plan which should, among other aims, revitalise the hill land without destroying its wildlife, agricultural value, or scenic beauty. Since ignorance is one basic reason for destruction, the importance of each of the sites that reached the *Development Plan* is outlined in Tables 7.1 and 7.2, and their location is indicated on Figure 7.1. Additional notes are provided on six particularly interesting and easily visited areas.

Since insufficient data were available, the selection of sites for listing was somewhat arbitrary, and several of value were unfortunately omitted. The main basis was botanical (see Table 7.1). There are, as yet, no sites

Table 7.1 *Basis for the listing of sites of special ecological importance*

Site		1	2	3	4	5a	5b	6	7a	7b	8a	8b	8c	8d	9	10a	10b	10c
1	Ayres buffer	—	•	•	•	•	•	•	•	•	•	•	•	•	•	—	•	•
2	zones	—	•	•	•	•	•	•	•	•	•	•	•	•	•	—	•	•
3	Maughold	[•]	•	•	•	•	—	•	—	•	•	•	†	—	•	—	—	•
4	N. Barrule	—	•	•	—	—	•	—	—	—	•	•	—	—	—	—	—	—
5	Snaefell	•	•	•	—	—	—	—	•	•	•	—	—	—	—	—	—	—
6	Slieu Dhoo	—	—	—	—	•	—	—	—	•	—	—	—	—	—	—	—	• !
7	Colden	•	•	•	—	—	—	•	•	•	•	—	—	—	•	—	—	—
8	Gob ny Calla	—	—	—	—	—	—	—	—	—	—	•	—	—	—	—	—	—
9	Skeirrup	—	—	—	—	—	—	—	—	—	—	•	—	—	—	○	○	•
10	Thistle Head	—	—	—	—	—	•	—	—	•	•	•	—	—	—	—	—	•
11	S. Barrule	—	•	•	—	•	•	—	—	•	•	•	○	—	—	—	—	—
12	Cass ny Hawin	•	•	•	•	•	•	—	•	•	•	•	—	—	•	—	—	—
13	Langness	•	•	•	•	•	•	—	•	•	•	•	S*†	—	—	•	•	—
14	Scarlett	•	•	•	•	•	•	—	—	•	•	•	†	—	—	•	•	•
15	Perwick/Pt. Erin	—	•	•	•	•	•	—	—	—	—	—	—	—	•	•	•	—
16	Calf of Man	—	—	•	•	•	•	—	•	•	•	•	S•	•	•	•	•	•
17	Lough Cranstal	—	•	•	•	•	•	•	—	•	•	•	—	—	•	—	—	—
18a	Laggagh	•	•	•	•	•	—	—	•	•	•	•	—	•	•	—	—	—
b	Gat e Wing	—	—	—	—	—	—	—	—	•	—	—	—	—	—	—	•	— !
c	Ballakinnag	—	•	•	•	—	—	—	•	•	•	•	•†	—	—	—	—	—
d	Curragh e Cowle	—	•	•	•	•	—	—	•	•	•	•	†	—	—	—	•	—
e	Bishops Dub	—	—	•	•	•	—	—	•	•	•	•	†	—	—	—	•	—
19	Ballachurry	—	—	—	—	•	•	—	•	—	•	Bats	—	—	•	—	—	—
20	Close e Kewin	—	—	—	—	—	—	•	—	—	•	—	—	—	•	—	—	• !
21	Nappin Ponds	—	•	•	•	•	—	—	—	—	•	•	—	—	—	—	—	—
22	Narradale	—	•	•	•	•	—	—	•	—	—	•	—	—	•	—	—	—
23	Glen Auldyn	—	•	•	•	•	—	—	•	—	—	•	—	—	—	—	—	—
24	Traie ny Halsall	—	—	—	—	—	—	—	—	—	•	—	†	—	—	—	—	—
25	Dhoon	•	•	•	•	•	—	•	•	—	—	•	†	•	—	—	—	—
26	Glen Roy	—	•	•	•	•	—	—	•	—	—	•	—	•	—	—	—	•
27	N. of Groudle	—	•	•	•	•	—	—	—	—	—	•	—	—	—	—	—	— !
28	Central valley Curraghs	—	•	•	•	•	—	—	•	—	•	—	—	—	•	•	—	• !
29	Ballamooar	—	•	—	—	•	•	—	—	—	—	—	S	—	•	—	•	•
30	Glen Maye	[•]	•	•	•	•	—	—	•	—	•	•	—	—	•	—	•	—
31	Dalby etc.	—	•	•	•	•	—	—	•	—	•	—	—	—	—	—	—	—
32	Billown	•	•	•	•	—	—	•	•	—	•	—	—	•	—	—	—	—
33	Ayres	•	•	•	•	•	—	—	•	•	•	—	—	•	•	•	•	•
34	Curraghs	[•]	•	•	•	•	•	—	•	•	•	•	•	•	•	•	•	•

• Factors on which inclusion depends (see key below).
[•] Species on which inclusion depends may be extinct.
○ Site destroyed.
S Bird sanctuary under Manx law.
† Site of geological interest.
* *Stenobothrus stigmaticus*, unique in the British Isles.
! Site requires reconsideration.

* KEY TO BASIS OF LISTING
1 Presence of plants not found elsewhere in the Isle of Man.
2 Presence of plants rare in Man; fewer than ten sites.
3 Presence of plants rare in this area of the island.
4 Number of plant species present.
5a Degree to which plant assemblage is representative of the habitat.
 b Degree to which plant assemblage is markedly unusual for the habitat.
6 Variety of plant species present.
7a Presence of unusual plant communities.
 b Presence of representative plant communities.
8 Presence of species, assemblages or habitats that are:
    a Rare in Man.
    b Rare in this area of the island.
    c Rare in the British Isles as a whole.
    d Rare in Northwest Europe.
9 Diversity (variety of habitats present).
10 Major bird uses:
    a Passage on migration.
    b Wintering.
    c Breeding.

Table 7.2 *Habitat types represented by the listed areas*

Habitat type	Listed areas
Ayres heath	1 2 33
Maritime heath	1 2 8 9 10 12 13 14 15 16 24 31
Hill heath	4 5 6 7 11 22 23 26
Wetlands*	13 14 17 18a 18b 18c 18d 18e 19 20 21 27 28 31 32 33 34 †
Open water	14 17 18a 18b 18c 18d 18e 29 32 33 34 ¶
Other	32 (inland limestone grassland)

* Some very small in area.
† Additional wetland areas not included in the initial listings are: 35: Guilcagh, Jurby 'Pollies', Loughcroute, Rhendhoo (already destroyed) and Tosaby.
¶ It would seem desirable to add Eary/Kionslieu and, on the grounds of local value, Tromode Dam. Additional northern ponds which might also be protected include: Braust, Glasgoe and the peripheral curragh ponds such as Ballacain.

listed on geological grounds, although both the volcanic area at Scarlett, and features such as the Chasms and the Langness Arches are included within potentially protected areas. A *Marine Nature Reserve* has existed round the south coast by a general agreement among divers, but has no force in law.

The absence of primarily agricultural land, apart from Ballachurry, should be noted. There are already a number of other private Bird Sanctuaries in addition to the long-established one that covers the whole of Langness. These should have Wildlife Refuge status. The Manx National Trust properties afford all the protection possible whilst retaining public access for ramblage, while the Calf is a Bird Sanctuary with a recognised Observatory.

The Ayres (areas 1, 2 and 33)

The Ayres conservation area lies at the northern extremity of the Island, roughly comprising the area seaward of the last maintained hedges. The western half of the

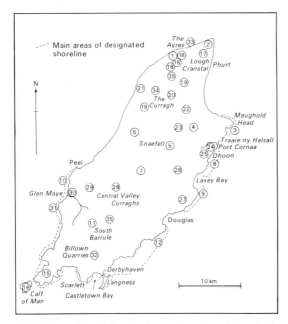

Fig. 7.1 Location of sites designated as of special ecological importance in the *Isle of Man Development Plan*

7.1 The spectacular cliffs of the south-west coast provide one of the most important breeding areas for sea birds

area is owned by the IOM Government, whilst the eastern portion is owned mainly by gravel companies. The Manx National Trust presently owns about a quarter of the proposed Manx Nature Reserve, but all of the Government-owned land is now managed by a joint committee, mainly for wildlife.

Vehicle access with parking is available off the A10 coast road at: Blue Point (NX393023), Ballakinnag/Rue Point (NX414030) and Ballaghennie (NX437033), and also at Point of Ayre (NX464046), reached by the A16 from Bride. A census of breeding birds has been published by Wallace (1972) and a plant species list by Garrad (1972).

The Ayres is an area of unique heathland which has developed on a thin layer of blown sand overlying shingle. It is now generally accepted that it was formed by a series of shingle spits that grew from the region of Blue Point, with some of the area coming into being after the first people established themselves (see Chapter 5). The lower inter-tidal zone is often fairly sandy, with pebbles at the top of the shore and typical, but not luxuriant, tideline vegetation. The shore is fringed by a narrow belt of Marram on blown sand. Flowering plants range from the endemic Isle of Man Cabbage (*Rhynchosinapsis monensis*), Sea Bindweed (*Calystegia soldanella*), Sea and Portland Spurges (*Euphorbia paralis* and *E. portlandica*) to somewhat oddly, Pyramidal Orchid (*Anacamptis pyramidalis*). Invertebrates include such moths as the Shore Wainscot (*Leucania litoralis*), which has caterpillars that eat Marram, and the Sand Dart (*Agrostis ripae*), the larvae of which feed on Sea Holly and Sea Rocket. Also present are the spider, *Arctosa perita* and various colour forms of the Common Field Grasshopper. The rear slope of the fixed dunes is covered by a mat of Burnet Rose, interspersed with various yellowheads, Common Storkesbill (*Erodium cicutarium*) and, in one locality only, the Dense-flowered Orchid (*Neotinia intacta*). This plant has a European range extending to Cyprus, Crete and the Canary Islands but is otherwise found in the British Isles only in Clare, Galway and Mayo. Regrettably, a series of dry springs combined with rabbit damage may have exterminated the colony, which did not flower during 1987. Other orchids present have included Fragrant (*Gymnadenia conopsea*) and Early Purple (*Orchis mascula*), the latter not known elsewhere in Man.

Common, Arctic and Little Terns have maintained breeding colonies here in the face of human disturbance, while offshore delights include diving Gannets and Grey, or Atlantic, Seals. There are a few passage and wintering waders and Mallard, Shelduck, Lapwings, Curlew and Oystercatchers also nest. The most recognisable species of the lichen heath are *Cladonia impexa*, *Hypogymnia physodes* and *Usnea articulata*. The last is typically corticolous but here grows on the ground, just possibly a habit it adopted during the last glaciation as a refuge off southwest Ireland. The interest of the area is considerably increased by the existence of a series of wet hollows, mimicking the slacks and lows of a true dune system. Ponds dug in these have been maintained since at least 1868, the publication date of the 25" Ordnance Survey plans. Common frogs, damselflies and a wide variety of flowering plants are present. During the last twenty years or so a series of much larger excavations near the Point of Ayre have also gone below ground water. Although domestic waste was used as a fill in one of these, the remaining ponds were swiftly colonised by *Chara* pondweed and flowering plants, such as willows and Marsh Marigold (*Caltha palustris*) and have proved popular with breeding birds, ranging from Herring and Black-headed Gulls to Mute Swans, Tufted Duck and the first Merganser ever to nest in Man. Another, more deliberate experiment has resulted in a small colony of Grass of Parnassus (*Parnassia palustris*), not known elsewhere in Man at the present day.

If you should visit this superb heathland, please observe the spirit of the management plan and do not take wheeled vehicles on to the heath, or shore, or unduly disturb breeding birds.

Maughold Head/Port Mooar (area 3)

Maughold Head, to the south-east of Ramsey, is a distinctive hill of Manx Slate separated from the upland massif by the pre-glacial Mooar valley and cut by the sea into steep and impressive cliffs. Parking is available in Maughold village and at Port Mooar, from which a coastal footpath leads, by courtesy of the landowner, to Maughold church.

The Maughold cliffs are home to a wide variety of cliff-nesting birds, including House Martins as well as the more usual gulls and auks. The vegetation is restricted, but a single oak near Stack Mooar may be a survivor of the wildwood. Port Mooar shore has a fluctuating maritime flora which has included *Spartina*, Sea Spurge (*Euphorbia paralis*), Sea Kale (*Crambe maritima*) and Sea Lavender (*Limonium vulgare*) as well as commoner species. The intertidal zone repays investigation better than might be anticipated.

Langness (area 13)

The Langness peninsula, originally an islet but now joined to Man by a tombolo, separates the bays of Castletown and Derbyhaven. Parking is available at the end of the lighthouse road and at Derby Fort.

The solid geology comprises mainly slate and Carboniferous conglomerate (Chapter 3), but the grassland has developed on a cover of blown sand, probably partly a relic of a once more extensive intertidal zone. The conglomerate rests uncomfortably on the slate, and is clearly demonstrated at the Arches where the slate has been tunnelled away to leave the bright red conglomerate precariously poised. This whole feature is probably a mining relic adapted to provide a sheltered harbour for a small boat, rather than a natural phenomenon. It is most easily located by following the cliffline round from the gateway on the lighthouse side of the car park.

On the outer portion of the peninsula there is fine maritime heath with Purple Milk Vetch (*Astragalus danicus*), Spring Squill (*Scilla verna*), Field Gentian (*Gentianella campestris*) and (sometimes) Dodder (*Cuscuta epithymum*). The main international wildlife importance of the area, however, is as the home of the only colonies of the grasshopper *Stenobothrus stigmaticus* known in the British Isles; please do not collect. The area is also of great value as a place to observe migration of both lepidoptera and birds. Castletown Bay/Derbyhaven is the main Manx wintering ground of waders and wildfowl, with quite large numbers of Mallard, Teal and Widgeon as well as occasional rarities. It has long been a bird sanctuary.

True salt marsh is absent, but the area below the Madoc Memorial/Langness farmhouse ruins has Sea Club Rush (*Scirpus maritimus*), Ragged Robin (*Lychnis flos-cuculi*) and Celery-leaved Buttercup (*Rununculus sceleratus*), while Chestnut-headed Sedge *Blysmus rufus* may be found nearer the mine. The vegetation of the gullies is also worth investigation. The wildlife is, of course, not static. The oldest Manx colony of *Spartina* grass has just vanished from the Castletown Bay side, while Sea Wormwood (*Artemisia maritima*) has appeared on one of the outer stacks, where it is almost invisible among the abundant lichens. Visit also the dwarf 'wood' of Sloes on the raised beach between the mine and the memorial.

Scarlett (area 14)

The headland of Scarlett, immediately south of Castletown is approached via Queen Street, to park near the quarry, or on foot from Poyllvaaish. The Manx Nature Conservation Trust visitors' centre is usually open 2–5 pm, Thursday, Saturday and Sunday.

This again is an area of noteworthy geology (see Chapter 3). The Stack is columnar basalt and there are outcrops of lava, volcanic ash and conspicuous dykes. The various pools offer a range of salinity each with appropriate

NATURE CONSERVATION

7.2 The Calf of Man, once the home of vast colonies of Manx shearwaters, is now a bird sanctuary and observatory maintained by the Manx National Trust

wildlife. In the case of the quarry this includes Fennel Pondweed (*Potamogeton pectinatus*), which is rare in Man, a pink woodlouse (*Androniscus dentiger*) and damselflies. The locally rare Strawberry Clover (*Trifolium fragiferum*) and Parsley Water-Dropwort (*Oeanthe lachinelii*) are present, together with more common plants such as Marsh Arrowgrass (*Triglochin palustris*), Brookweed (*Samolus valerandi*), Watercress (*Rorippa nasturtium-aquaticum*), Fools' Watercress (*Apium nodiflorum*) and Samphire (*Crithmum maritimum*). The last was pickled in barrels, exported and sold as health food in the days when winter diet featured much salt meat and fish.

The Calf of Man (area 16)

Lying just off the south-west tip of Man, this 616 acre island of Manx Slate (Figure 7.2) is now a bird sanctuary and Observatory maintained by the Manx National Trust. Access is by boat from Port Erin or Port St Mary and there are no major restrictions, although dogs, camping and the lighting of fires are prohibited. Limited hostel accommodation is available. Details can be obtained from the Secretary, the Manx Museum, Douglas (Douglas, IOM 75522).

In historic times, the Calf was home to a vast Manx Shearwater colony and the bird was named from here. As elsewhere in areas of Norse settlement, the colony was extensively culled for human food without diminution since the killing of the fat young, left by their parents to slim until they could fly from their nest burrows, merely replaced later natural losses. The incursion of rats, however, has decimated these burrow-nesting birds and there are now only a very few pairs, precariously protected by anti-rat campaigns.

The islet's wholly rocky coastline is being intensively surveyed as the groundwork of a future Marine Nature Reserve. There is a breeding colony of Grey (or Atlantic) Seals which has risen to its present numbers since the 1880s. Despite lending their name to several coastal features, such as Gob ny Rona in Maughold, seals had not bred in Man in later historic times.

There are also many sea birds and an unusually dense population of breeding Choughs. Part of the Manx Museum's flock of the native Manx sheep, the Loghtan, is permanently pastured here to maintain the short, ant-filled turf beloved of this handsome 'red-legged king of the crows'. The Observatory has a resident warden and assistant who are happy to explain and demonstrate their work when its pressures are not too great. Very large numbers of birds are ringed during the migration periods and at these seasons assistance is preferred to conversation.

In 1986, 5,471 birds of 79 species were ringed, 957 in April alone. During the life of the Observatory, 109,953 birds of 135 species have been ringed.

Ballaugh Curraghs (area 34)

History

The Ballaugh Curraghs are an area of wetland, of little agricultural value, between the villages of Ballaugh and Sulby on the northern plain (Figure 7.3). The area is thought to be the site of the proglacial Lake Andreas (see Chapter 4), and may have remained as an area of continuous wetland throughout the Postglacial (see Chapter 5). Limited parking is available at the Killane bridge, but the area is a wetland so please proceed with extreme caution and do not pick Common Wintergreen (*Pyrola minor L*).

The area is presently drained by the Killane and, in the north, by the largely canalised Lhen. A chain of ponds, including the Laggagh Mooar (area 18a), Curragh e Cowle (18d), Ballakinnag (18c) and other wetlands may mark another drainage route, their anti-

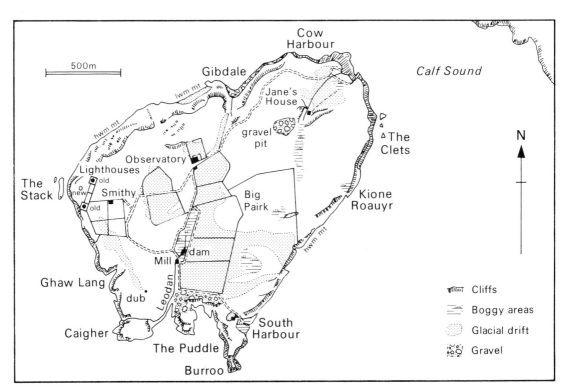

Fig. 7.2 The main features of the Calf of Man

7.3 Ballaugh Curraghs: A. The central track north from Ballavolley, with the Killane in its artificial channel along the west edge and the open water beyond. The man with the horse and cart is in front of the site of the Wildlife Park. The wide track is now an overgrown footpath and 3m high willow scrub flanks it almost continuously save where the channel side has been newly cleared. B. Open water fringed by Bog Bean. It is now impossible to see the distant view of the tower of Ballaugh church from within the main area of willow carr. Photographed about 1890–1900 by G. B. Cowen. Reproduced with the permission of the Manx Museum

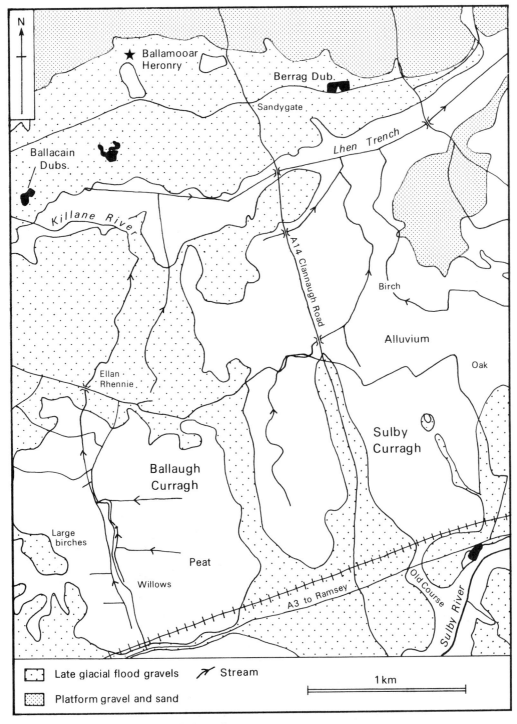

Fig. 7.3 Main features of the Ballaugh Curraghs

quity partly explaining their modern botanical interest. They may be assumed, like the Curraghs, to have been continuously wetland but, again like the Curraghs, the first (on the evidence of a painting by John M. Nicholson now in the Manx Museum) has become very much more scrub-covered than it was in the nineteenth century.

Wildlife

The Manx Gaelic word *curragh* is applied fairly generally to wetlands. Other notable examples of wildlife importance include Curragh Beg (also known as Curragh Pherick) north of Ramsey and Greeba Curragh and Curragh Glass which form part of area 28, the Central Valley curraghs (Figure 7.1). They are now fairly densely overgrown with tall willow scrub, with invading birch in 28 and 34. In fairly recent times the curraghs of the Lake Andreas basin were kept fairly open by a variety of uses including peat cutting, grazing and the cutting of hay and litter. In Medieval times the curraghs seem to have been extensively scrub-grown, the scrub probably providing charcoal for such industrial uses as iron smelting and working. Anxiety about limited vegetation changes may, therefore, be unwarranted. There was certainly less tree growth during the period 1850–1950, as evidenced by G. B. Cowen's photographs or the bird watching P. G. Ralfe's description, in 1897, of an open watery landscape. This parallels Lamplugh's (1903) view of Lough Cranstal (area 17) as 'an open grassy flat'. Recent aerial photographs, however, show there is still plenty of water and there seems no proof that the destruction, early this century, of dams of

7.4 Laggagh Mooar (the big miry place) by John Miller Nicholson, 31 December 1883. Reproduced by courtesy of the Manx Museum

the Killane and the Lhen mills had any great effect on drainage. A full hydrological and ecological study of the area is, however, long overdue.

At present there is an interesting range of communities, from the maintained open water of the drains, where bladderwort (*Utricularia*) seems to have been the main casualty, through the Bog Bean (*Menyanthes trifoliata*) mat to the Bog Myrtle (*Myrica gale*) swamp, which in turn gives way to the willow/birch scrub and the less friendly scrub of the reverted fields. The Bog Bean mat is best seen by crossing the first bridge to the west, south of the southern road bridge over the Killane. This also affords access to the Bog Myrtle area, which is notably wetter than that recently acquired by the Manx Nature Conservation Trust (MNCT) at Close Sartfield. Both areas have the lines of Royal Fern (*Osmunda regalis*), a generally abundant plant here, which are thought to mark the boundaries of old peat cuttings. Where there are wetter patches, plants of interest may include Round-leaved Sundew (*Drosera rotundifolia*), various marsh orchids (the Manx ones are *Dactylorhiza incarnata* ssp *incarnata* and *coccinea* and *D. purpurella*, with a range of misleading hybrids) and sedges. The area is rich in invertebrates such as damselflies, but its former variety of aquatic snails, adapted to Man's soft water, seems to have fallen victim to run-off from agricultural land. The majority of those recorded by Taylor in the 1900s could not be refound during work on the *Non-Marine Molluscs Atlas*. However, the Duck Mussel was refound at the Guilcagh in 1985.

The main interest of the area is the continuity of its vegetation with that preserved in the pollen record and as wetland. It seems probable that coppicing of the willow would enhance the small bird population and, as was shown at Loughcroute, revitalise the ground flora. For example, the meadow where Fragrant Orchid grew is now completely scrub-covered and this has been the fate of most of the former grassland east of the main track. *Deschampsia* tussock has taken over many areas including the MNCT property near Curragh Aspic, formerly the property of the Lord Bishop of Sodor and Man. An experiment in growing New Zealand Flax (*Phormium tenax*) as a fibre source has left it an established part of the flora and it is also present near Greeba. Devoted work by Mark Fitzpatric has revealed continuing value as a small bird habitat. Breeding species include Redpolls and there is a major winter Hen Harrier roost, as well as a variety of ducks.

The birch would also appear worthy of study in greater detail. It appears likely that Manx birch commonly attains a greater age than can trees in the drier parts of eastern England. A birch (*Betula pubescens*) felled in Lezayre, for example, had 125 rings of healthy wood, which is equal to the oldest recorded by Rackham (1980). The inter-breeding between species and sub-species, the distribution of seedlings and their survival and the other wildlife to which they play host could all be researched.

Invertebrates

Recent investigations have not generally produced results commensurate with the effort expended, and this is not wholly a reflection of lack of experience/expertise on the part of the collectors. Unfortunately, sufficient work has not been done to establish all the causes of our failure to re-find creatures recorded in the nineteenth century. It seems likely that two relevant factors are de-acidification of the water as a result of enhanced run-off from agricultural land and drastic habitat alteration caused by scrub growth. Recent pH tests (for acidity) on water drawn from the Killane between the Wildlife Park and the road bridge to the north have given figures of 6.5–7, whereas the indications are that a natural figure would be of the order of 5.5–6 (run-off on South Barrule, for example, is as acid as 3.5–4!). Detailed species lists are appended.

APPENDIX:
BALLAUGH CURRAGHS SPECIES LISTS

MOLLUSCS (F. Taylor's list of 2.9.1909, with some additions).

Aplexa hypnorum
Carychium minimum agg.
Columella edentula
Lymnaea palustris
*L. perega
L. truncatula
*Physa fontinalis
Pisidium milium
*P. nitidum
P. obtusale
P. pusillum

*P. subtruncatulum
*Planorbis albus
P. contortus
P. spirorbis
Punctum pygaeum
Sphaereum corneum
Valvata piscinalis
Vertigo antivertigo

Anodonta anatina DUCK MUSSEL
Lhen mouth, 1922–25, Guilcagh 1985.

Results of investigations 1969–87 suggest a loss of several species. One hour's fishing at SC363951, 28.7.69, added Pisidium hibernicum but otherwise refound only the species starred.

DIPTERA Flies.

Bibio marci
B. reticulatus
Bombylius canescens
Calliphora vomitoria
Cheilosia rosarum
Chirosia parvicornis
Chloromya formosa
Chrysopilus cristatus
Chrysopa relictus
Chrysotoxum restivum
Cinxia borealis
Dasyneura ulmariae
Empis chioptera
E. lividia
E. tesselata
Epistrophe eligans
Epitriptus cowinii†
Geosargus cuprarius
Haematopta crassicornis
Helophilus pendulus
Ischryosyrphus glaucius
Lathrophthalmus aeneus

Morellia hortorum
Neoascia podagria
Nephrotoma guestfalica
Opomyza germationis
Pamponerus germanicus
Phaonia basalis
Philonicus albiceps
Phytomyza ilicis
Rhamphomyia subcinerascens
Rhingia campestris
Sepsis violacea
Statinia marginata
Syrphidis ribesii
Tephrochlamys rubiventris
Thereva nobilitata
Tipula fulvipennis
T. lunata
T. vernalis
Tubifera intricaria
T. pertinax
Varichaeta consobrina
V. radicans

† Peculiar to Man.

96 THE ISLE OF MAN

ODONATA Damselflies etc.

The entire Manx list occurs in the curraghs.

Aeshnea juncea COMMON AESHNEA
Enallagma cyathigerum COMMON BLUE DAMSELFLY (mainly blue)
Ishnura elegans COMMON ISHNURA (blue near end of body)
Lestes sponsa GREEN LESTES old records from Ballacain dubs
Libellula quadrimaculata FOUR SPOTTED LIBELLULA
Pyrrhosoma nymphula LARGE RED DAMSELFLY
Sympetrum striolatum COMMON SYMPETRUM

HYMENOPTERA Bees and wasps

Agrothereutes abbreviata var. pygoleeucos
Andrena clarkella
A. denticulata
A. labilis
Bombus agrorum
B. lucorum agg.
B. muscorum
Crabo cribarius
Dolerus liogaster
Dolerus cothurnatus
Halictus calceatus
Lissonata bellator
Myrmica ruginodis
Parabates cristatus
Pontania viminalis
Psythyrus campestris
Spilichneumon occisorius
Trichoma enecator

ORTHOPTERA Grasshoppers

Omocestus viridulus COMMON GREEN

NEUROPTERA

Chrysops vulgaris
C. ventralis
Hemerobius lutescens
Micromus variegatus

TRICHOPTERA

Limnophilus auricula
L. flavicornis
L. elegans a species very rare in the British Isles was recorded in 1909

HEMIPTERA Bugs

Acanthosoma haemorrhoidale HAWTHORN SHIELD BUG
Capsus meriopterus a BROOM MYRID
Corixa praeusta a WATER BOATMAN
Gerris lacustris COMMON PONDSKATER
Hydrometra stagnorum WATER-MEASURER
Myzus cerasi
Peizodorus lituratus
Plagiognathus arbustorum
Subsigara fossarum
S. scotti
Velia currens

COLEOPTERA Beetles

Agapus bipustulatus
A. chalconatus
A. guttatus
Brychius elevator
Cantharis cryptica SOLDIER
Coccinella 7-punctata 7-SPOT LADYBIRD
Hydroporus incognitus
H. lepidus
H. lineatus
H. obscurus
H. palustris
H. pictus

Colymbetes fuscus
Copelatus agilis
Deronectes depressus
D. duodecimpustulatus
Dytiscus semisulcatus
Gyrinus caspius
G. marinus
G. natator
G. urinator
Haliplus flavicollis
H. lineatocollis
Hydroporus cephalus
H. discretus
H. erythrocephalus
H. gyllenhalli

H. planus
H. pubescens
H. striola
H. tesselatus
H. umbrosus
Hygrotus inaequalis
Ilybius ater
I. fuliginosus
I. obscurus
Laccophilus minutus
Melolontha melolontha MAY BUG
Noterus capricornis
Orectochilus villosus
Rantus bistriatus

LEPIDOPTERA

BUTTERFLIES

Aglaia urticae SMALL TORTOISESHELL
Anthocharis cardamines ORANGE TIP
Aphantopus hyperanthus RINGLET†
Argynnis aglaia DARK-GREEN FRITILLARY
Celastrina icarus HOLLY BLUE
Coenonympha pamphila SMALL HEATH
Lycaena phlaeas SMALL COPPER
Maniola jurtina MEADOW BROWN

Nymphalis io PEACOCK
Pararge megera WALL BROWN
Pieris brassicae CABBAGE WHITE
P. napi GREEN-VEINED WHITE
P. rapae SMALL WHITE
Polyommatus icarus COMMON BLUE
Vanessa atalanta RED ADMIRAL
V. cardui PAINTED LADY

† Extinct

This is effectively the whole Manx list and it is extremely unlikely that other species occur other than as casuals or introductions. This caveat applies, in particular, to blues and fritillaries.

MOTHS

Abraxas grossulariata MAGPIE
Acleris hastiana
Acronicta rumicis
Adaina microdactyla
Agrochola lota
Agrostis segetum TURNIP
Allophyes oxyacanthae
Amphipoea lucens
Ancylis geminana
Anticlea derivata
Apocheima pilosaria
Argyresthia pruniella
Autographa gamma SEE Plusia
Bactra furfurana
Biston betularia peppered PEPPERED
Brotopha smilis
Cabera exanthemata
Caloptilia stigmatella
Cerura vinula PUSS
Cidaria fulvata

Euphroctis similis YELLOW-TAIL
Eustrotia uncula
Haplodrina blanda
Hepialis sylvina ORANGE SWIFT
Hydriomena furcata ab. sorditata
H. ruberata
Lomaspilis marginata
Macrothylacia rubi FOX
Malacosoma neustria LACKEY
Mesoleuca albicillata BEAUTIFUL CARPET
Neofaculta ericetella
Noctua pronuba LARGE YELLOW UNDERWING
Nonagria typhae BULRUSH WAINSCOT
Ochropleura plecta FLAME SHOULDER
Oligia fasciuncula
O. latrunculus
O. strigilis
Opisthograptis luteolata BRIMSTONE
Opsibotis fuscalis
Orthosia stabilis COMMON QUAKER

Cilix glaucata CHINESE CHARACTER
Clepsis consimilana
Coleophora albicosta
C. albidella
C. alticolella
C. glaucicolella
C. muripennella
Drepana falcataria PEBBLE HOOK-TIP
Ectropis bistorta
Elachista argentella
E. subalbidella
Eligmodonta ziczac PEBBLE PROMINENT
Ennomos alniaria CANARY-SHOULDERED THORN
Eptinotia nisella
E. ramella
E. subocellana
Erannis defoliaria MOTTLED UMBER
Eulithis pyraliata
E. testata
Euphyia bilineata YELLOW SHELL
Eupithecia pygmmaeata MARSH PUG
E. subumbrata
Ourapteryx sambucaria SWALLOWTAIL
Perizoma albulata
Phalera bucephala BUFF TIP
Phigalia pedaria
P. pilosaria PALE BRINDLED BEAUTY
Phlogophora meticulosa ANGLE SHADES
Phyllonorycter viminiella
Plusia gamma SILVER Y (immigrant)
Plutella porrectella
Rheumaptera hastata
Rivula sericealis
Smerinthus ocellata EYED HAWK
Spilosoma lutea BUFF TIP
Stenoptilia bipunctidactyla GREY WOOD PLUME
Sterrha lactata WAVE
Teleiodes notatella
xanthia icteritis
X. togata
Xanthorhoe montanata CARPET
Xylena exsoleta SWORD GRASS
Zygaena trifolii 5-SPOT BURNET

PART ONE: THE PHYSICAL ENVIRONMENT

Conclusion

The physical environment of the Isle of Man should not be viewed as independent of the Island's cultural and economic development. Man's insular status, position near the centre of the northern Irish Sea and diverse landscape have had a strong influence on its inhabitants from the earliest times. Nor is the role of the many generations of Manx people in altering their physical environment to be underestimated.

The first Mesolithic hunter-gatherers must have arrived on a forested island with a rich supply of food ranging from the fruits and game of the forest to the fish, shellfish and sea birds of the shore. Although there are no local outcrops of flint, ample pebbles occur in the glacial drift and they would have been easily collected from beaches. Perhaps of all Man's inhabitants these were in greatest harmony with their environment and did little to alter it. Their successors, the Mesolithic users of heavy-blade flint tools, probably began the alteration of the Manx landscape by the use of fire to drive game.

The destruction of the native Manx wildwood must have accelerated markedly with the introduction of farming by the Mull Hill and Ronaldsway Neolithic cultures. Even before the first metal tools arrived the Manx people would have irreparably altered their physical environment. Most of the potential arable land that did not require elaborate drainage may already have been cleared of trees, whilst in the uplands the burning of woodland and scrub had probably initiated the development of peat. The diminution of the woodland continued during the Bronze Age and Iron Age, with wood being used for construction and to produce charcoal. When the first Norsemen arrived they would have viewed a landscape largely denuded of trees and very different from that encountered by the first Mesolithic seafarers.

For the Norsemen, however, the Isle of Man's physical environment had two great virtues; its position controlled the trade routes of the northern Irish Sea and it possessed a natural harbour at Peel. Peel became the capital of the Norse Kingdom and the influence of the Manx 'Vikings' is still writ large upon the landscape, in the pattern of sheading, parish and farm boundaries, as well as in the Manx culture and constitution.

After the demise of the Norse Kingdom the physical environment continued to influence historical development. In particular, the position of the Manx uplands, producing problems of transport and communication, accounts in large part for the historic division between the north and south of the Island. In pre-reformation times this was reflected in the dominance of religious and secular control. Even today administrative and cultural divisions persist.

Whereas until the nineteenth century Man's physical environment favoured a fishing and farming economy, with the industrial revolution the presence of extractable minerals made mining important. The scars can still be seen on the landscape, particularly around St Johns. The use of water power to pump one of the mines involved altering catchment hydrology and has bequeathed us the magnificent Lady Isabella wheel at Laxey. At this time the absence of fossil fuel emerged as an important element of Man's physical environment and has remained so. An extension of the Cumberland coalfield beneath the northern plain of the Island now seems unlikely and so relatively high fuel costs will probably remain a feature of the Manx economy.

In Victorian times, Man's proximity to the

industrial north of England combined with its insular status, beautiful scenery and beaches, encouraged perhaps the greatest alteration of the Manx landscape since Viking times: the invasion of tourism. Manx glens were bought and planted with flowers and trees (with the result that they became 'crammed with rampant lovers' (Herbert, 1909, p. 13). The transport network was extended, including the construction of the tourist railway from Laxey to the summit of Snaefell. The topography of the east coast necessitated the construction of the electric tramway from Douglas to Ramsey. Tourism has declined and Man can hardly hope to compete with the sunshine of the Mediterranean for the return of the mass market. However, the diverse beauty of the Manx landscape, combined with a unique history and culture, still produce a venue which has much to offer.

The finance sector is now booming and the Manx economy is continuing to improve. The Island is attracting ever more affluent immigrants and there will undoubtedly be a demand for more development of rural sites. It is to be hoped that whilst encouraging economic development, the Isle of Man Government will not lose sight of the fragility of their greatest asset, the physical environment.

PART TWO

HISTORY, ADMINISTRATION AND POPULATION

Introduction

Few authors have attempted rigorous definitions of the concept of place. Those who have, agree that the physical characteristics of the landscape are crucial. Part One of the book has therefore concentrated upon this facet of the personality of the Isle of Man and has shown how the unique appearance of present-day landscapes is the product of a variety of geological and geomorphological processes. Any definition of place has, however, to acknowledge that the physical landscape is important not only in itself but also as the crucible within which man's activities are forged. The conclusion to Part One has already hinted at the way in which man's activities on the Isle of Man have been shaped by the physical environment and its range of opportunities and limitations. Parts Three and Four look in more detail at those activities, the way in which they have changed and the way in which they have left an indelible imprint upon the landscape. These elements of place combine to form a simple working definition of the concept, summarised by Relph (1976) in *Place and Placelessness*:

> These three components of place . . . the static physical setting, the activities, and the meanings—constitute the three basic elements of the identity of places. (Relph, 1976, p. 47).

Equally though, Relph felt that there is an additional or fourth element to the idea of place. This, he suggested, is the spirit of a place, or its personality:

> Obviously the spirit of a place involves topography and appearance, economic functions and social activities, and particular significance deriving from past events and present situations—but it differs from the simple summation of these'. (Relph, 1976, p. 48).

Part Two of the book tries to address this intangible and transient element. It describes the shared past of the Islanders and the cultures that have, at varying times, dominated the Island. It provides an account of the changing constitutional position of the Island not only in relation to its distinctive internal arrangements, but also vis à vis its evolving relationship with the United Kingdom and supra-national bodies like the European Community. And it shows how the Island's population has been forced to respond to environmental limitations and more recently state intervention on demographic policy. In each case, the authors indicate how these broader considerations are either crystallised or lend meaning to particular localities on the Island. Chapter 8 itemises some of the more important archaeological sites and interprets their meaning. Chapter 10 describes the changing relationship between town and country, the shifting balance of power between Castletown and Douglas and the symbolism of Cronkbourne industrial village. At a different scale, Chapter 9 provides an account of the ceremony at St Johns, the importance of that place and the significance of its physical setting.

Clearly such an account cannot hope to capture the full richness of Manx identity. To do this, readers must visit and experience the Island first-hand, for it is only then that they might capture its essence.

CHAPTER 8
History
David Freke

Prehistory

The present appearance of the Isle of Man is the result of an interaction between Man and the environment which has been operating from the earliest period of human occupation on the Island. The peat in the early levels of Lough Cranstal contains charcoal which is considered to indicate that early settlers on the Island were burning the forest to enhance their hunting and gathering activities (Tooley, 1978). Even earlier evidence for the presence of groups of these Mesolithic (c.8,000–4,000 BC) peoples can also be found in the form of scatters of flint waste from tool making. Quantities of material have been recorded from around Peel, Port St Mary, and on the northern plain. One site, Cass ny Hawin near Ronaldsway, has produced hazel nuts and flint waste typical of the very tiny tools known as microliths which have their nearest parallels in southern Scottish and Cumbrian assemblages (Woodman, 1978). The only other *in situ* find of similar flint is at Peel Castle, where in 1982 Peter Gelling revealed a hollow containing over 2,000 waste flakes while excavating a post-mevieval gun battery (Smart, 1986). Even at this earliest stage of the Island's community there seems to be evidence of insularity, with a hollow-based form of flint point which is peculiar to Man (Woodman, 1978).

A later type of flintwork characterised by a heavier blade has been found at both coastal and inland locations, in particular Ginger Hall and other sites in the north and Glen Whyllin (Woodman, 1978) on the west. These later assemblages have close affinities with Ulster groups, perhaps indicating that the source of the influences on the Island was changing at the end of the Mesolithic period. There is some evidence for Neolithic innovations being brought into the Irish Sea area from the south and west, thence to spread into the north-west of Britain (Smith et al., 1981), and it is important to establish if there were flourishing contacts between Ireland and Man already in place to facilitate this diffusion.

The first farmers of the Neolithic are most obviously visible through their burial monuments on the Island, which link the Manx communities with the wider distribution of similar structures throughout western Europe (Henshall, 1978). Two early groups may be discerned, a distinction based upon the forms of their burial cairns and on their pottery styles. One has been called the Castal yn Ard group after the dramatic megalithic tomb of that name near Maughold, a three-phase structure with a semi-circular forecourt and five chambers at the west end, and a heap of fire-reddened stones at the east. The very large uprights around the forecourt, the chambers and the kerbstones of the two phases of the cairn can still be seen, although some of them were re-erected following the excavations in 1932–35 (Fleure and Nealy, 1936). The similar tomb called King Orry's Grave can be seen on the hill above Laxey. It consisted of an elaborate east end with a portal, forecourt and single chamber and 60 metres away to the west, now separated from it by a road, another forecourt and double chamber. The ends appear to have been

8.1 Meayll Circle overlooking Port Erin. The site consists of a circle of six Neolithic lintel graves. Each was about six feet long, partly buried, and was entered through a passage facing out of the circle

linked at a later date by a long cairn, now destroyed (Megaw, B, 1938). Two other simpler chambered tombs, even more damaged, can be found at St Johns—the Kew—and at Ballakelly, Santon. The pottery used by this group has very coarse fabric and large baggy shapes and has been found in the tombs and on probable settlement sites in many areas of the Island.

The second group of early farmers, called the Meayll Hill group, seem to have co-existed with the Cashtal yn Ard people as their pottery, a much finer ware with a greater variety of shapes and decoration, is sometimes found on sites together with the coarser wares. The megalithic tombs of the Meayll group may be best represented by the very fine passage grave on Meayll Hill itself. The monument comprises six pairs of chambers, each pair entered through a passage at right angles to them to form a 'T' shape, the whole group arranged in a circle. Excavations undertaken in 1911 (Kermode and Herdman, 1914) and 1971 (Henshall, 1978) have shown that it was originally covered by a cairn, with possibly a primary cist burial in the centre. The form of this monument is unique, and can be proposed as an example of an insular modification of the widely distributed passage grave type. Another tomb which has produced Meayll type pottery was recently excavated prior to its destruction by quarrying at Ballaharra, St Johns (Cregeen, 1978). It is clear that the Isle of Man participated in the Neolithic megalith culture, whose grandest expression in north-west Europe is the Boyne Valley group in Ireland. In the Isle of Man the only surviving decoration comparable with the magisterial sequence in the Boyne Valley is a spiral cut on a stone at Ballaragh. Other parts of Britain are also in contact with the Island at this period and the importance of the western seaways is demonstrated by Manx affinities with the Cotswolds monuments (Henshall, 1978).

A later Neolithic (c.2200–1899 BC) hut site was discovered and excavated at Ronaldsway in the late 1930s (Bruce et al., 1947; Moffat, 1978). This consisted of a rectangular house with occupation debris, including characteristic forms of pottery, flint scrapers and stone axe heads. About half of the petrologically examined Ronaldsway-type axes were made of stone from nearby Ballapaddag, but other polished axes from the Island, possibly earlier in date, derive from Cumbria. The notable difference between Ronaldsway-type axes and other polished axes is the treatment of the butt end, which was deliberately roughened by well-defined pecking. It is reasonable to suggest that here is another insular development, especially as there are no such axes from elsewhere in Britain and there is no clear evidence for imports of other axes into the Island at this period (Garrad, 1978a). Another diagnostic artifact from Ronaldsway is the 'hump backed scraper', a variant of the common Neolithic form. The two basic types of Ronaldsway pottery are either made of a very coarse fabric or of finer material, but both types can have either a bevelled or a thickened rim form and all are more or less straight sided and round bottomed. This pottery too appears to be an insular development.

Pottery of Ronaldsway type interpreted as occupation debris has recently been recovered from two sites, one in the north and one in the south, both above the 200 metre contour line (Garrad, pers. comm.), which suggests that upland areas were being exploited by the late Neolithic. A recent summary (Smith et al., 1981) concludes that a wetter, cooler climate than that experienced since the retreat of the ice began to affect north-west Britain in the early Neolithic. This, together with the interference of human communities on fragile upland ecosystems may have initiated upland peat formation (Moore, 1975). Northern Irish sites suggest a delay between Neolithic activity and the onset of peat formation. In the Isle of Man the work of Russell on the peats on the southern slopes of Snaefell gave a carbon 14 date of 2865±45 BP for their origin, well into the Bronze Age (Russell, 1978, p. 48).

The viability of the dates from the onset of peat formation around the Irish Sea suggest that human interference was a decisive factor in the pace at which this waterlogging developed, and the Manx evidence, such as it is, indicates that the uplands were not exploited for browsing until very late in the Neolithic with the consequent waterlogging occurring later still in the Bronze Age. This accords with an assessment of the archaeological evidence for increasing arable at the expense of pasture in Britain as a whole in the late Neolithic to early Bronze Age period (Bradley, 1978, p. 52).

After an initial 'Beaker' phase, represented on the Isle of Man by very few artifacts, the Bronze Age in the Island, as elsewhere in Britain, seems to fall into early and late

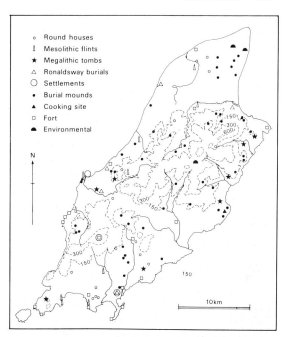

Fig. 8.1 Selected prehistoric sites. Some burial mounds and round houses may be Viking or early Christian in origin (see text)

periods, differentiated by changes in burial ritual, artifacts, social organisation and climate. The pre-war notions of a 'Beaker Folk' migrating into Britain bringing a knowledge of metal working and new funerary rituals is now generally rejected, the beaker and its associated objects now being seen as the visible trappings of a rapidly transmitted cult (Longworth and Cherry, 1986, p. 41).

In the early Bronze Age the funerary rite changed from the use of communal cemeteries to cairns and earth mounds—barrows— covering primary central burials, although secondary interments then often took place, sometimes with an enlargement of the mound. The primary burials were sometimes placed in stone cists, visible at St Johns in the road-side adjacent to the Tynwald mound (itself a probable burial mound) and at Balladoole inside the circuit of the later defences (Bruce, 1974). There are many mounds and cairns on the Island which may be Bronze Age burial structures, often in prominent locations on the crests of hills or prominences where, like their Neolithic counterparts, they were probably visible from the settlements, but others may be in lowland locations. Fine upland examples are to be seen on either side of the main road south west of Dalby Mountains, and there are several lowland mounds in the areas around Bride and Andreas in the northern plain.

The early Bronze Age from c.1800–1000 BC was warmer and drier than today, with estimates of temperatures varying from 2–3°C to 1–1.5°C higher (Tinsley, 1981, p. 211), and elswhere in Britain this period is associated with considerable evidence for the extension of upland farming. The stone field boundaries which covered much of Dartmoor, for instance, appear to be the result of a single planned act acround 1300 BC. The only evidence for similar land use in the Isle of Man is the existence of upland barrows from which we may infer associated upland settlements. Open air cooking sites have been excavated on good farm land (Cubbon, 1964 and 1972–74) but they remain the only physical evidence for settlement in the early Bronze Age so far identified on the Isle of Man.

Like the funerary monuments the Bronze Age metalwork from the Island exhibits no insular traits (Davey, 1978). Bowen (1972) considered that the early material shows close affinities with Ulster assemblages, and he suggested that the Island's early Bronze Age trade and contacts generally were orientated to Ireland and Galloway, while Davey stressed the nearly synchronous introduction of new late Bronze Age technology into southern England and the Isle of Man. It is probable that there was bronze production on the Island using the mineral sources at Bradda and Langness by the later period, and two late Bronze Age founders' hoards have been discovered (Davey, 1978, p. 230).

In contrast to the earlier periods the Isle of Man appears to be more integrated with its neighbours in the Bronze age, there being no insular forms of field monument or artifact, which may suggest that here, as elsewhere in Britain, there was increasing pressure on the community. Corroborative evidence for this suggestion seems to be offered by the 70 or so huts in the major hillfort on South Barrule (Gelling, P, 1970a), a striking example of the need for communal defence in the late Bronze Age. Other British hillforts begin to appear at about the same time and they are interpreted as evidence of increasing tension. This may be linked with late Bronze Age climatic deterioration and movement of peoples which combined to put pressure on resources, and it is at just such times that the Isle of Man's strategic location would become an element in local Irish Sea power struggles.

The Iron Age (c.500 BC–AD c.500 in the Isle of Man, there being no Roman invasion) saw the construction of coastal promontory forts at Close ny Chollagh (Gelling, P, 1958), Cronk ny Merriu (Gelling, P, 1952), Cass ny Hawin (Gelling, P, 1957), Langness and up to 18 more around the Manx coast. Together with the inland fortified homesteads at Ballanicholas (Gelling, P, 1966) and Castleward and the defended hilltop sites at Cronk Sumark, Maughold Head, Burroo Ned, Chapel Hill and the later defences of South Barrule, the Iron Age presents a picture of both communal and individual defence. The recent excavations at St Patrick's Isle (Freke, 1985a, 1985b, 1987, forthcoming) have unearthed a pre-Christian village, and its inconvenient location must have been chosen for security reasons. One large hut was a granary, and a large grain store on a secure site suggests an accumulation of staple provisions for financial and political control (Gent, 1983). The village on the islet may have been a chieftain's stronghold.

Three very large round huts were excavated by Dr Gerhard Bersu (Bersu, 1977) in the 1940s at Ballacagan and Ballanorris. They were up to 30 metres in diameter with concentric circles of load-bearing posts, and are considered to be a local variant of the Scottish crannog (Gelling, P, 1978). These structures seem to have their origins back in the third century BC and the finds suggest a continued occupation into the first century AD and possibly even longer (Gelling, P, 1978). Those investigated by Dr Bersu were all in low-lying marshy areas, perhaps chosen for their inherent defensive qualities, but there are other round houses in higher, better drained locations, such as Cashtal Lajer, Manannan's Chair and the Braaid, although none of these has proven by excavation to be Iron Age in date. There seems to be continuity in Manx cultural life through from the early Iron Age to what elsewhere in Britain would be the Roman period, and the results of excavations by Gelling at Port y Candas and Kiondroghad on round houses in marshy situations very similar to those excavated by Bersu, suggest that the tradition of the round house persisted even into the late first millenium AD on the Isle of Man.

The predominance of defended or inaccessible locations for the known settlement sites is clear, and argues for an unsettled social situation, like that in Britain as a whole. The cross ridge dyke above Sulby reservoir may mark out an Iron Age territorial boundary. There is not the evidence for immigration in the centuries BC, however, as there is in the south of England (Cunliffe, 1974) and as Bowen argues for the Scottish Isles (Brown, 1972, p. 68).

The late Iron Age and early Christian period

Sir Cyril Fox pointed out that British contact with continental Europe has been channelled either across the Straits of Dover and the North Sea or through the western seaways, and which of these is favoured at any particular period depends upon the stability of the relationship between Europe and Britain. In times of disturbance and movement of peoples, the western seaways become active, and in times of relative peace the eastern routes revive and the importance of the western decline. The later Roman period was just such a settled time and consequently there is little evidence of activity and change in Man. There is, however, the possibility that tribal movements in advance of the Roman invasion established a loose tribal hegemony in the Irish Sea region, with similar tribal names appearing in Ireland, Wales and Northern England. This evidence is very ambiguous, however, and may be the result of quite unconnected tribal names or scribal errors.

In the early years of the Roman advance the Irish Sea zone was an important strategic arena and Agricola in AD81–82 considered invading Ireland, carrying out intelligence gathering operations on ports and the number of troops required (Breeze, 1982, p. 46). In the event he took his fleet and armies to south west Scotland, possibly from their winter quarters in Chester, and left Ireland, and Man, undisturbed. The presence of a Roman fleet around the shores of Scotland required by Agricola's activities, though, would have extended the pax Romana to Manx shores at the end of the first century AD. The promantary fort at Close ny Chollagh, and the marshy locations occupied by Ballacagen, Ballanorris and Ballanicholas, all defended or defendable sites, were abandoned and there is no evidence for new defences to replace them.

Continuing unrest in Scotland in the second century, leading to the Roman withdrawal to Hadrian's Wall, may have encouraged movements around the Irish Sea. The western forts of Hadrian's Wall and Moresby, St Bees, and Ravenglass forts along the Cumbrian coast were all strengthened during the second century presumably to counter attacks from Ireland and Galloway. The introduction of duns (small defended sites) and souterrains (defended sites with underground chambers and passages) into southern Scotland argues for some movement from the west. The Isle of Man must have been involved in these movements but little archaeological evidence remains except the general similarity of the excavated structures and artifacts to those found in contemporary Ireland, Scotland and Wales. Part of the

Ronaldsway early Christian site has been interpreted as a souterrain (Laing and Laing, 1984-87).

Another line of contact is to be found in the myths and legends of the pre-Christian era in which the Island is linked to Wales, Scotland and Ireland through heroes such as Mannan, a personification of the sea, Conchobar, a King of Ulster, Culainn, a Manx smith, and the war chief Finn. These stories are not to be accepted as historical narrative, but their general premise assumes a Celtic Irish Sea cultural region, although Alcock puts their origin back to the first Celtic migrations of the Iron Age and throws doubts on the unity of the post-Roman tribes around the Irish Sea (Alcock, 1970, p. 55).

The pre-Patrician Christianity of the Celtic west outside the boundaries of the Roman Empire had developed through missionary contact from the eastern Mediterranean desert Fathers. Their disciples brought an ascetic Christianity based on isolation and meditation in contrast to the local community-based church of the Roman occupation which was re-introduced into southern Britain in AD597 by St Augustine. The collapse of Roman Britain in the early fifth century had the effect of stimulating the sea routes to the Celtic west which encouraged contact with southern Gaul and the hermetical tradition of Christianity there. This tendency continued in the sixth and seventh centuries despite the earlier efforts of both St Ninian at Candida Casa (Whithorn in Galloway) and St Patrick in Ireland who were operating in the region as agents of the Roman Church to correct the wayward doctrines of the ex-Roman *foederati* (native tribal mercenaries). It took several centuries for the western Celtic church and the Roman church to come into line. The south of Ireland had already adopted the Roman rites in the early seventh century, but despite the ruling of the Synod of Whitby in 664 the northern Irish, and by implication the Manx, retained the increasingly untenable Celtic forms until 716.

The Manx evidence for this Early Christian period is principally in the unsurpassed collection of carved stone crosses on the Island, and some Saints' dedications. The two strands of evidence come together at Maughold, named after a minor Celtic Saint Machaoi, who founded the monastery of Nendrum on Maghee Island in Stranford Loch and whose other dedications are to be found in Galloway, Kintyre, Ulster and the Clyde estuary (Bowen, 1972, p. 87). Maughold seems to have been the principal pre-Norse religious community on Man, and the Parish has produced the largest collection of early Christian cross slabs in the Island, many of them from the churchyard which is believed to mark the extent of the monastic precinct. Some crosses from this period on the Island bear inscriptions in the Ogham alphabet which is to be found also in southern Ireland, west Wales and Cornwall. Ogham is a version of Latin letters adapted to simple stone and wood carving techniques used to write in the old Irish language. It seems to have appeared by the fifth century and continued into the seventh century, and is sometimes accompanied by an inscription in Roman letters in Latin. Some of the Manx stones show the links Man had with its neighbours. A stone from Ballaqueeney, Rushen, now in the Manx Museum, is dedicated in Ogham to Bivaidu, a member of a tribe living near the River Boyne in Ireland. Two Latin inscriptions on early crosses at Maughold include one to Branhui who is credited with bringing water to the monastery, and another to Guriat, a north Welsh king. Other fine early crosses can be seen at Kirk Lonan and Onchan. The Calf of Man has produced a crucifixion scene with Irish iconography expertly executed in slate which is interpreted as an altar frontal from a lost monastic community on the Calf.

The Isle of Man has over two hundred keeill (chapel) sites, many of which are likely to be on the sites of pre-Norse chapels. The stone or sod ruins which can now be seen in the Manx landscape are likely to be later Norse structures, and several have early Christian crosses built into their walls. But many of them may be on the sites of the original keeills, to judge by the number of early, or at least simple, cross slabs found in or near them and the early Celtic saints after whom almost all of them are named (Megaw, B, 1978, p. 297). It has been suggested that there may once have been a keeill on every treen (an early administrative land division comprising between 200 and 400 acres), but careful analysis throws doubt on this (Bruce, 1968, p. 74). There can be little doubt though that most of the keills served the local farm community, and are located on farm land.

Some sites are more elaborate, surrounded by precinct walls with a tiny circular cell just outside for the incumbent. The locations of some of these 'hermitages' have clearly been chosen for their spectacular topographical characteristics like Lag ny Keeilly, half way down a precipitous slope into the sea on the western coast, and Spooyt Vane sited above a dramatic waterfall near Kirk Michael. These locations confirm the special value attributed by the Celts of the west to natural phenomena, in addition to the austerity learnt from the desert Fathers. It is demonstrated also by the veneration for wells and springs (chibbyr in Manx), which are frequently associated with holy sites. This outlook may be seen as a survival of the pantheism of the pre-Christian era, and forms part of the unique contribution of the celtic west to the fabric of Christianity.

Associated with many early church sites are cemeteries with characteristic stone lined graves called 'lintel graves' because of their large capping stones. These can certainly date from before the Norse period as indicated by the lintel graves disturbed by the Viking boat burial at Balladoole, but the stone-lined burial technique seems to have continued in use occasionally until quite recent times. Also the label 'lintel grave cemetery' may be misleading in that not all graves in such a burial ground need be stone-lined. Because the acid soils of most of the Isle of Man tend to destroy all traces of ancient corpses, including the bones, detecting wooden coffin burials, or interments with no coffin at all, requires the use of modern archaeological excavation techniques, and many early burial grounds have been uncovered accidently by the plough or by other modern earth-moving activities, where only the very obvious stone slabs of the lintel graves are likely to be noted. It is possible that some zones in an early cemetery or some periods of use may have involved lintel graves exclusively as at Glentraugh (Garrad, 1978b), but the recent excavations at Peel Castle have revealed the complexities which might be expected in a cemetery with a long period of use serving a mixed local population.

The population at this time seems to be shorter in stature, but the available evidence is so limited that until more well preserved skeletal material is forthcoming this can only be a provisional suggestion (Garrad, 1978b, p. 248). The people lived by farming and fishing and inhabited round houses not much different in form from those of their pre-Christian forebears (Gelling, P, 1977). An example can be seen at Port y Candas near Ballacraine where an excavation by Gelling revealed a circular house on a mound in a boggy area very similar to the situation of the large Iron Age houses at Ballacagen. The round structure at the Braaid was thought by the excavators (Fleure and Dunlop, 1942; Gelling, P, 1964) to represent a pre-Norse house incorporated into the later Norse farm there, but the site remains very enigmatic. The only village site of this period is at Ronaldsway, now under the airport. In the late 1930s lintel graves and round huts were excavated underlying later Medieval metal-working debris. The site produced cross slabs and a beam balance of pre-Norse date, and the concentration of prehistoric, early Christian, Viking and Medieval occupation in the area of Ronaldsway clearly demonstrates the importance of the site, presumably because of the proximity of a sheltered anchorage and a rich hinterland (Megaw, B, 1938–40; Neely, 1940; Laing and Laing, 1984–87).

The division of the farmland into treens and

quarterlands is generally acknowledged to date to this period also. An analysis of keeill sites in the parish of Kirk Michael indicates that there tends to be one per treen and that they are peripherally sited in relation to quarterland boundaries. Enough to establish that this relationship is beyond doubt pre-Norse in date. The excavation and survey at Keeill Vael, Druidale, in 1979–80 revealed evidence for upland exploitation in the pre-Norse period, with a small keeill built into an earlier circular structure, interpreted by the excavator as a hut (Morris, 1981; 1983). This tends to corroborate suggestions by a number of workers that there was a system of upland migration in the pre-Norse period which utilised marginal land between the 160 and 300 metre contours (Megaw, E, 1978). They can be identified by the place name element 'Eary' which means 'summer pasture'. They are at the margin of farmland but within the enclosed land and many are still farmed today. This group of settlements seems to be distinct from the highest settlements, the 'sheilings', which occur typically above the 300 metre contour on unenclosed open moorland and which may be Norse in date (Gelling, P, 1962–63).

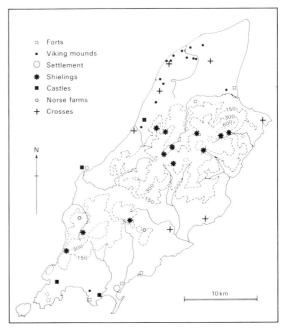

Fig. 8.2 Selected early Christian, Norse and Medieval sites

The Norse raids and settlement

At the end of the eighth century the Norse adventurers reached the Irish Sea. Irish annals record the sacking of Inis Patraic in Dublin Bay in 798, the first mention of Scandinavians ('Lochlannaibh') although Iona had been attacked in 795 and an entry in 794 records the devastation of 'all the islands of Britain by Gentiles'. Various factors lay behind the Scandinavian expansion, including a system of inheritance which whittled down the amount of land available to each son, the political unification of Norway under King Harald Findhair which encouraged unruly chieftains to look further afield to achieve the cultural ideal of personal heroism and daring, the market forces which led to international trade in Scandinavian furs, amber and other highly valued commodities, with the consequent development of their highly efficient ships.

The initial half-century of Scandinavian presence in the Irish Sea was essentially Viking, a word which should retain its connotations as a 'job description' for a pirate or raider. The raids on Scotland, Ireland and England were carried out for the purposes of gaining wealth in the most efficient way available, and the frequent recorded attacks on rich monasteries and churches should be seen as being 'pro-loot' not 'anti-Christian' in motivation. Although there are no records of attacks on the Isle of Man in the early eighth century, we can reasonably assume that the monastery at Maughold (the scene of a later Norse raid) was ransacked and the coastal communities of the Island harassed in the same way as adjacent Irish, Welsh and Scottish areas. The sandy beaches around the north of the Island with the rich agricultural hinterland of the northern plain would have been particularly vulnerable to the raiders' depredations. It is also very likely that the Vikings appreciated the strategic location of the Isle of Man in the middle of the north Irish Sea, astride the sea routes which we have seen were so important for trade and politics in the region. The sheltered harbours around the Manx coasts would have proved invaluable to them as bases from which to raid their neighbours.

The transition from raiding to settlement in the Isle of Man is as obscure as the impact of the raids themselves. From elsewhere we know of the Scandinavian practice of 'nesstaking' (building a dyke across a headland) as a preliminary to full invasion, and there is a possible candidate for such a dyke on Langness. Most Scandinavian evidence from the Island, however, is firmly related to the settlement itself, through place names, personal names, political institutions, the settlement sites and finds of artifacts in graves and elsewhere. The analogy with Ireland and Scotland suggests that settlement was established by 830 and that a favoured approach was the dynastic marriage. By the middle of the ninth century there was an identifiable group of mixed blood warriors in Ireland called the *Gall-Gaedhil*, or 'foreign Gael', and a similar situation can be envisaged for Man. Later there certainly are mixed marriages as evidenced by Norse and Gaelic names on cross slabs of the tenth century, but the most advantageous periods for such a strategy would be during the initial settlement when marrying local heiresses would greatly facilitate the peaceful acquisition of land and wealth.

The Norse incomers were pagans and the Isle of Man has an impressive series of burial mounds which are generally ascribed to the first settlers. There are over twenty mounds on coastal ridges or headlands which may be Norse period burials and the brilliant excavations by Bersu in the 1940s, together with earlier work by Kermode and Bruce, have resulted in a reasonably clear picture of this facet of the period (Bersu and Wilson, 1966; Kermode, 1930; Bruce and Cubbon, 1930). Three boat burials have been identified; at Balladoole, Knock y Doonee and Cronk yn Howe, and two other mounds at Ballateare and Cronk Mooar produced coffin burials. The rich array of goods discovered in these graves include swords, spears, axes and the shields as well as horse harness and in one case blacksmith's tools. An elaborate burial ritual is hinted at by quantities of cremated animal bones and a human sacrifice at Ballateare and the possibility of human sacrifice at Balladoole. These mounds with their weaponry and farm gear have been interpreted as establishing family claims to newly acquired land, although recent attempts to demonstrate a statistically valid relationship between property divisions and mounds in Jurby have proved inconclusive (Fletcher and Reilly, 1988).

At Balladoole the positioning of the Viking boat had actually displaced earlier lintel graves while at Cronk yn Howe a keeill was built over the mound. This continuity of ritual use can be seen again at Jurby churchyard, where a probable Norse burial mound is sited within the faintly discernible early cemetery boundary bank. Pieces of Scandinavian gear have been recovered over the years from a number of early Christian cemetery sites and the recent Peel Castle excavations have a burial ground in use from pre-Norse to late Medieval times apparently without a break. Included amongst the hundreds of interments were eight burials with grave goods, although only in two did these amount to more than dress ornaments or fasteners.

One of the exceptions was the richest accompanied woman's burial of the Norse period yet discovered outside the Scandinavian homelands (Holgate, 1987). A middle-aged woman who suffered from a vitamin D deficiency had been buried in an elaborate lintel grave with three knives, a small set of shears, an antler comb, a work box or pouch with needles in it, a cooking spit, a charm pendant and a necklace of over seventy glass, amber and jet beads. She was clearly a woman

8.2 A view of St Patrick's Isle at Peel with the man-made causeway to the left and the life-boat station to the right. The size of the curtain wall is obvious from this shot, as are the prominent siege-tower and cathedral

of some status in the community, but she lacked the oval brooches which would indicate that she was wearing Scandinavian dress. She may have been a native of Man or one of the other Gaelic countries around the Irish Sea, an example of the Norse policy of marrying heiresses, or she may have been a Scandinavian who adopted celtic dress. Her grave, although elaborate, was recognisably a lintel grave and it was aligned east-west in conformity with those around it. The accumulated evidence for Peel, Balladoole, Jurby and elsewhere suggests a high degree of tolerance between the Gaelic Christian and the Norse pagan communities, at least by the time of the settlement.

The ninth century was a period of raiding and settlement which gave the Scandinavians control of the sea and which established Scandinavian communities in the surrounding lands and islands. The subsequent centuries saw the rulers and chieftains of these groups jockeying for power, and the Isle of Man's strategic location made it both a target and a stronghold. We learn from the *Orkneyinga Saga*, written several centuries later but probably incorporating much historical memory, that raiding parties would overwinter on the Isle of Man before sallying forth again in summer campaigns. The Norse defences of the Island are still very little understood, however. The three small refortified Iron Age promontary forts at Cass ny Hawin, Cronk ny Merriu and Close ny Chollagh have been shown by excavation to contain Norse-style long houses (Gelling, P, 1957; 1952), and the watch site at Vowlan near Ramsey was occupied in Norse times (Bersu, 1949). These can have been no more than coastal lookout stations, perhaps under central control, but they underline the importance of the Irish Sea as a strategic factor in local politics and trade.

The control of the sea also lies behind the fortification of St Patrick's Isle on the west coast. Recent excavations inside the late Medieval castle defences have revealed a massive stone dump rampart on the eastern edge of the islet, the only side not naturally protected by cliffs, and although precise dates are still awaited, stratigraphically the rampart falls between the early Christian cemetery and the later Medieval defences (Freke, 1989). It is tempting to see this as the Isle of Man's principal Norse stronghold, defending the sheltered harbour of the River Neb and the central valley route between the west and east coasts of the Isle of Man. Later Norse kings of Man were living on St Patrick's Isle and the *Chronicles of the Kings of Man* frequently record voyages ending at Peel and meetings being held there. Magnus Barefoot landed at St Patrick's Isle in 1098, and the Norse kings Godred and Olaf died on the islet in 1187 and 1237 respectively.

The Iron Age hill fort at Cronk Sumark may have been refortified as there is a striking similarity between its layout and the re-used sites of Dunadd and Dundurn in Scotland and Dinas Emrys in Wales. An obvious function for such a site at Cronk Sumark would be the watch and defence of the fertile northern plain and its sea approaches. Such inland defended sites seem to have been administrative in function, rather than concerned with personal security, as the known Norse farm sites are undefended. The lowland farm on rich agricultural land at the Braaid is undefended as is the contrasting marginal settlement at Doarlish Cashen excavated by Gelling (Gelling, P, 1970b). The metal working on a low mound in a marsh in the seventh to eighth centuries appears to have been abandoned and used as a dumping site in the Norse period.

The pagan Norse seem to have been converted to Christianity by the late tenth century, in a process which was probably very gradual. One of the accompanied graves on St Patrick's Isle contained a coin of Edmund (c.950). The Isle of Man saw a flowering of Norse carving on cross slabs of the tenth and eleventh centuries, some of which show an appreciation of both Christian and pagan motifs, like Thorwald's cross at Andreas which depicts Odin and Fenris at Ragnarok on one side and a figure holding aloft a book and a cross while trampling a serpent on the other. On other crosses graphic episodes of the Sigurd myth are interwoven into the interlace, and dragons writhe around hunting scenes and runic inscriptions. The characteristic ring chain motif found on crosses throughout the Island are claimed by Gaut, son Bjorn, to be all his work, according to the runes on fine crosses at Kirk Michael and Andreas, although the claim to have made them 'and all in Man' may be exaggerated, and the ring chain motif itself is not unique to Man. The Isle of Man has at least twenty-six stones with runes cut on them, a very high number when compared with thirty-three from Norway itself and only seven from all the other possessions outside the Scandinavian homelands. A tentative explanation for this is given by Page (1983), who suggests that the interaction of the strong western celtic stone carving tradition and the Norse rune stone tradition together with the requirements of a church prepared to accept records in any form, combined to encourage runic memorials. The Isle of Man again seems to have benefited from a fruitful interaction of indigenous and alien cultures.

The Norse Kingdom of Man and the Isles

The history of the Isle of Man becomes more visible though not necessarily any less complex with the coming of Godred Crovan, the 'King Orree' of Manx legend, a local survivor of Harald Hadrada's army defeated at Stamford Bridge in 1066. In 1079 at the battle of Sky Hill near Ramsey he defeated the Manx at the third attempt and welded together an extensive island kingdom which encompassed the western Scottish isles. After his death on Islay in 1095 there was a period of civil war in Man which was to the advantage of Magnus Barefoot, King of Norway, who arrived in 1098 having apparently subdued the Orkneys and Galloway en route. He subsequently used the

Isle of Man as a base from which to harass the coasts of Wales and Ireland, coming to grief eventually in an ill-starred attempt to invade Ireland in 1103.

After a brief period of confusion the descendants of Godred Crovan ruled the Isle of Man until Magnus, the last Norse King of Man, died in 1265 and Norway gave the Island to Scotland. In the reign of Godred II (1153–87) the Manx kings lost control of the islands of Mull and Islay, retaining only the Lewis and Skye groups, and increasingly the Isle of Man found itself between warring factions representing Scottish, English and Norwegian interests. Norway finally bowed out of the contest with the death of the last Norse king, leaving Scotland and England to dispute the ownership of the Isle of Man for the next three-quarters of a century until 1333 when Edward III granted the Island to William de Montecute, first Earl of Salisbury.

It is clear that during the Norse period Scandinavian administration and cultural customs became established on the Isle of Man. The present Court of Tynwald is a descendant of the 'Thing' of the early Scandinavian settlers, the traditional Norse procedure for settling disputes and grievances. However, the extent to which inhabitants of the Isle of Man became assimilated into Norse culture has been the subject of considerable debate, turning on the interpretation of place-name evidence (Megaw, B, 1978; Gelling, M, 1978). The majority view now considers that a Gaelic-speaking proportion of the population persisted throughout the period of the Norse domination, and that the Norse themselves had become Gaelicised by the end of the thirteenth century. There is little doubt that Norse settlers were in positions of control, though the thirteen Gaelic names out of a total of forty-six cut in runes on Norse crosses show that Celtic origins did not automatically relegate people to a socially deprived sub-group, and the marginal Norse farm at Doarlish Cashen indicated that the settlers were not all in prime positions.

The Norse period population followed a farming economy which maximised the potential upland husbandry, and the climatic amelioration of the eleventh and twelfth centuries encouraged the increased exploitation of higher land through transhumance. Cattle would be driven to upland pastures—shielings—in the summer, the people living in small turf walled huts. More than fifty groups of mounds around the 300 metre contour have been located and mapped, principally by Peter Gelling (1962–63) whose excavations at the site above Block Eary on the north slopes of Snaefell produced a penny of Stephen dated to 1135–41. The large number of huts—up to thirty-seven—in some of these groups suggests a long period of use, and Peter Gelling's excavations recovered evidence for cheese-making, weaving and stock control. No date can be advanced for their abandonment except that post-Medieval commentators were unaware of the practice of transhumance and the meaning of 'Eary', so it is possible that the later Medieval climatic deterioration made shielings untenable and led to the consolidation of the earlier and lower 'Eary' pastures into regularly farmed land (Megaw, E, 1978).

Although the submission of the church in Ulster and Man to Roman discipline occurred in 716, the Viking incursions seem to have delayed the adoption of the full Roman parochial system as late as the twelfth century. The keeills continued in use and the monastery of St Maughold retained its pre-eminent position probably until St German's Cathedral on St Patrick's Isle was established in the late twelfth or early thirteenth century. Before this there were a succession of bishops starting with the Norseman Roolwer in the eleventh century. The site of their headquarters is unknown. King Olaf 1 (1113–53) invited the monks of Furness to the Island in 1134 and the monastery they founded was sited at Ballasalla, where substantial ruins of their Cistercian house can still be seen. The abbey became a focus for the consolidation of the Roman church with the monks granted the right to elect the bishop and given the responsibility of spreading Roman rites and administrative arrangements to the Isles.

The ultimate siting of the cathedral on St Patrick's Isle may have had a dual motivation in that it diverted power away from Maughold and also acknowledged the sacred character of the islet. St Patrick's Isle already had three churches and the celtic-style round tower and had been used since pre-Norse times as a burial ground. The recent excavations revealed the third church to add to the two still visible as ruins, and established the presence of the extensive cemetery, which continued in use with the building of the cathedral until the last century. The location of the original parish churches of both St Patrick and St German on the islet suggests that the organisation of the Isle of Man into parishes came late, and had to accommodate the already existing churches by the uncomfortable expedient of drawing the parish boundary across the islet between them.

Bishop Simon (1227–1247) is credited with building St German's Cathedral, although the present structure is obviously the accumulated result of several building episodes. It is still very unclear for what portion of the cathedral Bishop Simon was responsible, as there is evidence of a smaller building under the chancel, and the nave, now without its south aisle, has a different floor level from the crossing and the chancel. The crypt has been thoroughly remodelled, and the tower and transepts were battlemented probably in the late fourteenth century. The excavations have shown that there was domestic occupation of the area to the north of the cathedral from the twelfth century, replaced by a dormitory block and hall for lay priests or vicars chorale in the fifteenth century. The bishop himself built a palace near Kirk Michael at Bishopscourt. The present tower house dates from the fourteenth century and has been substantially added to in later periods.

The ceding of the Isle of Man to the Scots by Norway in 1266 ushered in almost a century of dispute between Scotland and England in which the Manx took the brunt of the warring factions. Their own claims foundered at the battle of Ronaldsway in 1275 when the Manx army under Godred son of King Magnus was defeated by the Scots. Scottish dynastic weakness at the end of the thirteenth century allowed Edward I of England to take possession of the island and it became a royal gift first to Baillol, then Anthony Beck the Bishop of Durham, then to Piers Gaveston the favourite of Edward II, before Robert Bruce intervened to restate Scottish claims in 1313. His army landed at Ramsey and proceeded south to besiege Castle Rushen in Castletown, which surrendered after a month. The castle was reputedly virtually destroyed, and Bruce granted the Isle of Man to Thomas Randolf, Earl of Moray. The lack of strong government left the island open to opportunistic pillage, as an attack by Irish freebooters in 1316 showed. Scots and English claims were pursued through attacks on each other's forces on the island in 1329 and 1388, and trouble can be inferred from truces agreed in 1318, 1328, 1333, and 'protection money' paid to the Scots in 1343. The island's distress was no doubt an encouragement to the French who in 1377 also exacted money in order to prevent them burning the Manxmen's houses.

During this period a succession of Scots and English Lords were granted the Island by their respective monarchs. Sir William de Montecute was granted the Island in absolute possession in 1333, and it was inherited by his son, the second Earl of Salisbury, in 1344. In 1392, exercising his absolute right of ownership, Sir William Montecute II sold the Island and its crown to a supporter of Richard II, Sir William le Scorpe, who on the accession of Henry IV in 1399 was beheaded, leaving the Island in the hands of the English crown again. Henry granted the Island to Henry de Percy, Earl of Northumberland, but after the Percys' rebellion in 1403 it was seized by the crown and in 1405 handed over to Sir John Stanley, on the service of rendering two falcons to Henry and to the future monarchs of England at their coronation in token of homage. A degree of stability then ensued for several centuries of almost unbroken Stanley rule.

The life of the islanders during this period of dispute was undoubtedly very hard, the depredations of outsiders being compounded by the rise of the power of the church in the Island, which although it also suffered, especially in 1316, steadily increased their taxes and tithes. In 1291 tithes were introduced on fish, and the products of merchants, smiths and other artificers, in 1334 visitation dues were introduced, and a punitive 'smoke tax' on hearths was imposed in 1302 as a result of Manx dissension over the questionable appropriation of the churches of St Michael and St Maughold by Furness Abbey. Within the church itself there was a struggle for control being fought between the monastic powers and the Pope, the Pope slowly establishing his ascendancy with the bishops to go to him for their consecration from 1348. Before this date the Bishop of Sodor and Man was consecrated by the Archbishop of Trondhjem in Norway, a legacy of the Norse kingdom. The bishop amassed a considerable proportion of the land, comprising ninety-nine farms and seventy-seven cottages, and he held his own courts, with his own prison in Peel Castle. Other lands in the Island were held by Furness Abbey, Rushen Abbey, and by abbeys and priories in Bangor, Sabhal, St Bees and Whithorn.

The work of the Cistercian monks of Rushen Abbey on their own holdings and their pressure on their tenants must have enhanced the productivity of the land, however, through their administrative efficiency and agricultural acumen. A number of factors worked against a thriving agriculture though,

8.3 Bishops Court, built by Bishop Simon in the 1230s, was the ecclesiastical centre of the Island for many years. It housed the church's own court and was even equipped with gallows. The present building, which largely dates from the eighteenth century, is now privately owned

8.4 A period view of Castle Rushen showing the seventeenth-century additions which are even now inhabited

including political unrest, exploitation by landlords, worsening climatic conditions in the late middle ages and the system of land holding. The land was traditionally held by the king and the barons, and the people had no right of tenure, a circumstance which did not encourage agricultural improvements from the grassroots. The worst effects of such a system, however, were avoided by the gradual assumption of tenurial rights by the people in this period of political flux, but its reimposition later had unpleasant consequences.

Medieval farming seems to have been frequently disrupted by warfare and raiding, but some features of the Manx landscape were well established. The mountain hedge or felldyke divided the open moorland pasture from the fields, and although it is not mentioned in statutes until as late as 1422, the description of the bounds of Rushen Abbey, now dated to the late thirteenth or early fourteenth century (Megaw, E, 1978), mentions 'an old wall known to the locals along the sloping sides of the mountains' above Laxey (Broderick, 1978, f54v). Farm boundaries and field boundaries, if later disputes are any guide, were much less well defined, often consisting of 'field marks' rather than physical barriers. Fences or walls seem to function more as stock control devices than property divisions until the post-Medieval period. There is still a debate about the date at which the severe lack of trees began to cause shortages and problems on the Island. Magnus Barefoot in 1098 constrained the men of Galloway to provide wood for his Manx forts, but against this evidence is the account of Godren Crovan hiding three hundred men in the woods on Scacafell in 1079.

In the thirteenth century the centre of power moved from Peel Castle to Castle Rushen, reflecting a different defensive posture. Peel is essentially a defence against sea-borne attack, its isolation is a measure of protection for the garrison against the sea raid, but it is vulnerable itself to attack from the heights of Peel Hill. It relies upon a friendly populace. Its defences have recently been the subject of detailed archaeological investigation which have shown that a norse stone dump rampart on the low eastern edge of the islet was heightened with a turf bank in the Medieval period. It is likely that both these ramparts had a timber stockade either along the top or as a revetment. The earliest use of the name 'Peel' (which originally meant 'stockade') is in a charter of 1399, by which time there were stone towers at Peel Castle, but Margaret Gelling suggests that it may have originally been applied in the late thirteenth century in its original sense (Gelling, M, 1989). The ramparts and stockades do not appear to have encircled the entire islet, as attempts to locate them on the north by Bersu in 1947 (Wright, 1980–82) and on the south by Freke in 1986 (Freke, 1987) led both to the conclusion that the present fifteenth century stone wall is the earliest circuit there. Understanding the defences is complicated by the fact that there has been some erosion of the islet in the vicinity of the east end of the Cathedral, which in the late Medieval and post-Medieval period led to some remodelling of the ramparts and walls in this area (Freke, 1983).

As outsiders wrestled for control of the Isle of Man a more feudal stronghold was required to defend them and their garrisons against an uncertain hinterland and roaming

8.5 The fourteenth-century Monk's Bridge over the Silver Burn adjacent to Rushen Abbey. It was a packhorse bridge on the old Monk's Road said to run from Rushen Abbey to Bishops Court. The Abbey was an outstation of Furness Abbey in Cumbria

8.6 Castle Rushen, taken from the main square in Castletown. The exterior curtain wall is in the foreground, the central keep with its very early clock is behind it. The castle was once the administrative and military centre of the Island. Subsequently it became a prison, and part of it now houses the Registry of Births, Deaths and Marriages

bands of marauders. Although Castletown does not have the advantage of such a good natural harbour as Peel, it being both rocky and exposed, it does command some of the best farm land on the Island and there is reasonable access to the shelter of Derbyhaven. The castle appears to have been commenced in the twelfth century and by the end of the Norse period had become the headquarters of the kings, the last Norse king, Magnus, dying there. In the account of Robert Bruce's attack on the Island in 1313 the Scots landed at Ramsey and apparently by-passed Peel Castle to besiege Castle Rushen, perhaps an indication of Peel's decline. By 1377, after the addition of the gatehouse tower and the east tower and the raising of the keep, Castle Rushen was strong enough to withstand the French, who resorted to threats against property. The curtain wall was added in the late fourteenth century and the keep raised again to enable the garrison to fire over it. The function of the motte and bailey castle at Cronk Mooar north of Port St Mary is unclear, but such small castles were a response to unsettled political conditions in feudal England, and they may be in the Isle of Man too.

The Earls of Derby and the Dukes of Atholl

The gift of the Island to Sir John Stanley in 1405 led to over three hundred years of relatively stable government with the Stanleys as Kings of Man until 1504 and then as Lords of Man until their line failed in 1736. Their rule was interrupted for sixteen years at the end of the sixteenth century between the fifth and sixth Earls of Derby and again for nine years in the seventeenth century during the upheavals of the Civil War. Sir John, the first Stanley ruler (1405–14), did not visit the Isle of Man, but his son Sir John II (1414–32) took a great interest in its affairs, and began to reform its administration, writing down and codifying its laws, curbing the power of the church barons and restoring the ancient constitution. In the next hundred years visits by the Stanleys were rare: governors effectively ruled on the Stanleys' behalf as religious changes in England eventually led to the dissolution of Rushen Abbey and the Nunnery just outside Douglas as well as the Friars at Ballabeg in Arbory in 1540. The fourth Earl of Derby (1572–93) visited the Island several times but not until 1627 and the accession of James, the seventh Earl, 'the Great Stanley' as the Manx called him, did a Stanley actually reside on the Island. He also took the unprecedented step of appointing a Manxman, Edward Christian, as deputy governor, an appointment which was used by Christian ultimately to advance

8.7 A period wood-cut of Castle Rushen at Castletown

Manx interests in grievances about tithes and land tenure.

At the outbreak of the Civil War in 1649 Edward Christian was instructed to raise a militia on behalf of the absent James Stanley, an army which he then incited to rebellion. Lord Derby returned and astutely out-manoeuvred Christian at a meeting he called at Peel Castle to hear the complaints of the people, afterwards arresting and imprisoning Christian as ring-leader. Although the Earl placated the people with reforms of rents and tithes, he exacerbated a dispute which was to dog Manx affairs for another half century. All land in Man was owned by the Lord or the crown and Manx farmers held their land as the Lord's tenants-at-will, but in time through custom this had become almost a system of fixed tenure. Attempts to re-establish the original rights of the Lord over the disposition of land began at the end of the sixteenth century, but it was the seventh Earl who brought the dispute to a head by enforcing a system of land leasing to replace the traditional 'straw tenure'. His introduction of a three-generation (twenty-one-year) leasehold for land effectively limited inheritance and ownership, to the dismay of farmers who had grown accustomed to regarding as permanent their access to land which technically they held from the Lord. In the late seventeenth century the resulting insecurity of tenure led many to abandon their farms to turn to the more lucrative practice of smuggling. The land tenure problem was resolved by the Act of Settlement in 1704, although the related disputes over the enclosure and rights pertaining to the common land of the mountains was not settled until 1866.

The seventh Earl rallied to the royalist cause

8.8 A beautiful period view along the earlier promenades of Douglas. It shows the characteristic horse trams and the face that urban expansion had yet to spread beyond the Villa Marina. The bay frontage is now completely built-up

in 1651 but the small force of Manxmen he took to Lancashire was overwhelmed at Wigan, the Earl escaping to join the king's forces, only to be defeated at Worcester, tried in Chester and executed at Bolton. On the Isle of Man the news of his capture encouraged William Christian—'Illiam Dhone'—a relative of Edward Christian, to rebel with the militia he had been entrusted to use against the parliamentarians. He took all the castles except for Peel and Rushen, and offered to surrender the Island to Colonel Duckenfield who was leading the force sent to capture the Isle of Man. On hearing the brutal news of her husband's death, the doughty Countess Charlotte eventually surrendered Castle Rushen but the rebels got no better deal from their new rulers in relation to their complaints, and ten years later, after the restoration of the monarchy in England and the Derbys in Ireland, William Christian was executed at Hango Hill for his part in the rebellion in an act of ill-concealed revenge by Charles Stanley, the eighth Earl.

The years of the later Stanleys are marked by the incumbency of Bishop Wilson (1698–1755), whose doomed attempt to re-establish the ascendancy of the church are redeemed by his undeniable identification with the lives of his flock. The tenth Earl of Derby (1702–1736) died intestate and the Island passed to the second Duke of Atholl, whose family ruled either as Lords from 1736 to 1765, or as governors from 1793 until 1830. The British Crown was keen to halt the illicit trade carried on in the Island and persuaded the third Duke of Atholl (1764–65) to sell the Island for £70,000. The fourth Duke succeeded to the title in 1774 and began to agitate for an increase in the sum paid by the crown for the rights of the Atholls in the Isle of Man. He was mollified by being made governor-general in 1793, and he built the newly restored Castle Mona in Douglas as his Manx residence. His initial popularity was eroded by his habit of nepotism and when in 1825 his nephew Bishop Murray levied an outrageous potato and turnip tithe there were riots. The Duke was eventually paid £417,000 by the British Government for his rights in 1828. The deal gave the revenues from the Isle of Man taxes to the British with no effective representation, which led to a period of neglect of capital projects and services on the Island.

CHAPTER 9

The Isle of Man constitution
Robert Quayle

In 1979 the Isle of Man celebrated the Millennium of Tynwald, its parliamentary assembly. The date was chosen in a somewhat arbitrary manner as there was no evidence, documentary or otherwise, to suggest that an event occurred in 979 AD that was readily identifiable with the festivities in the Island one thousand years later; what was beyond argument was the fact that the celebration marked a remarkable survival, not just of a form of consultative assembly but also of a nation whose identity had endured the passage of years, as well as the attentions of larger and more powerful neighbours.

To have done that was ample justification for a spectacular party and after the inevitable carping of the critics (the local term is 'Manx crabs') had been quelled, that was just what took place.

Origins

The Vikings seem to be best remembered for the attributes that identify present day football hooligans—marauding bands of gratuitously violent travellers intent on pursuing their particular anarchic form of entertainment, principally rape and pillage. Fortunately, thanks to the work of the modern Viking specialists, their more remarkable qualities have also now been recognised—their bravery, their incredible seamanship and their feats of exploration centuries ahead of better known successors such as Columbus. What still remains largely unknown, however, is their developed sense of government and the need to create a form of organisation, involving quite a sophisticated form of consultative process in the countries they colonised.

Most people attribute the birth of modern European, even world, democracy to the Parliament of Simon de Montfort and because of that, Westminster is affectionately described as the 'Mother of Parliaments'. Tynwald, and its sister assemblies throughout the Scandinavian world can lay claim to be the collective 'grandmother' of parliament or, since to the writer's knowledge no actual connection has been established between them and the first Westminster parliament, 'great aunts'. For it is in the 'Ting' or assembly of the people (which is at the root of the Althing, the Folkting, the Storting, the Logting and the Tynwald) that the common Viking influence is identified in the parliamentary assemblies of Iceland, Denmark, Norway, the Faroes and the Isle of Man. Some can trace a direct link to the original Viking assembly—Iceland can produce documentary evidence that its Althing first met in 932 AD—but even it has to concede to the Isle of Man that there was a very substantial break in continuity; the Isle of Man cannot produce such written evidence of Tynwald's pedigree, but there are grounds for believing that its origins may well pre-date the Althing by as much as 100 years, and it has continued with breaks that can be measured in, at the most, decades since that time.

The common philosophy at the root of these great institutions is that the people should know the law under which they are ruled, and in turn should be able to bring their

9.1 This building, found adjacent to Castle Rushen in Castletown, was the home of the House of Keys between 1710 and 1874. The House of Keys and Legislative Council now meet in Douglas

grievances to their ruler, and obtain justice. Thus the earliest accounts of Tynwald refer to such components: the public declaration of the law by the ruler, who would sit, surrounded by his officers, in some public exalted place; the people gathered in some natural amphitheatre to listen; and their right to approach the ruler in order to seek justice, which was meted out then and there. If one compares the geographical qualities of Thingvollir in Iceland, the meeting place of the Althing, and the present and probably most regularly used site of Tynwald (St John's) in the Isle of Man, at each the artificial small mound on which the ruler sat was but a small prop against the larger backdrop of hills behind the natural escarpment below, which made the chosen site a natural focal point.

Today's Tynwald Ceremony, held at St John's each year on 5 July, or Old Midsummer's Day, embodies all the principal components of those early Viking assemblies. Firstly the coming in full panoply of state of the ruler and his principal officers to the chosen place: in the earliest days presumably the Viking Chief would have arrived with his followers from Peel or some other place where his fleet was moored—today the Sovereign or her Lieutenant Governor arrives in more sedate style, but still with armed retainers in the form of a Ceremonial Guard of Honour. This is followed by a form of religious observance. Now it is an act of corporate Christian worship and dedication of the Island and its people. Then it would have been some pagan ritual involving Odin and his fellow gods of Viking mythology. Thereafter the Lord, the principal Officers of Church and State, and the members of the Legislature process from the place of worship (now the Royal Chapel of St John the Baptist) to Tynwald Hill, said to be made up from soil from all parts of the Island. The declaration or promulgation of the law then takes place from the Hill, although it is now only a summary of the previous year's new legislation, albeit still read both in English and Manx. The people then have a chance to present petitions for Redress of Grievances—with today's sophisticated judicature these have to fall within a limited and clearly defined range, but in bygone days the ruler's judgements were summarily executed, in some cases with considerable but then acceptable levels of brutality (i.e. the rolling of witches in barrels down the adjacent hill called Slieu Whallian). Once it was all over the participants hastened off to a great banquet. The Vikings certainly knew how to enjoy themselves, and even if their modern day successors retire to the more restrained environs of a local hotel, one can sense the ghosts of Viking revellers looking on with approval.

The transformation from Viking assembly to modern Tynwald Day Ceremony presided over in recent years by HM the Queen, and other members of the Royal Family, and attended by Heads of State such as the President of Iceland and a future Vice-President of the USA, reflects the history of the Island during the intervening period and the development of its unique identity.

9.2 Tynwald Hill at St Johns. Tynwald still meets here once a year, on 5 July, to promulgate the laws of the previous year. Members process from the church on the right, to the Hill, which is made of turf from each part of the Island. The Island's war memorial can also be seen, as can the very edge of the arboretum which was opened in 1979 to celebrate the Millennium.

Development

As is inevitable in a small island community surrounded by much larger and more powerful neighbours, the history of the Island is inextricably bound up with events outside its shores.

The Viking influence which merged with the local celtic culture during the tenth and eleventh centuries was relatively short-lived, the Island proving an inevitable pawn in the power struggles which took place between the kingdoms of Man and the Isles (Manae et Insulae) and Scotland. A number of bloody battles took place, fortunately not all within the Island, but eventually the Island kingdom faded away, its only legacy being the phrase 'Sodor and Man' perpetuated in the title of the Island's bishop: Sodor, of course, being the latter day derivation of 'Sudreys' or the southern isles of Western Scotland. The period of Viking control ended in 1266 (other than in the enduring survival of organisation and cultural heritage) with the ceding of the Island to Scotland. The Scottish influence, too, was short-lived, as Scotland was more preoccupied with its larger neighbour south of the Border and with a degree of inevitability, sovereignty over the Isle of Man passed in 1333 to Edward III, King of England. The Island seems to have been alternately a nuisance and a plaything for its masters; at times ignored, at other times passed as a gift to temporary favourites such as Piers Gaveston. But eventually, in 1406 the subsidiary sovereignty of the Island was granted to the Stanley family, successively Earls of Derby. Their interest in the Island and its inhabitants (apart from the extraction of revenue) depended on their activities elsewhere, but they ruled as 'Lords of Man' for a period of some three hundred years, often managing their Island

Kingdom through Governors or other trusted officers of their household. From time to time the Island got caught up in events going on in the wider world; during the English Civil War the Earl of Derby made the mistake of being on the losing side, and lost not only his head, but also his Island Kingdom, despite a spirited defence by his Countess against a Parliamentary force led by Colonel Fairfax. With the Restoration in 1660, the Stanley family's fortunes revived, and retribution was quickly meted out to one William Christian (or Illiam Dhone) who had seen the temporary discomfiture of the Stanleys as an opportunity for a resurgence in Manx Nationalism. He remains a potent symbol of Manx patriotism.

After the death of the 10th Earl, the kingdom passed through the female line to the 2nd Duke of Atholl. The tenure of the Atholl family was relatively short-lived as, despite their inevitably larger resources, relations between Scotland and England were not as settled as they might have been, and in 1765 the third Duke was subjected to the eighteenth-century equivalent of compulsory purchase, being obliged to surrender his kingdom to the English crown. The ostensible reason was the increasing irritation of the English Customs and Excise service with the activities of smugglers—not so much Manx smugglers who obtained their contraband in the Island—as those whose trade with the West Indies and further afield was prospering. However, the present Duke has suggested that the acquisition was inextricably bound up with the process of reducing the power of the Scottish nobles in the light of their questionable loyalty to the Crown during the Jacobean uprisings of 1715 and 1745.

Whatever the reason, with the passage of the Isle of Man Purchase Act in 1765 the Island entered into a black period of its history with the remittance of all revenues collected in the island to London, and virtually no investment in the Island, its infrastructure and its inhabitants. It could have been worse—there were moves to annex the Island and make it part of an English county such as Cumberland. These were fiercely and successfully resisted by representatives of the people drawn from the House of Keys and led by its Speaker Sir George Moore.

During the period of the Island's history in which it was prey to the activities of its larger neighbours, Tynwald had not fallen silent. Indeed there are a number of recorded meetings of Tynwald during which it carried out its legislative functions, although it probably met more frequently in its now defunct role as a judicial assembly. However, the period also witnessed the growth of the popular component, the House of Keys. The exact date of its birth is as unknown as the true origins of its name. Opinions differ as to whether this derives from 'Keise', the Manx word for 'chosen', or from 'Kaire-as-feed', the Manx for 'twenty-four'. Whatever the answer by 1765, indeed long before, twenty-four representatives of the people had met together periodically as spokesmen for the people and as a sounding board for the Lord or his representative. They had no executive authority—that was vested in the Lord, and his Council formed of the senior officers of church and state—but the indispensability of their support was recognised in the part they played in consenting to legislation desired by the Lord.

It is somewhat tortuous to describe them as representatives of the people, other than as self-appointed, since in those days the House was self perpetuating. On the death or resignation of a member the names of two candidates were put forward for appointment by the Lord himself—a process analagous to the appointment of Bishops in the United Kingdom today—and inevitably these men tended to be drawn from the larger, more powerful families on the Island. However tendentious their claim, they did hold themselves out as representing the resident population of the Island and were clearly accepted as such both by the people and the Lord and his officers.

The bleak period which followed the Revestment was made even more unpleasant for the Manx people by the singularly tactless, even foolish, appointment of a subsequent Duke of Atholl as Governor of the Island. Apart from building himself a very gracious mansion in Douglas (fortunately still preserved as an hotel), he made it his object to enhance the paltry sum his father had received for his interests in the Isle of Man by milking the revenue of the Island even more comprehensively than the Government of Westminster was doing. Eventually his excesses became too great to tolerate, even for Westminster, and after an investigation by a Commission of Enquiry he was offered a once and for all settlement of any outstanding claims he might have against the Island, and a single ticket back to Scotland. To avoid a repetition of a representative taking upon himself a role to which he was not entitled, the Duke's successors have never been more than Lieutenant Governors with appropriate constraints on their power and authority.

The modern period

In 1863, the Island received Mr Henry Brougham Loch, its newly appointed Lieutenant Governor, and it is his arrival which ushers in both the modern period of Manx history and the constitutional development which was inextricably bound up with it.

Loch had already distinguished himself in farther flung parts of the world. He had known the inside of a Chinese prison during the Boxer Rebellion and it was clear that he had enlightened and far-sighted views on how this small Island kingdom should develop. In his 19 years on the Island, before he moved on to govern the new state of Victoria in Australia and a peerage, he stamped his mark on the Isle of Man and earned a place in its history as the architect of a modern constitution.

Whether he was familiar with the adage 'No taxation without representation', or not, is something we will probably never know, but several things must have struck Loch on his arrival. The Island was run down and its infrastructure inadequate and neglected. All taxes raised on the Island were automatically remitted to the United Kingdom. The Branch of Tynwald which was ostensibly representational was anything but. And it had recently experienced a rebuff by having an arbitrary act of petulance in imprisoning a local newspaper editor for contempt, quashed by the English Privy Council. The solution to these problems was as pragmatic as it was far sighted. The Island would have the right to raise taxes to be retained in the Island for its own benefit, and in exchange, there would be elections to the House of Keys on the basis of the suffrage then obtaining across the water. So in 1866 the first full elections to the House of Keys were held, and of the 24 members returned, over half were former members of the House. This probably proved very little, save that the men who were given the vote detected an identity of interest with the men who had been in the House of Keys all along.

But the Island had embarked upon the path towards modern constitutional development and, despite attempts to restrain the process from time to time, that process has continued inexorably onwards.

It is essential to realise that the steps taken in 1866 were only small and hesitant. The Island still remained subject to the whim of the United Kingdom Government at Westminster, and the Government of the Island was still firmly in the hands of its duly appointed representatives. However, the next one-hundred-and-twenty years saw a gradual process of devolution, a process which is by no means fully completed today. For ease of reference this process is divided into three distinct areas, all of which combine under the generic term 'constitutional development'.

Constitutional development

Legislative reforms

The changes made in 1866 were so fundamental that they presaged a period of some fifty years of relative inactivity whilst Tynwald caught its breath. This was only punctuated by periodic extensions of the suffrage, even though attempts to give women the vote in 1881 were frustrated by a United Kingdom Government that had still to be convinced on the subject. A move to abolish the death penalty at the same time suffered a similar fate—ironically the Island has not made any attempt to abolish it for a second time.

The catalyst for change was the relative lack of influence of the popularly elected House of Keys, whose limited reformist zeal was circumscribed by the Legislative Council, the other Chamber of Tynwald, now recognisably carrying out the functions of a legislative body as opposed to just an advisory body for the Governor. Not only could the Governor veto its more radical aspirations, but he often did not have to resort to that sanction, as his job was done for him by the Legislative Council. This comprised exclusively ex-officio members of greater or lesser worth, ranging from the law officers (Deemsters and Attorney-General) through ecclesiastical potentates (Bishop, Archdeacon and Vicar General) to minor functionaries whose very titles would have inspired W. S. Gilbert, and whose duties had ceased to have any real relevance. Needless to say, these potentates of Island Society—'the lilac establishment'—were in no hurry to be divested of their power and influence. It took the explosive mixture of an autocratic Governor, Lord Raglan, a popular demagogue from Liverpool, Sam Norris, and ultimately the First World War to bring about the changes in the composition in the Legislative Council.

The Report of the MacDermot Commission—a Constitutional Commission appointed as a result of the local unrest expressed by such heady action as the throwing of sods of earth at Lord Raglan—recommended certain changes which were introduced in 1919. These included the replacement by nominees of the Governor of some of the defunct ex-officio members and the election by the House of Keys, itself, of all the Legislative Council members (other than the Bishop and Attorney-General) for a fixed term, presently five years. Along that route the Legislative Council lost its ultimate veto over any legislative or other changes approved by the House of Keys, a veto being replaced with a delaying function similar to that exercised by the Upper House at Westminster. One of the final stages in that saga saw the Governor being removed as Presiding Officer of the Legislative Council, fittingly enough in 1979, although he still presides when both Chambers meet together in Tynwald.

The actual process of enacting legislation follows closely the pattern of Westminster with each 'Branch', as the House of Keys and Legislative Council are called, giving a bill three readings and, after further legislative formalities, submitting the same to the Sovereign for Royal Assent, a final step which can never be taken as being a foregone conclusion. One further unique hurdle remains. Unless that Act is read out, or 'promulgated' at the next meeting at St John's of Tynwald, it ceases to have effect. Tynwald, the combined assembly of the Branches, meets regularly each month to consider matters of policy, finance and administration, in particular subsidiary legislation.

Constituencies

The Isle of Man has been divided, traditionally, into Sheadings, and these are in turn sub-divided into Parishes. The latter, although not necessarily the same as the ecclesiastical sub-divisions, are more understandable than Sheadings, a term probably derived from a Viking division of the Island into 'ship-divisions'—areas responsible for the provision and manning of a ship in time of emergency. It was these units that formed the basis for the division of the Island into constituencies for election to the House of Keys and these have remained the same until the present day, with the occasional variation in boundaries due to shifts in population. The only change of significance to the method of voting for members of the House of Keys (which takes place every five years) has been the change from the traditional first past the post system to a form of proportional representation based on the single transferable vote. Used for the first time in the 1986 General Election and having narrowly (one vote) survived an attempt to return to a more traditional form of voting, the system is thought to be more suitable for the many constituencies which return more than one member.

Traditionally members have been elected on their personal attributes and not because of any party affiliation. This generalisation is, of course, subject to exceptions created by the Manx Labour Party and in more recent times by Mec Vannin, the nationalist party. However, contemporary developments discussed later may, in the writer's view, lead to more widespread changes.

Judicial

The Island, as befits a separate kingdom, has an independent bar and judiciary. Its judges are called Deemsters, its stipendiary magistrates are called 'High Bailiffs', and its members 'advocates', indicating a fused profession. In every other respect the Island's legal system resembles its larger neighbour, with the Island's legislation embodying the best of United Kingdom legislation and eschewing the rest (in theory at least). Originally Tynwald exercised a more judicial and legislative function, its meetings in the latter role being far less numerous than a Court. However, with its development as a Legislature and a growing recognition of the undesirability of involving the Judiciary in the passage of legislation, there has been a phased removal of the Law Officers from the Legislative Council, leaving only the Attorney-General with a seat and a voice, albeit without a vote.

Executive

The exact status of the executive has created more misunderstanding than any other element of the Isle of Man Government, particularly in recent years. The term 'the Government of the Isle of Man' has elicited many different interpretations, but one popular misconception must be laid to rest: Tynwald is not, and never has been, the Government of the Isle of Man any more than Parliament at Westminster has been the Government of the United Kingdom.

In 1866 the Government was clearly the Lieutenant Governor who needed Tynwald approval for the exercise of certain of his functions; but he was unmistakably 'the Government'. The process of devolution started in a hesitant way with certain executive functions being delegated to 'Boards' composed of a number of members, including some drawn from the membership of Tynwald. As this process gathered speed, more and more executive functions were delegated to Boards, which became known as Boards of Tynwald, with functions as diverse as responsibility for Airports to Tourism through Highways, Harbours, Health, Education, Forestry and Social Security. Despite this, the Governor retained his executive functions for all areas not devolved to Boards, but like them, he was dependent on Tynwald for the resources he needed to carry out his duties. Most significantly, until 1961 he retained control of the purse strings, acting as Chancellor of the Exchequer until the creation of the Finance Board, but even then retaining a significant influence on the exercise of that Board's very considerable powers.

At the height of its development, the Board system absorbed the talents of all members of

Tynwald, and a few more beside, as each Board comprised at least three members, and in the case of education, as many as twenty-nine. The majority of members found themselves serving on at least two Boards, and the level of cooperation and coordination between the respective Boards was not always as great as might be imagined. Indeed there is at least one example of a Board indulging in litigation against another.

The ability to coordinate effectively the activities of Boards decreased by the same factor as their independence increased through the devolution of the Governor's powers. In recognition of this, as well as the desire to spread the Governor's executive role, a consultative body was created called 'Executive Council'. In many ways this took on the function exercised by the House of Keys in the early days of its life, being a sounding board for the Governor. Moreover, it enabled him to use the phrase 'after consultation with Executive Council' which could justify any action in which democrats would argue that the people should have a say.

Gradually Executive Council took on a life of its own. It became more and more like a Cabinet, particularly after dispensing with the physical presence of the Governor himself on all but a limited number of occasions. The trend was further accentuated when members were elected according to their responsibilities rather than their personal merits. The latest step along the road has been indissolubly linked with the continuing but probably most radical change in the Government's composition for the past 100 years.

The Ministerial System

For the very reasons referred to above, and because of the perceived need for a more efficient and effective form of Government to meet the needs of a modern-day world, pressure grew over the early years of the 1980s for the introduction of a ministerial system. An important step along the way was the rationalisation of the Board system effected through a reduction of both the numbers of Boards and of the numbers of Tynwald members serving thereon. The changes in nomenclature from Board to Ministry, and from Board Chairman to Minister were less significant than the subsequent change which devolved upon the Minister the ability to reach a decision on his own authority, notwithstanding opposition from his colleagues on that ministry. That more than any other, in the author's view, has hastened the evolution of a more widely recognised Government/Opposition confrontationalist form of government as opposed to the traditional consensus form of Government to which the Island has been accustomed. Traditionally there were no rigid divisions, but members voted according to their individual consciences (and to their Board loyalties) rather than any party or philosophical 'whip'.

But the process is not yet complete—the Ministerial system headed by a Chief Minister elected by Tynwald and a group of Ministers chosen by him (but approved by Tynwald)—is but three years old and its weaknesses as well as its strengths are still being assessed in the crucible of experience.

Constitutional relations

As is apparent from this account of the shaping of the modern Isle of Man Constitution, the Island's relationships with its adjacent neighbours and in particular the United Kingdom have been a major factor in that process. However, the Island is not, and never has been part of the United Kingdom. That great English jurist, Lord Justice Coke described it as 'The ancient and absolute kingdom of Man', and the separate and sovereign nature of the kingdom is reinforced by a number of legal judgements, particularly one relating to the Countess of Derby in 1598. The most recent restatement of the position was made in the Royal Commission on the Constitution (the Kilbrandon Report) in 1971. Its precise, and unique, status is that of a Crown Dependency, in other words a possession not of the United Kingdom Government but of the Crown. What, in practice, does this mean? The English Crown has always been the sovereign power in the land, the Lordship of Man being subsidiary thereto, each holder acknowledging (traditionally by the presentation of a pair of falcons at each Coronation) that the title was held in fief. With Revestment the title theoretically lapsed, but the Manx have retained the use of the title 'The Queen, Lord of Man' even though it is not part of the Royal Title. As ultimate possessor of the territory, the Sovereign is responsible for its good government, and appoints a Lieutenant Governor to maintain and preserve the Crown's interest. In practice that responsibility is exercised on the Sovereign's part by one of her ministers, traditionally the Home Secretary and his department.

Although Tynwald is a sovereign parliament, Parliament at Westminster claims a right to legislate for the Island, a claim which is upheld by the substantial quantities of legislation from Westminster which has been extended to include the Island in its scope. However, by convention, that power is exercised only with the consent of Tynwald and the modern practice is for Tynwald to legislate for the Island itself, even if such legislation mirrors similar United Kingdom legislation. The convention has not been put to the test in recent years, although there have been occasions when one sensed it was imminent, as for instance in the 1970s over the matter of judicial corporal punishment—still on the statute book of the Island despite being in contravention of the European Convention on Human Rights, to which the Island is subject.

By convention also, the Island Government, whilst fully competent on all matters domestic, leaves all matters appertaining to foreign affairs and defence to the Sovereign's Ministers. Thus the Island has no army (Manxmen serve in the British Armed Forces as the Armed Forces Acts are extended to the Island), no ambassadors, no passports (other than an amended version of a British passport) and is not party on its own account to any international treaties or conventions. In the latter case, the United Kingdom becomes a party on the Island's behalf, at times something of a mixed blessing, as for instance when the United Kingdom found itself brought before the European Court over the Island's use of judicial corporal punishment, a practice which the Island steadfastly refused to renounce (even though it has ceased to be used in practice). Thus the Island is not a member of the Commonwealth, but by virtue of Britain's membership, participates in a number of its activities, particularly on the Parliamentary level. The Treaty of Accession to the European Community is one example where the Island occupied a status in its own right. In 1971 the Island was faced with the choice of full membership on the back of the United Kingdom, or complete separation which could only have been brought about by independence. In truly independent fashion, along with Jersey and Guernsey who found themselves in a similar quandary, a special relationship was negotiated embodying what, to the Islands, were the strongest attractions whilst eschewing some of those Community features, such as harmonisation of taxes and the common agricultural policy, which could have brought rapid doom to the Island's unique economic base.

The future

The advent of the 1990s finds the Island in a continuing period of change and development. The welcome prosperity from the financial sector may have brought temporary security, but such prosperity can be transient, and the real security from a well defined and universally respected and efficient governmental structure is still developing. The spectre of 1992 and its effect on the Island and similar semi-independent states of Western Europe is still something of an unknown quantity.

Of one thing a Manxman can be sure—just as the motto of the Isle of Man is 'Quocunque Ieceris Stabit' or roughly translated, 'a Manxman always lands on his feet', so the Manx are confident that uncertainty brings with it opportunity and just as the Island has survived the attentions of its larger neighbours during the last Millennium, it will do that for the next, preserving all that is best and adding to it when the need arises.

CHAPTER 10

Social demography

Vaughan Robinson

Introduction

Three elements determine the demographic course of any region: fertility, mortality and migration. The relationship between the first two of these has, of course, been studied in detail and even formalised into the well-known demographic transition model. This proposes a broad pattern of demographic change for a society undergoing economic development from a pre-industrial state to what is now described as a post-industrial state. Although the model has obvious weaknesses, it is of considerable didactic value with its neat periodisation of demographic change into the four phases of high stability, early expansion, late expansion and low stability. In the first of these phases, birth and death rates are both high, but subject to considerable fluctuation because of famine, war or disease. With the onset of economic development death rates fall sharply, although birth rates remain largely unaltered. The population increases rapidly as a consequence. In the late expansionary phase the birth rate also falls until it is close to the death rate, which has by now stabilised. In the low stable phase birth and death rates are in close accord and population increase is sharply reduced. Some commentators have even added a fifth stage based upon the experience of several Western European nations. In this post-industrial phase, the birth rate actually falls below that required to replenish the population which consequently declines in numbers and becomes progressively more aged. However, one of the critical shortcomings of the demographic transition model is that it envisages societies as being essentially closed or self-contained. Emigration and immigration are largely ignored and homeostasis between population and resources can only be maintained through the regulation of births and deaths. In the real world this is not true. As the Isle of Man clearly demonstrates, population can be exported at times when resources are limited and imported when the need arises. Indeed the Isle of Man also underlines another weakness of the model in that it excludes direct state intervention in population control. As later sections of this chapter show, the state can consciously manipulate not only the size of a population but also its age structure and social composition. The Isle of Man thus provides us with a fascinating natural laboratory in which to try out basic demographic concepts. The chapter is divided into three major sections. The first provides a broad account of changes in the Manx population prior to the late 1950s. It concentrates upon the shifting balance between births and deaths as a response to the changing economic and social circumstances under which the Manx population had to live. The second section focuses on the issue of emigration and the way in which this phenomenon helped maintain the balance between population and resources. The third section describes the development and impact of state intervention in population matters.

The Manx population through history: its size and characteristics

For the period before the eighteenth century our knowledge of the demography and social conditions of the Isle of Man is both limited and somewhat impressionistic. Moore's (1900) monumental history of the Island provides the best source, although Kinvig (1975) provides a complementary account.

There is little direct evidence of the way of life extant in the Isle of Man during the celtic period. Several factors, however, suggest that it closely paralleled that of Ireland. If this was the case society would have been tribal with each tribe closely connected to its own territory. The tribal chief was effectively the landowner and his subjects became tenants-at-will. During the winter all land was held in common and the unfenced pasture was used by all. In the growing season some land remained as common pasture whilst the remainder was divided annually and allocated in strips to individual tribe members. In exchange, tenants offered dues to the tribal chief either in the form of money, hospitality, provisions or service during times of war. Those tenants who fell down in their obligations could be coerced by that portion of the tribe that remained in servitude to the chief. Farmers would grow oats and raise cattle and they lived in stone and earth houses which were grouped into villages. Society contained several gradations down from the chief through the nobility, the freeman with property (e.g. cattle), the freeman without cattle to the unfreeman.

Society would have remained essentially tribal during the Scandinavian period with a chief presiding over his family, followers and servants. Tenants would owe fewer feudal dues, with the freemen obliged only to fight for their leader. Arable land was still redistributed annually but pastoral farming had become transhumance. Animals would be taken up into the mountain pastures in

10.1 Still one of the most characterful parts of Douglas, the North Quay. It was here that the settlement was first located, alongside the Dhoo and Glass rivers

summer-time from whence turf would also be extracted ready for use in winter. Women were responsible not only for milking the cows, but also for spinning and weaving flax and wool. Men would combine their agricultural duties with fishing. Extended families tended to share the same home which most frequently consisted of several small buildings arranged around an open quadrangle. Society remained stratified with, for example, freemen enjoying considerable control of their own affairs. The hierarchical nature of society would also have been reflected in the use of language. Norse was the official language and was used by the masters whilst Gaelic remained the tongue of the servants, conquered celtic slaves and women. From the eleventh century onwards the structure of secular society was overlain by new power relationships arising from the introduction of Christianity. By 1257 the Bishop controlled his own courts and estate and claimed one-third of the tithe.

The period 1266–1405 was one of continual changes of sovereignty. Again little direct evidence has survived which might indicate the social and demographic conditions of the period. It is known that the alternating Scottish and English rulers of the Island sought to extract as much as possible from its native population. The Island, according to Moore (1900) was impoverished and plundered. Although the church and some members of the established ruling class retained their importance, land ownership effectively reverted to the king. Some was granted to barons whilst the remainder was rented directly to tenants. The church, too, increased both its own landholdings and the tithe that it extracted from the population. Moore notes that it is not clear whether the Black Death visited the Island in 1348 but, if it did, its impact was limited. He also notes that the progressively increasing value of labour forced a change in status for the serfs who became free labourers in the early part of the fourteenth century.

The period of the first Stanleys, which straddled the fifteenth and sixteenth centuries, has left a limited legacy of direct evidence about the people of the Isle of Man and their lifestyle. Moore (1900) cites the writings of Bishop Meryck (1577) and Blundell (1648) as being particularly informative. The evidence suggests that, at least in the early period, there was a recovery from the poverty of Scottish and English rule although feudal obligations became more extensive. Subjects were expected to provide free corn, herring, beef and turf to the Lord of Man, they had to help repair his forts and domestic properties and they had no choice but to train for service in his militia. Watch duties were also obligatory. The economy, too, was closely regulated with taxes on fishing for herring and the import or export of any goods. Farmers were allocated to particular mills (owned by the Lord) and strict penalties were imposed on those who dealt with another miller. Individuals could also be 'yarded', i.e. taken into the Lord's service compulsorily for a nominal wage. In exchange for these feudal obligations the Manx people enjoyed some stability and peace. Moore (1900) cites the wages and prices current in the 1420s and concludes that many Manx lived in considerable relative comfort. Meryck and Blundell both describe the upper classes in the Isle of Man as being direct parallels to their Lancastrian counterparts: they lived in well-built and handsome houses sited on their own country estates. Townsfolk lived in two-storey houses. Even the poorest owned two such properties whilst shopkeepers might own three. Conditions were less favourable in rural areas. Some were thought to be comfortable, but the typical property of the 1640s was still a stone and clay, one room, thatched cottage which housed not only the family but also their livestock. Even so, Sir Spencer Walpole wrote of the common man:

> the inhabitants of Man at this period enjoyed considerable advantages over those of England and Ireland . . . Except for the duties to his lord, his church, and his country, the Manxman enjoyed comparative liberty. The soil brought him the little which was necessary for his frugal existence, and the rich though fluctuating harvest of the sea supplemented the food which he drew from the land . . . And thus the little Island after a rude fashion flourished. (cited in Moore, 1900, p. 296).

One word of Walpole's which is of considerable significance is 'fluctuating'. Economic conditions still depended upon the vagaries of the weather and the toll of intermittent disease. Thus whilst 1641 and 1649 were noted for their scarcity and poverty, 1642–48 and 1650 were good years with bumper harvests. Population was still being held in check by the availability of natural resources as it is in any pre-industrial society. The total population of the Island probably numbered no more than 12,000 people of which the vast majority were rural inhabitants. The most important town—Castletown, with its courts, Governor and Key officers—would not have exceeded 750 people in size. Douglas was the second town because of its port function. It is recorded as having 730 residents in 1511, although this seems rather unlikely given that it contained only 50 houses clustered around the North and South Quay. Peel was mentioned only as the site of a sheading court, and Ramsey was the smallest of the 'towns'. The latter also emphasised even the 'urban' dwellers' dependence on nature, since in 1630 it was almost totally destroyed by high tides which washed away land and buildings. Lastly, the period was also noted for its strict regulation of emigration and immigrants. No-one was allowed to leave the Island without an official licence and sailors caught transporting illegal emigrants stood to forfeit their ships. Although the Isle of Man was popular with immigrants in the seventeenth century, these aliens were treated very much as second class citizens. If they committed any crime on the Island the penalty was death, they were unable to pass on their property to their offspring, they were more heavily taxed and on at least one occasion some of them suffered compulsory repatriation.

The later period of Stanley rule and that of their Atholl successors (1660–1765) appears to have been one of greater economic hardship. At the end of the seventeenth century wages were depressed by legislation to the levels current in 1609. Corn and cattle prices were low and farmers were therefore less well-off. And labourers were little better than serfs. Emigration and disease, however, produced

10.2 Traditional cottages as occupied by the early farmer-fishermen on the Island. These particular properties are at Niarbyl, on the west coast

progressive labour shortages and as a result labour laws were slackened and wages rose naturally. By the late 1720s economic conditions had consequently improved somewhat, although there was a further deterioration by 1765. Again conditions indicate the fragility of progress, the dependence on nature and the vulnerability of the population. 1737 was marked by many deaths from cold and fever. 1739 saw a drought, poor crops and many families made destitute. The winter of 1739–40 was particularly harsh, whilst 1741 witnessed much hardship and even starvation. In contrast 1742 produced a good harvest, although this was followed by a bout of contagious fever which killed whole families. Poor people still lived in sod cabins and survived on a diet of herring and potatoes. They burnt gorse and turf for heat and cooking. Whilst seventeenth century accounts stress that the Isle of Man was a poor country, Sacheverall put this into a broader context with his observations that whilst:

> there are few that can be properly said to be rich, so neither are there many that can be said to be miserably poor.

Moreover the Island had:

> fewer beggars in proportion than in any other nation (cited in Moore, 1900, p. 407).

The period of late Stanley and Atholl administration was the first for which we have actual census data. The first, almost certainly inaccurate, census was held in 1726. At that time the population was thought to number 14,426 of which 2,530 were urban. Douglas was by now the largest town with a population of 810 followed by Castletown (785), Peel (475) and Ramsey (460). The second census was actually a series of estimates collected from local clergy. This gave a 1757 population of 19,144 of which 4,146 lived in towns. Again Douglas was the leading town (population of 1814) with Castletown (915), Ramsey (882) and Peel (805). Conditions in the towns were obviously poor despite their relatively small size, since a report of 1701 describes them as filthy. Douglas was clearly a cosmopolitan centre, having a population in 1681 which was over one quarter 'foreigners and Frenchmen'. The lenient debt laws which made the Isle of Man a haven for debtors after 1737 also added to Douglas' population. Many of the new arrivals spent a good deal of money and this made Douglas not only the most populous but also the richest town on the Island. Indeed as early as 1721 townsfolk of other settlements were complaining of the growing primacy of the port whilst farmers bemoaned the fact that they were tied to other fairs and could not, therefore, bring their goods to market in Douglas. Despite Douglas' commercial power, Castletown remained the administrative and social centre of the Island. Moore (1974) describes how the town hosted the House of Keys and how it was the place of residence for many of the élite. The richest families had up to five servants and they enjoyed the benefits of balls, musical evenings and poetry readings. Many of the town's finest houses were erected towards the end of the eighteenth century. Ramsey had still not recovered from the tragedy of 1630 and there is evidence of a continued threat from the sea until 1721.

The Revestment of 1765 marked the start of another phase in Manx history which persisted until the 1830s. In the early part of this period (1765–92) economic conditions deteriorated as society struggled to adjust to the end of the smuggling industry. Over 1,000 people emigrated in the one year during which Revestment took place, and poverty was widespread amongst those who remained. Prices rose rapidly but incomes remained fixed. Respite arrived from 1793 onwards with the posting of an increased contingent of soldiers to which could be added a growing number of English or Irish immigrants attracted by the relatively low cost of living on the Island. Both groups boosted domestic demand, created the need for more labour and pushed up wage rates. This new wealth was reflected in a growing population which

10.3 North Quay in Douglas. One of the few parts of Douglas which pre-dates the period of popular tourism. Proposals to redevelop the harbour as a marina were rejected in 1989

rose from 27,913 in 1772 to 40,081 by the time of the first official census in 1821. It was also reflected in the landscape with considerable new building taking place in both the towns and country. Moore (1900) describes how most of the turf and thatched cabins were replaced by stone cottages which slate roofs. This was particularly the case in the south of the Island where building materials were more plentiful (Herbert, 1909). The towns benefited, too. Bregazzi (1964) notes that Douglas was practically rebuilt between 1800 and 1830. The Red Pier and Fort Anne Jetty (Brown, 1964) improved port facilities. The two new thoroughfares of Duke Street and Strand Street were added. Oil street lighting was installed in limited quantities from 1829 onwards. The 1834 Douglas Waterworks Act made provision for piped drinking water for the first time. The Duke of Atholl erected his new residence, Castle Mona, out of town. And as a result Douglas became the new centre of the Isle of Man (Norris, 1904) not only for the local aristocracy and the retired army and navy officers, but also for the increasing number of summer visitors. The permanent population rose accordingly, climbing from 2,411 in 1784 to 6,054 in 1821. Castletown, too, benefited from this new prosperity (Moore, 1974), with new homes, the new Assembly Rooms, and other symbols of civic pride. Population rose from 1,318 in 1784 to 2,036 in 1821. Contrary to many of these visible signs, the economy was not yet stable or wholly robust. The end of the Napoleonic Wars and the return of demobbed soldiers and sailors created problems which were compounded by poor harvests and the iniquities of the Corn Laws. Serious rioting occurred in 1821 because of the lack of grain for sale, and the 1820s as a whole witnessed an economic depression. Social concomitants included emigration, a rise in crime and increasing numbers of beggars. Population totals faithfully reflected these problems and between 1821 and 1831 the population rose at a much reduced rate (40,081 in 1821 to 41,751 in 1831). Moreover, throughout the entire period 1765–1830 the mortality rate remained at a high, if fluctuating, level. Disease still made periodic incursions resulting in the local decimation of the population. Smallpox swept parts of the Island in 1765, 1772–73, 1780 and 1799. In the first of these years 48 out of every 100 people in German parish and Peel town were killed by the disease. Ballaugh parish lost 28 in 1764–65, 33 in 1772–73 and 22 in 1799 to the same cause.

The period 1835–95 saw a rapid transformation of the Manx economy brought about by the growth and development of the tourist industry and the final coming to fruition of the mining sector. Manufacturing, too, was stimulated by the other economic developments and expanded accordingly. The fiscal reforms of 1866 also made their contribution. This economic metamorphosis (which is discussed more fully in the introduction to Part IV and in Chapter 14) produced profound demographic changes of three types. Firstly, improvements in sanitation, hygiene and personal incomes contributed towards declining levels of mortality. Whilst epidemics were not yet totally banished they did become less frequent in the latter part of the century. Moore (1900) mentions epidemics of smallpox in 1839, 1851 and 1864–66, of typhoid in 1837 and 1864–66, and of cholera in 1849 and 1853. But he makes no mention of any repetition of these attacks after 1866, and we are left to conclude that initiatives such as the introduction of mains water sewerage in Douglas in the 1860s had reduced the population's vulnerability to disease. This is also reflected in the death rates recorded during the period. By the 1860s, mortality had fallen to 19 per

10.4 A general view of Port Erin showing the way that the nineteenth-century hotels clustered along the bay front. The row of cottages behind the beach, marks the site of the earliest modern settlement, located near a source of freshwater

thousand and even with the twinned epidemics of 1864–66 the rate was less than 27 per thousand. In contrast, the birth rates that we have for the period show no decline. Between 1851 and 1860 the rate was 28 per thousand and in the following decade it actually rose to 30 per thousand. Clearly the changing relationship between fertility and mortality would have produced a significantly higher rate of population growth. As a result, the second major demographic change of the period was the rapid expansion of the Manx population. As Table 10.1 demonstrates, the Island experienced a sharp growth in its population from nearly 42,000 in 1831 to nearly 56,000 in 1891. This represents an increase of one-third, despite the set-back of the commercial depression of 1847–51 and high levels of emigration in the period 1851–61. Indeed at times the annual growth rate almost reached 1.5%. However, because of the uneven distribution of this growth, the Isle of Man also experienced a third demographic trend, namely urbanisation. Improvements in agricultural profitability did much to hold the rural population on the land but after 1851—when the rural population reached its peak—numbers fell away sharply. Between 1851 and 1891 the population found outside the four main towns fell by 9,500 or 27%. Indeed this fall would have been even greater were it not for the influx of miners and their families into the rural areas of Foxdale and Laxey. The Lonan district, for example, which contained Laxey saw its population rise by 1,817 persons between 1831 and 1871, an increase of 95%. However, it was really the major towns that experienced the most rapid growth in their populations. In Peel's case this was the result of the successful professionalisation of the fishing industry, but others benefited from the new tourist industry. Douglas was foremost amongst these. The establishment of the Steam Packet Company (see Chapter 12) in 1830 and the progressive extension of harbour facilities (Brown, 1964) allowed the number of visitors to double between the 1840s and the 1860s.

The majority of these looked to Douglas for their holiday needs and the town grew accordingly. Norris (1904) described Douglas as the 'Naples of the North' and enthused over its 1,000 boarding houses, its two miles of promenade and its 'Palace' dance hall. The contemporary writer Cummings (1848) provides another glowing description:

> In the foreground we have the town of Douglas stretching along the south-western edge of the bay . . . The more elevated localities have in later years been seized upon by the better classes for their habitations and for the lodging-houses of strangers, and a new town has thus rapidly grown up of a more respectable character, and this from its position being more conspicuous than the more ancient one, happily impresses the visitor . . . with a very favourable opinion of the spot'. (Cummings, 1848, p. 13).

He goes on to describe Castle Mona, Villa Marina and the Falcon Cliff. However, the improvements to the town were not simply cosmetic and Bregazzi (1964) provides a more down-to-earth account of the way that Douglas was becoming a healthier place in which to live. He notes the new hospital (1885), the new West Baldwin Reservoir (1905), the public abbatoir (1894), the introduction of gas lighting (1835), the new cemetery (1899) and the clearance of parts of the old town (1895). In due course Douglas became the Island's capital and was incorporated as the first Manx borough. Not surprisingly all these developments were not only a response to population growth, but also stimulated it. The population of Douglas, which had been only 6,774 in 1831, climbed to 12,511 by 1861, 15,719 by 1881 and 19,525 by 1891. Although other towns were also growing, it was the rapid increase in Douglas' population which was responsible for the progressive urbanisation of the Manx population. Douglas itself contained 35 per cent of all Manx residents by 1891 (up from 16% in 1831), and by that year the four main towns between them rep-

Table 10.1 *The Manx population in the period 1831–91*

Year	Total	Absolute change	% change	% in Douglas	% in 4 towns	Rural population
1831	41751	—	—	16	29	29437
1841	47986	+6235	+14.9	18	32	32819
1851	52387	+4401	+9.2	19	33	34933
1861	52469	+82	+0.2	24	39	31846
1871	54042	+1573	+3.0	26	44	30303
1881	53558	−484	−0.9	29	48	27742
1891	55608	+2050	+3.8	35	54	25408

resented over half the Island's population (Table 10.1).

As on the mainland, a combination of economic prosperity and the contemporary desire to innovate, produced a series of urban experiments in the Isle of Man. Some of these were more speculative than others. One envisaged the erection of a 370 foot high tower on the North Quay in Douglas. The tower would have six floors and would house a circus, a skating rink, a concert hall, a dance hall, a variety theatre, a restaurant, a bazaar and an open observation deck. The capacity of all these facilities would have enabled 40,000 tourists to be entertained simultaneously and a high level suspension bridge would have provided access to Douglas Head. Despite the purchase of land, and the subscription of £25,000 by local notaries, the remaining £75,000 was never forthcoming from the public and the idea never materialised. Elsewhere in Douglas, and on an entirely different scale, residents were experimenting with cooperative neighbourhood management. Derby Square was erected in the late 1830s for professional and upper middle class people. By 1854, however, the area had suffered from vandalism and an Improvement Society was formed by local residents. They levied a subscription and spent the resulting capital on the installation of gas lamps, the provision of new paths and the repair and painting of railings. The following year all residents agreed to adopt a common colour scheme for external decoration and shortly afterwards they employed a caretaker/groundsman and installed security gates which were locked each evening one hour after sunset. The society was obviously successful since it did not relinquish its management role to Douglas Town Council until 1948. And finally near Tromode, to the north of Douglas, there was an even more comprehensive urban experiment in the form of the planned industrial village of Cronkbourne. The village contained forty-two cottages when it was completed in 1850, having been planned and funded by the Moore family who owned the nearby sailcloth factory. It not only provided an economic return for the Moores, who charged their workers rent for living there, but it also afforded them a degree of social control. There was no public house within walking distance and there were 39 rules and regulations to which residents had to adhere. Failure to do so resulted in fines. Enlarging the holes in the gas burners led to a one shilling fine, for example, whilst failing to clean the road and gutter outside a cottage cost three old pence. In exchange for this social control, though, the residents received considerable benefits. There was a sick fund, a reading room, a wash room, a school, and street and domestic lighting. Analysis of 1861 census data for the village shows a youthful population (average age 24 years) drawn from the nearby area and also from Ireland. Average household size was 5.4 persons and women were clearly very fertile. By the 1881 census overall numbers were still about 250 people but there had been considerable turnover of residents and the village's age-structure was more diversified.

Elsewhere in the Island other towns were less fortunate than expanding Douglas with its innovatory initiatives. Castletown, in particular, failed to benefit from the tourist or industrial booms and its population remained resolutely around 2000 people from 1831 right through to 1891. Despite no longer being the Island's capital, it remained an area of residence for the upper classes and a seat of learning. The 1851 Census showed it to possess nine teachers and 345 scholars, and there were known to be five schools sited there in the latter part of the century.

It might seem unusual to group together the years 1896–1945 as one period when they began with the very first motor cars and ended with jet aircraft. Nevertheless, despite the astounding technological advances, this half century was essentially one of specialisation and retrenchment for the Isle of Man. In the introduction to Part IV a somewhat finer division of the period will be argued but this is not really relevant here. What is of relevance is that the mining industry experienced terminal decline, fishing increasingly passed into non-Manx hands and the manufacturing sector, which was heavily dependent upon both of these two enterprises, also experienced severe recession. Indeed after the First World War the tourist industry itself began to experience difficulties (see Chapter 16). The demographic and social response was predictable. Whilst the death rate continued to drift downwards from its 1898 level of 19.4 per 1,000, the birth rate fell sharply from 25.6 per 1,000 in 1898 to 15.1 per 1,000 in 1931. In fact it fell so sharply that from the end of the First World War onwards deaths have generally exceeded births in the

10.5 Cronkbourne Village on the outskirts of Douglas. The only example of a model industrial village on the Isle of Man, it was built in the nineteenth century to house workers at the nearby sailcloth factory

Table 10.2 *The Manx population in the period 1896–1945*

Year	Total	Absolute change	% change	% in Douglas	% in 4 towns	Rural population
1891	55608	+2050	+3.8	35	54	25408
1901	54752	−856	−1.5	39	57	23434
1911	52016	−2736	−5.0	41	57	22155
1921	49078*	−2938	−5.6	*	*	*
1931	49308	+230	+0.5	39	56	21592
1939	52029	+2721	+5.5	38		

* the 1921 Census was taken in late June and included significant numbers of tourists. The adjusted figure for residents only is used here.

Island (Corran, 1977). The shortage of work brought about by retrenchment, even in the tourist industry, created a severe problem of winter unemployment in the 1930s and forced the introduction of public works schemes. With nearly 2,000 registering as unemployed each winter, emigration again became important and population loss from this source when combined with natural loss produced a declining total population. As Table 10.2 shows, the Island's population did not fall consistently or at a steady rate but the net effect was the same. Between 1891 and 1921 the population declined by 6,530 people, or 12%. Over the whole period, the loss was 3,579 people or 6%. Once the decline in the tourist industry had become established, population loss became a feature of most parts of the Island. Despite new piers, promenades and a doubling of the number of residences (600 in 1891, 1238 in 1939) Douglas experienced population loss in absolute and relative terms. By 1911 Peel's population was barely 60% of what it had been in 1881, and Castletown saw a reduction in the number of its residents to below 1,800. The rural areas, too, saw a continuation of their population loss with a 15% fall between 1891 and 1931.

Set against this demographic gloom the Second World War provoked a somewhat unexpected reversal of trends. Construction activity provided work for local and imported labour. Military bases required the introduction of a sizeable corps of military personnel. And, lastly, the Island was selected as the site for the internment of enemy aliens. Whereas the 23,000 male internees held during the First World War had been kept in a single camp at Knockaloe near Peel (with its own facilities and railway) in the Second World War camps were opened in a variety of places. Chappell's (1986) account of the camps makes fascinating reading. He describes the various camps in Douglas, Peel, Ramsey and Port Erin, their inmates and the lives which they led for the six years of the war. At their peak in August 1940 the camps held 10,024 male internees and a further 4,000 women and children. The latter were accommodated in hotels and guest houses in Port Erin and Port St Mary which were effectively sealed off for the purpose. The camps not only contributed to the Island's economy—the first camp at Mooragh Promenade in Ramsey had a daily standing order for 500 pounds of bread—but also to its demography. Chappel mentions, for instance, the fifteen internees now buried in the Jewish section of Douglas cemetery, the suicides, the post-war marriages between internees and locals, and the births that occurred in the married quarters at Port St Mary.

The close of the Second World War signalled a temporary respite from long-term decline and, for almost half a decade, the Isle of Man experienced relative prosperity. In conjunction with the inevitable baby boom after the return of military personnel, this prosperity brought about a short-lived increase in the Island's population. The 1951

Table 10.3 *The Manx population in the period 1945–61*

Year	Total residents	Absolute change	% change	% in Douglas	% in 4 towns	% in 8 towns
1951	54024	—	—	37	53	67
1961	47166	−6858	−12.7	40	56	70

Census recorded some 54,024 residents, a level that had not been seen since 1901. However, the optimism and prosperity were short-lived and by the early 1950s the Isle of Man had returned to its previous course of specialisation and retrenchment. True, Government had begun to intervene more forcefully in economic matters, but prior to 1961 their initiatives were primarily designed to shore-up the existing structure, and regain powers from Westminster and the Lieutenant Governor. The 1950s were thus a period of demographic setback. Table 10.3 summarises some of the changes. The total resident population fell by almost 6,900 people. Winter unemployment reappeared as a significant problem: over 1,000 men were registered as unemployed in the winters of 1952 and 1953 and even by 1959 the number still exceeded 700. Government Winter Works Schemes were introduced to ameliorate the worst economic and social corollaries of unemployment but, despite these, emigration soon increased. Some individuals undertook seasonal mobility, spending the winter months working in the sugar-beet fields of Lincolnshire. Others left for good. In fact net loss through emigration accounted for some 84% of the inter-censal population decline between 1951 and 1961: a net total of 5,788 people left the Island. These emigrants must have been drawn disproportionately from rural parts of the Isle of Man because the 'four towns', and Douglas in particular, increased their share of the insular population over the decade. Although emigration was the chief drain on population, deaths continued to exceed births and the margin in fact increased in the 1950s. Whereas in 1951 the birth and death rates had been in temporary parity (15.1 and 15.7 per thousand respectively), once the post-war fertility boom faded this was no longer the case. Fertility fell to 13.8 per thousand by 1961, whilst the death rate rose to 17.8. Even without emigration, the Island would have experienced a fall in its population of 1,070 people or 2% over the decade. Since this deficiency of births was exacerbated by the emigration of the youngest and most fertile elements of society, the Isle of Man quite speedily came to possess an ageing population. The proportion over retirement age rose from 17% in 1951 to 22% ten years later. By 1961, then, the Isle of Man's economy was not only in disarray but the Island was plagued by the social and demographic problems which so often beset marginal regions.

Population emigration: a key element in social and demographic change

Before discussing State intervention in Manx demographic matters during the 1960s, it is worth looking in a little more detail at the issue of population emigration. The demographic transition model assumes an effectively closed system in which population and resources are kept in homeostasis by the continually adjusting inter-relationship between births and deaths. In reality this is a gross over-simplification since emigration and immigration are more immediate and direct responses to changes in the resource base. Indeed, in the Isle of Man emigration was *the* prime safety valve releasing population pressure in the 1820s, 1830s, 1840s, 1920s and 1950s. Whilst not denying the centrality of natural change—particularly in earlier historical periods—it would appear that the significance of emigration as an element in demographic dynamics has been considerably underplayed in the literature. This conclusion is further reinforced when one takes account of the potential *social* impact of migration, not only upon the individuals involved, but also on the donor and receiving societies. It is these issues which are amplified in the section that follows.

Emigration appears to have been a problem for the Isle of Man for much of its recent history. Moore (1900) notes that legislation was in force as early as 1422 to prevent unlicensed individuals leaving the Island. Moreover, similar laws were still known to be in existence in 1609 and even at the start of the eighteenth century. Even so, by that time emigration was already an acknowledged problem. An Act of Tynwald in 1713 admitted that many men and women left the Isle of Man at the age of 16, never to return. The Act described the supposed effects of emigration:

> this Island is no better than a nursery for other places, and the useful servants going off, and but a few left besides such depraved, useless or inactive people who are rather a burden than any real service to the Island. (quoted in Moore, 1900, p. 398).

Most of this movement would simply have been to mainland Britain, but early evidence also points to Manx involvement in long-distance international migration. Accounts exist of Manxmen in Barbados in 1650 and of a small colony of Manx in Jamaica in the 1760s. Myles Standish, the military leader of the Pilgrim Fathers, was also reputed to be from Lezayre. He was, of course, one of the first European settlers in North America. Although individual migration was thus a limited feature of both the seventeenth and eighteenth centuries, it was not really until the nineteenth century that we have well documented accounts of substantial international emigration. Since that time, there have been five major periods of movement overseas, although a steady stream of emigrants also left the Island in the intervening years. Each of these major exoduses has shared the same essential cause, namely economic depression, but in some cases this was exacerbated by other factors.

The emigration of 1821 to 1831 is, strangely, one of the best documented. There is some disagreement about the pre-eminent 'push' factor that stimulated emigration. The *Manx Advertiser* took the line that economic problems were to blame. Farmer-fishermen were suffering from the decline in herring catches. Income from this source had always been essential to supplement that from subsistence farming which was still heavily reliant upon the elements. A good harvest and successful fishing allowed the crofter to survive. Without the latter, many were forced into debt and had to use their smallholdings as collateral. As the debts rose many had little choice but to sell their land, and some used the balance to finance emigration. The *Advertiser* also bemoaned the lack of work for labourers in the north of the Island. The *Manx Sun*, however, felt that these were background issues only. Their editorials through 1827 repeatedly blamed the ecclesiastical tithes and when these were extended to the potato crop in 1826 many crofters were simply forced from the land. One of their leader articles on 27 May 1827 provided a detailed account of why tithes had to be the prime cause of emigration over and above the economic depression. Emigration during this period was sizeable. The *Sun* reports that 1,200 people had left the north of the Island in March and April 1827 alone, out of a resident population of 12–15,000. Both papers carried reports of regular departures of 50 or 100 people to Liverpool, whilst the *Sun* of 5 April notes that so many crofters were leaving the rural parts of the Island that farm prices had become artificially depressed.

Push factors in themselves are often not sufficient to stimulate permanent migration. Opportunities must also exist elsewhere. During the late 1820s and 1830s these were to be found in North America. Moreover, visits by speakers who had already travelled there and the letters of early arrivals soon provided a fund of information which potential emigrants could tap. Several diaries and letters of the pioneer migrants still survive and these provide graphic accounts of the journey across the Atlantic. They also tell us that families often chose to settle in localities that resembled their homelands. The Cleveland area proved particularly attractive to Manx migrants because it was at the end of a transport corridor, it possessed glen-like scenery and it enjoyed red clay soils similar to those found on parts of the Island. It was for these reasons that the settlers of 1827–28 put their roots down there and colonised eight villages on the low plateau adjacent to Lake Erie. Thomas Kelly and others of the pioneers soon sent reports back to fellow-islanders who were encouraged to join them. Chain migration of relatives and friends ensured the steady expansion of the Manx community in the Cleveland area. Kinvig (1955) records that there were 3–4,000 Manx and their descendants in Cuyahoga County by 1883, and that by 1950 this figure had grown to 25–30,000. A 1964 account estimated that the community had grown further to the point where 40–50,000 people of Manx origin lived in Northern Ohio with 10,000 in Greater Cleveland itself. Commentators were not averse to describing Cleveland as the Isle of Man's second largest town and when wealth was also taken into consideration it was undoubtedly the Manx people's première city. Not surprisingly, in view of the strong culture and traditions of the Manx, their shared experiences and the persistence of their language, the community enjoyed a high degree of institutional completeness. Among these facilities were a Manx area in the City graveyard, the headquarters of the North American Manx Association, a school, the Mona's Relief Society, and particular neighbourhoods dominated by Manx residents.

A similar account could be provided of later migratory periods, but the essence was the same. In 1847–61 emigration was stimulated by the failure of the potato crop and encouraged by the discovery of gold in the United States and Australia. In the late 1880s and 1890s the decline of the mining industry in Laxey and Foxdale was the motive force for leaving the Isle of Man whilst the Gold Rush in the Klondike area of Canada provided an attractive destination. In the 1920s and 1950s, a lack of economic opportunity turned people's thoughts to emigration whilst Canada, Australia and New Zealand welcomed new white settlers even to the extent of offering assisted passages.

The different historical peaks of emigration also shared other common features. First, in each period there were people for whom migration marked the threshold of enormous success and wealth. Joseph Mylchreest of Peel was one such person. He left the Island in the 1850s and had found his way to New South Wales by 1860. There he became involved in mining operations before moving on to New Zealand, British Columbia, Bolivia, Peru and Chile. In 1876 he moved on again to South Africa where he was to make his fortune in the diamond industry. By 1888 he was the largest individual owner of mines in South Africa, but this did not prevent him retiring back to Peel where he acted as the town's benefactor and, later, MHK. Second, although the migrations spread over a 130-year period, there do appear to have been particular parts of the island where migration was always more prevalent. An analysis of a sample from the Manx Museum's card index of migrants reveals that the parishes of Maughold, Onchan, German, Malew, and Rushen seem to have contributed many migrants. Bride, Jurby, Braddan, Marown and Arbory were also well represented whilst Ballaugh, Lezayre and Santon yielded none. Whilst these relative contributions undoubtedly reflect the populations available for emigration, they also indicate the strength of chain migration, drawing relatives and friends from the same village or locality. Chain migration also underlies the third characteristic of all the nineteenth- and twentieth-century emigrations: the concentrated nature of settlement in the receiving country. Figure 10.1 thus maps the distribution of members of the North American Manx Association in 1979. It reveals the agglomeration of members in particular states such as Ohio, Illinois, California and Virginia.

At a different scale Figure 10.2 shows the distribution of the Manx-born in England and Wales at the time of the 1951 Census. Ravenstein's laws would lead us to expect concentrations in nearby areas and in great centres of commerce. Figure 10.2 bears this out. It also underscores the veracity of the notion of distance decay with concentrations in Chester, Liverpool, Birkenhead, Wallasey, Bootle and Manchester. The third shared feature of all the major periods of Manx emigration was that many of the participants continued to retain strong links with their native culture and even country. Intermarriage and continued use of the language were early indicators of this. Later, symbolic ethnicity (Gans, 1979) replaced these core features as witnessed by the flowering of Manx Associations, the celebration of Tynwald Day, the ritual homecomings, the birth of the World Manx Association, the biennial Conventions of the North American Association, and the rekindled interest in the writings of T. E. Brown. As a past President of Mona's Relief Society commented:

> The Manx are more or less clannish people. Although they have assimilated easily because on the Island English as well as their own language is spoken, they will stay in their own organisations. They are Americans, but also very proud of their Manx ancestry. (quoted in the *Manx Courier*, 13 November 1964).

The same might also be said of the Manx in Canada, Australia, New Zealand, South Africa and Hong Kong.

The brief account above demonstrates that an analysis of the changing social demography of the Isle of Man would be incomplete without the inclusion of material on emigration. Emigration allowed a rapid adjustment of the home population to available resources. However, it also had a profound effect on those involved, the communities they left and those they chose to make their new homes.

State intervention in the Island's demography

During the 1950s the Government of the Isle of Man incrementally regained control of key powers which had previously been exercised by the Crown. This paved the way for a more interventionist style of Government in which the economy and demography of the Island were managed so as to achieve policy goals. Prime amongst these goals was the need to revitalise and diversify the economy so as to stem the post-war haemorrhage of population and the consequent ageing of the Island's remaining residents. Although, as we have seen, emigration was the historic direction of population movement, the Isle of Man had in recent times proved attractive to certain groups of immigrants. These included the mainland debtors, retired military personnel after the Napoleonic War, and the Lancashire folk who moved to the Island in the wake of the First World War to become hoteliers and landladies. Even so the decision to embark upon a new policy of stimulated immigration was bold, far-sighted and without precedent. The crux of the policy was to create a new tax regime which would be attractive to mainland residents with considerable resources. Death duties were abolished as was surtax, and income tax reduced to a level of 21.25%. The strategy was clearly successful. Between 1958 and 1966 the Island attracted 6,363 New Resi-

148 THE ISLE OF MAN

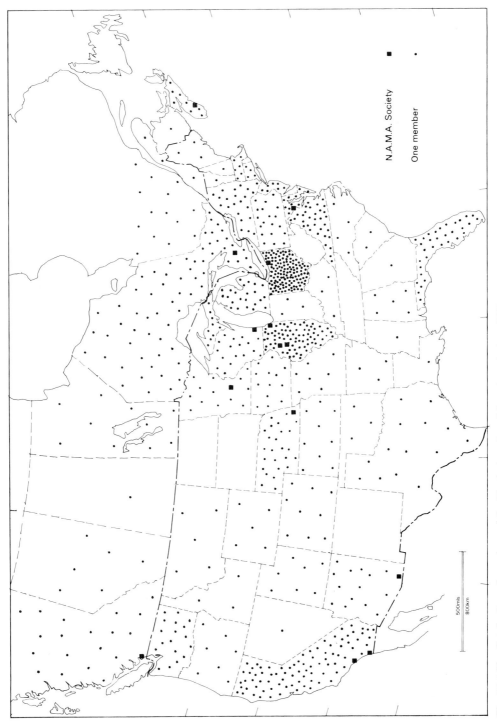

Fig. 10.1 The distribution of members of the North American Manx Association, 1979

SOCIAL DEMOGRAPHY 149

dents, more than enough to compensate for outmigration of Manx-natives and the loss of population through natural change (see Table 10.4). As a result the population of the Island increased by almost 5% (Table 10.5).

Despite, or perhaps because of, the apparent numerical success of the New Residents policy it was not long before it was criticised. In particular a belief developed that it was turning the Island into an offshore retirement home and that this was producing an inherent population imbalance. As a result, a Commission was established by Tynwald in March 1967 to investigate the issue of the imbalance of population. It took note of a variety of submitted documents including one from Mec Vannin, the Manx nationalist party. This particular document encapsulated the fears of many local residents albeit in a more politicised form than they would support. They argued:

Mec Vannin, as an avowedly nationalist and patriotic organisation, cannot help but stress that we are rather less concerned than the orthodox politician with the questions and rather more with the prospects of our survival as a nation, as a distinct Celtic community in its own right and with its own identity and independence. If we continue to attract new residents with plenty of money, the economic and financial problem is eased; but it will not help to preserve our indepen-

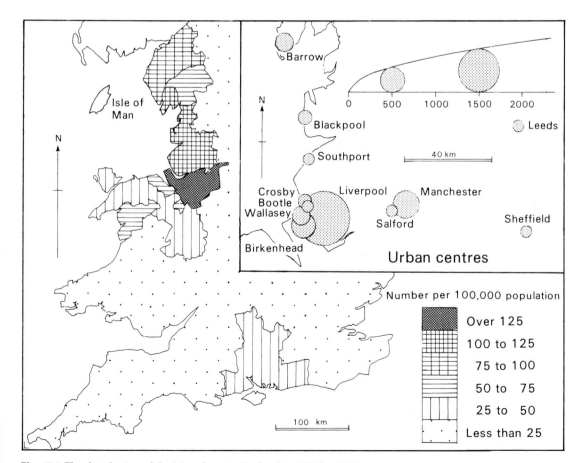

Fig. 10.2 The distribution of the Manx-born in England and Wales, 1951

150   THE ISLE OF MAN

Table 10.4 *Components of Manx population change, 1961–86*

Period	Absolute change	Births	Deaths	Natural change	Net migration	Immigrants	Emigrants
1961–66	+2146	3445	4114	−669	+2815	6006	3191
1966–71	+3916	3630	4391	−761	+4677	7910	3233
1971–66	+7268	3894	4972	−1078	+8346	10563	2217
1976–81	+4183	3586	5009	−1423	+5606	9097	3491
1981–66	−397	3525	4924	−1399	+1002	n/a	n/a

Note: births and deaths relate to calendar years, other data are April to April.

Table 10.5 *The Manx population in the period 1961–86*

Year	Total residents	Absolute change	% change	% in Douglas	% in 8 towns	% retired
1961	47166	−6858	−12.7	40	70	22.4
1966	49312	+2146	+4.5	39	71	23.7
1971	53228	+3916	+7.9	36	72	24.5
1976	60496	+7268	+13.7	33	71	24.4
1981	64679	4183	+6.9	31	71	23.8
1986	64282	−396	−0.6	32	72	23.8

dence and national integrity; in fact, quite the contrary. (quoted in Isle of Man Government, 1968, p. 74).

Other submissions concerned more practical aspects of the issue, including the supply of and demand for property, labour supply and the financial repercussions of an ageing population. The Commission, however, argued strongly in favour of the continuing immigration of wealthy retirees and suggested that the real problem was the loss of local people due to a shortage of economic opportunities. They stressed the importance of creating new, permanent manufacturing jobs and also of maintaining planned expansion of the housing stock. What is also of interest about the report is that the discussion is set within the context of movement towards a target population of 75,000, a figure first officially suggested by the Governor in June 1966. This effectively meant the planned addition of some 25,000 to the 1966 population of 49,312.

The 1968 Report clearly did little to allay fears or defuse the issue, for it was followed by a further investigation and report in 1973 (Isle of Man Government, 1973). This again noted the dichotomy of opinion on the Island between those wishing to see further population growth through immigration, and those wishing to protect the quality of life of the existing residents. The latter regarded New Residents as a potential burden on the Island's welfare services and a diluting influence on the Manx way of life. The Select Committee was able to study quantitative aspects of the issue using new data from the 1971 Census. This showed that the population had continued its strong growth with intercensal increases of 3,916 people and 8% (Table 10.5). The number of immigrants was also up sharply to 7,910, but the negative contribution of natural change had also increased whilst emigration had remained roughly constant. New data also showed that the fear of New Residents becoming immediate welfare dependents was misplaced. Indeed the largest percentage of arrivals were aged between 20 and 24 years old and immigration had actually contributed to a reduction in population imbalance by 1971.

The Select Committee viewed the New Residents policy as a way of broadening the population and adding new vitality to it. They recommended no major changes in Government policy but agreed that the temporary cessation of the positive promotion of the New Residents scheme was wise. They also considered that New Residents should be obliged to invest some of their resources in Government securities which might then be used to provide housing for indigenous residents.

The issue surfaced again at the end of 1978, and in January 1979 a further Select Committee was asked to look into the associated questions of population growth and immigration. They reported in 1980 and again they were aided in their deliberations by the availability of new census material for the year 1976. The Committee summarised the population issue as follows:

> The population debate can be divided into two broad categories; on the one hand it is argued that there has been sufficient economic growth and that further growth of the population and of the economy should be restricted and the quality of life should be the prime policy consideration. The other side contends that the standard of living is an integral part of the quality of life, that the Isle of Man has a low standard of living and can easily support further growth of population. (Isle of Man Government, 1980, p. 5).

In more detail, they listed possible adverse effects of uncontrolled immigration and population growth. These were the creation of a two-tier society; the loss of Manx culture; pressure on the countryside and leisure resources; pressure on the welfare and education services; a shortage of housing; and excessive demand for energy.

Superficially at least, the 1976 Census gave support to those afraid of overpopulation and a dilution of Manx Culture. Between 1971 and 1976 the Island's population increased by over 7,000 people, a percentage rise of 13.7 points (Table 10.5). Over 2,000 people emigrated from the Isle of Man and a further 1,000 people were lost through natural change. Immigration, on the other hand, rose by one-third (Table 10.4) and it became obvious that Manx-born people no longer constituted a clear majority of the Island's population. The Select Committee looked beyond these basic data, however, to test some of the common beliefs about the new arrivals. They pointed out that 'New Residents' was a misnomer since the definition of this term also included Manx-born people returning to the Island after a period of residence overseas that exceeded five years. Secondly, they noted that immigration had lowered the Island's average age, not raised it. Thirdly, New Residents were over-represented in the workforce relative to the local population and could therefore hardly be described as welfare-dependent. Fourthly, they occupied jobs for which locals did not yet possess the appropriate skills. And fifthly, the New Residents were making a greater than average contribution to Government resources. Given these facts the Committee recommended that immigration should not be halted but that controls might be instituted to allow its manipulation. They accepted that 75,000 was a feasible and not undesirable population target but that any such targets should be constantly reviewed and should not become ceilings. Finally, the report did sound a warning that if immigration was to continue, economic growth must be guaranteed to prevent the creation of a two-tier society with in-built tension and resentment.

By 1982, when the Manx Government made its next pronouncements upon population policy, economic growth could not be guaranteed since the Manx economy was in deep recession. Unemployment levels had risen two-and-a-half-fold since 1980, and 1,617 people had registered with the employment exchange in January 1982. The 1981 Census showed that emigration had increased by 5,790 since 1976 and the Island's loss through natural change was continuing to mount. Even immigration had receded, and as a result population growth was reduced sharply to 4,000 people and 7% (Table 10.5). Perhaps because of this change in circumstances the Government felt it necessary to defend its strategy of attracting New Residents and restate its desire to see continued population growth. As they put it:

> The relatively high proportion of new arrivals in the growth industries and in senior positions illustrate the extent to which the Island's economy relies upon the importation of skills and experience to encourage development. If Tynwald is committed to a policy of improving living standards and expanding job opportunities then it must avoid introducing unnecessary restrictions on the immigration of people whose presence would be of long-term benefit to the Island. [We] . . . believe that the Isle of Man has the room and resources for the continued expansion of population and [we] consider that restrictions upon immigration must not be allowed to inhibit economic development. (Isle of Man Government, 1982, p. A2).

The same strong belief in the necessity for continued immigration characterised the second report of the Select Committee on Population Growth and the Control of Immigration (Isle of Man Government, 1985). The Committee commented that immigration had declined sharply in the early 1980s and that the 1981–83 level was only one quarter of the annual rate recorded in the period 1976–81. They investigated the wisdom of expanding the Island's population to 75,000, and concluded that this could be achieved economically and that it would produce three clear benefits. These were: an increase in trade and turnover; a reduction in the unit costs of providing public services; and an expansion of available employment opportunities. The Committee therefore recommended 'a co-ordinated programme . . . of population expansion' directed at attracting the wealthy and those who are still economically active (although the two groups may overlap).

When the Manx Government made its next statement on population issues the economic context had changed yet again. By the time the policy document *The Development of a Prosperous and Caring Society* was published in 1987 the Isle of Man had moved out of recession into a period of economic prosperity. Despite this, the document confirmed what now appears to be a considered Government policy that seeks population growth through the positive encouragement of selective immigration. Growth is, however, clearly not being sought at the expense of quality of life since the document makes numerous mentions of the latter as well as the concept of social equality. Moreover the Government has signalled its determination to become actively involved in the management of population by funding a renewed awareness campaign. This seeks actively to attract New Residents and was the backbone of a new Residents policy that was successfully operated between February 1985 and June 1987.

The population of the Isle of Man in the 1980s

In 1986 the Isle of Man had a resident population of just over 64,000 people. In the five years between the Censuses of 1981 and 1986 the Island therefore experienced population loss for the first time since the 1950s. The loss was small. The number of residents fell by 397, or 0.6% (see Table 10.5). Table 10.4 shows the individual components which together produced the shift from population growth to population loss. Natural change barely altered, but the positive contribution made by net migration fell by 82%. Clearly, despite the Government's revitalised New Residents policy, the economic recession has had an impact upon people's willingness to settle in the Island. Economic recovery will undoubtedly change this and it will be surprising if the 1991 Census does not again record a growing population. The statistics for fertility and mortality underscore the dependence of the Island upon migration. Table 10.6 illustrates not only that the Isle of Man has experienced falling birth and death rates in the post-war period but also that the latter has always exceeded the former. Few other European countries are yet in this position but it is now widely seen as a characteristic of states that have moved into a post-industrial phase. The example of these countries requires an extension of the demographic transition model. The Isle of Man also typifies a post-industrial society in other respects; one of these is its age structure. Table 10.7 indicates the aged nature of the Island's population. This reflects not only the greater longevity of people as a result of medical advances but also the character of earlier migrations. The tendency towards smaller families and career couples choosing childlessness accounts for the diminishing number of children and young adults.

Analysis of the industrial profile of Manx employment also illustrates the increasingly post-industrial nature of insular society. Table 10.8 highlights the fall in employment in the primary sector, the relatively large proportion of the labour force supplying consumer services and the dramatic rise in employment within the business and finance sectors. The latter now account for one in five of all workers whereas in 1951 this figure was one in eleven. Table 10.9 indicates the repercussions of this restructuring of employment opportunities upon social class. Almost half the Manx population now have occupations categorised as white collar.

The spatial distribution of the Isle of Man's population has also undergone some change in the post-war years. Earlier parts of this chapter described the progressive urbanisation of the Manx population in the nineteenth century. This trend continued in the Isle of Man and elsewhere in the UK up until the Second World War. Post-war, however, mainland Britain has experienced progressive 'counterurbanisation' with population flowing back out of the major cities to smaller market towns and eventually to remote rural areas. There is some evidence that a parallel trend developed in the Isle of Man but at a much later date and on a far smaller scale. Urbanisation continued unabated until 1981 and, as Table 10.10 shows, the majority of the Island's 'towns' grew at a rate which exceeded the national average. Port Erin grew by 19%, Onchan by 17%, Castletown by 13% and Peel by 12%. Other areas actually lost population and these were often the more remote rural districts such as Patrick, Bride, Jurby and Maughold. After 1981 the pattern seems to have changed somewhat. Some of the 'towns' experienced sharply reduced population growth: Port Erin's growth fell back from 19% to 2%, whilst Onchan's dropped from 17% to 2%. Others actually experienced population loss: both Peel and Ramsey had reduced populations. Elsewhere in the Island some of the remote rural districts continued to lose population, but other rural districts were clearly more attractive to those leaving the towns. Michael, Ballaugh and German exemplify these new rural commuter areas.

Indeed, when all the twenty-six districts of the Isle of Man are ranked according to their population change over the period 1976–86, it is not the established centres of population that have experienced the most rapid growth. Rather it is the new suburban dormitories for immigrants and Douglas commuters (e.g.

Table 10.6 *Changing fertility and mortality in the Isle of Man, 1951–86, per 1,000*

Year	Birth rate	Death rate
1951	15.1	15.7
1961	14.1	18.2
1966	13.0	17.1
1971	15.1	17.4
1976	11.9	16.1
1981	11.6	15.1
1986	11.0	14.8

Table 10.7 *The age structure of the Isle of Man's population, 1986*

Age last birthday	% of males	% of females	% of total
0–4	5.5	5.0	5.2
5–9	6.0	5.3	5.2
10–14	7.2	6.3	6.7
15–19	8.0	6.9	7.4
20–24	7.5	6.4	6.9
25–29	6.0	5.7	5.9
30–34	6.1	5.6	5.8
35–39	7.4	6.8	7.1
40–44	6.3	5.9	6.1
45–49	5.7	5.3	5.5
50–54	5.3	4.9	5.1
55–59	5.1	5.7	5.4
60–64	5.8	6.4	6.1
65–69	5.3	5.9	5.6
70–74	5.2	6.1	5.7
75+	6.9	11.3	9.2
TOTAL	99.3	99.5	98.9
n=	30,782	33,500	64,282

Note: figures do not sum to 100% because of 'not stateds' and rounding.

Table 10.8 *Changing employment by industrial sector of Manx residents, 1951–81*

	% of labour force		
	1951	1971	1981
Agriculture, forestry and fishing	11	6	5
Manufacturing:			
food, drink and tobacco	3	2	3
engineering	4	5	7
textiles, clothing and footwear	3	3	2
Other; mining and quarrying	3	3	3
Construction	13	12	11
Gas, electricity and water	3	2	2
Transport and communication	9	8	9
Wholesale distribution	14	4	3
Retail distribution		12	10
Insurance, banking, finance and business services	2	3	6
Professional, educational, medical and scientific	7	12	14
Tourism, catering, entertainment and miscellaneous	20	19	18
Public administration	8	5	6
TOTAL	100%	96%	99%
n=	23,257	22,133	25,864

Table 10.9 *Social class of economically active residents of the Isle of Man, 1981 (%)*

Social class		Total population	Males	Females
I	Professional and managerial	3.3	4.9	0.8
II	Other white collar	23.2	23.0	23.8
IIIN	Junior non-manual	21.5	12.3	36.4
IIIM	Skilled manual	24.9	35.0	8.6
IV	Semi-skilled manual	16.8	14.8	19.9
V	Unskilled	6.0	6.2	5.6
TOTAL		95.8%	96.2%	95.1%
n=		27,564	17,027	10,537

Note: figures do not sum to 100% because of 'not classified' and rounding.

Onchan +19% and Port Erin +22%), the commutable villages (Michael +31%) and the environmentally attractive rural areas (German +21%, Michael parish +17%, Andreas +17% and Arbory +14%). The relative and absolute losers have been the older settlements (Douglas +2%, Laxey +3% and Ramsey +8%) and the less attractive rural areas (Jurby −9%, and Bride −12%). A more detailed description of intercensal change can be found in Figure 10.3.

Finally, in view of the importance of net immigration to the Isle of Man and also the centrality of New Residents to Government demographic and economic policy, it is worth considering the characteristics of this demographic group. Fortunately the 1981 Census provides us with detailed information on those people renewing residence or taking it up for the first time after 1961. They are shown to be predominantly (82%) of UK or Irish birth, although 9% were actually born on the Island itself. Fifty per cent had prior experience of extended residence on the Isle of Man, whilst 19% had a Manx spouse. Twenty-two per cent had associations with the Island through their parents. The largest single reason for migrating to the Isle of Man was for work although a significant number came to retire. The former group account for the higher proportion of migrants aged between 30 and 39 years whilst the latter are associated with an over-representation of migrants in the over 60–64-years age-group. Those who arrived between 1961 and 1971 were more likely to be white collar workers than were local residents whilst this was even more true of the post 1971 arrivals. The vast majority of New Residents had come to the Isle of Man from the UK but other previous countries of residence included Eire (2%), Australia (0.7%), South Africa (0.7%), Gibraltar and Malta (0.6%), Canada, Kenya and Hong Kong. The pattern of their settlement in the Isle of Man is shown by the percentage which New Residents form of district populations. This can be found in Figure 10.4. It illustrates how New Residents have been attracted to rural parishes and villages rather than towns. Douglas was the district with the lowest proportion of New Residents whilst Peel, Castletown and Onchan were also relatively unpopular. Both the parish and village of Michael, on the other hand, had around half their populations of New Resident stock. Maughold, Port Erin, Arbory, Ballaugh and Rushen had also attracted considerable proportions of New Residents.

Additional information on the New Residents can also be gleaned from a 1986 survey sponsored by the Isle of Man Government and undertaken by MORI. This revealed that the principal sources of information about the Island available to potential migrants were

Table 10.10 *Changing population distribution 1976–86*

	Resident population			% change	
	1976	1981	1986	1976–81	1981–86
Towns:					
Douglas	19897	19944	20368	+0.2	+2.1
Castletown	2788	3141	3019	+12.7	−3.9
Peel	3295	3688	3660	+11.9	−0.8
Ramsey	5372	5818	5778	+8.3	−0.7
Villages:					
Laxey	1242	1257	1279	+1.2	+1.7
Michael	437	532	574	+21.7	+7.9
Onchan	6395	7478	7608	+16.9	+1.7
Port Erin	2356	2812	2868	+19.3	+2.0
Port St Mary	1525	1572	1610	+1.3	+2.4
Parishes:					
Andreas	949	1113	1115	+17.3	+0.2
Arbory	1414	1651	1610	+16.8	−2.5
Ballaugh	655	681	745	+4.0	+9.4
Braddan	2095	2520	1804	+20.2	−28.4
Bride	428	413	378	−3.5	−8.5
German	866	1029	1051	+18.8	+2.1
Jurby	640	616	582	−3.8	−5.5
Lezayre	1339	468	1362	+9.6	−7.2
Lonan	1129	1126	1139	−0.3	+1.1
Malew	2116	2080	2054	−1.7	−1.2
Marown	1230	1290	1281	+4.9	−0.7
Maughold	763	759	755	−0.5	−0.5
Michael	402	445	470	+10.7	+5.6
Onchan	414	401	341	−3.1	−14.9
Patrick	1078	1077	1044	−0.1	−3.1
Rushen	1274	1341	1360	+5.3	+1.4
Santon	397	427	427	+7.6	0
TOTAL	60496	64679	64282	+6.9	−0.6

relatives, business associates and friends. The prime attractions of the Island were thought to be personal taxation, the low crime rate, the landscape, and the relaxed pace of life. Sixty per cent of respondents felt their expectations had been exceeded; particular features that were mentioned here were the friendliness of the local people, the natural beauty of the Island and the pace of life. Disappointments included shopping facilities, the ease and cost of transport to the Island, the cost of living and a sense of isolation.

Conclusion

The Isle of Man represents an interesting natural laboratory for demographic concepts. Its demographic history stands comparison with the idealised demographic transition model. It has experienced the effects of substantial emigration, and those people who have left have conformed to many of the regularities in patterns of settlement and maintenance of ethnic identity recognised amongst other immigrant groups. More

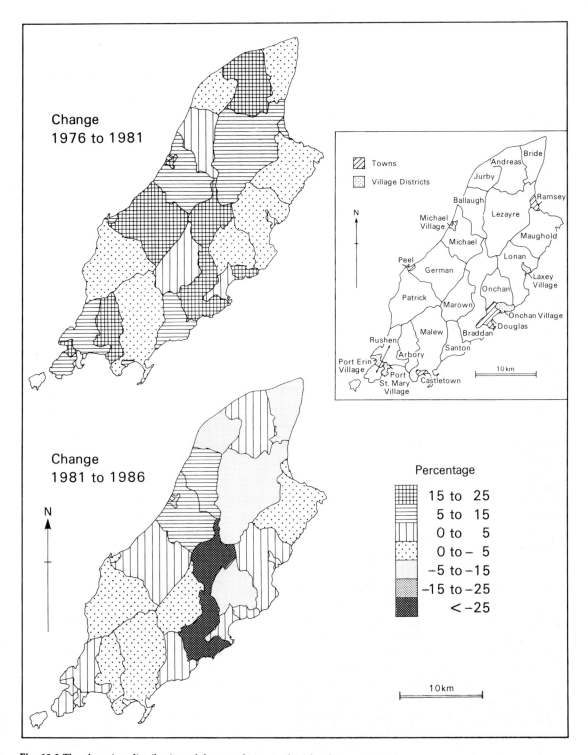

Fig. 10.3 The changing distribution of the population in the Isle of Man, 1976–86

recently, the Isle of Man has entered a new post-industrial phase in the development of its demography. There are no precedents for this, and as yet few parallels. Coupled with this, the Government of the Island has sought consciously to intervene in demographic matters through the setting of targets and the New Residents policy. The latter is itself relatively unique in the Western World. At a time when many developed countries are doing their utmost to curb immigration, the Isle of Man is deliberately stimulating it. Moreover this policy seeks to attract not only the wealthy but also working people. Again this is in direct contrast to many other nations which feel that their working population is already too large for the available employment opportunities and are therefore trying to limit new entrants.

10.6 A recent development on the Island; these are homes in the Saddle Mews retirement village on the outskirts of Douglas

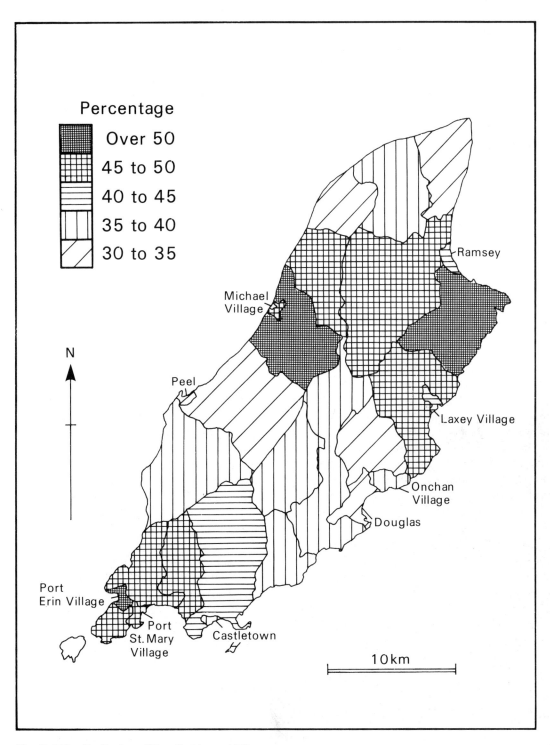

Fig. 10.4 The distribution of New Residents, 1981

PART TWO: HISTORY, ADMINISTRATION AND POPULATION

Conclusion

From the very earliest periods of man's habitation of the British Isles right down to the Second World War, the Irish Sea has always held a special strategic significance. In the earlier periods it acted as the main conduit for trade between the Mediterranean and the northern nations and more recently it dominated the vital transatlantic routes from Britain's western seaboard. The Irish Sea has thus always been an important thoroughfare but, in addition, it has been the boundary zone between adjoining cultures and peoples. Like any other boundary, it has responded to the dominance of any one particular power and at different times the Irish Sea has been counted as part of the sphere of influence of a variety of cultures.

Sited as it is, at the centre of the north Irish Sea, the Isle of Man has long been a focus for the dense networks of interaction which Bowen (1970) showed existed within the Irish Sea as early as the Bronze Age and which encouraged Mackinder (1902) to liken the region to the Mediterranean. Not only was the Island strategically important for the control of the region but it also represented a prized possession in its own right, courtesy of the fertile agricultural lands of the coastal plains. As a result, the Isle of Man has rarely been left to its own devices for very long, becoming a matter of contention amongst its neighbours in periods of political or economic stress. As Chapter 8 indicates, within the historic timescale the Welsh, Irish, Scottish, English and Scandinavian interests have, at various times, felt obliged to control the Isle of Man in pursuit of wider territorial or military ambitions. These powerful influences have not only shaped the present landscape (and archaeology) but they have also had a profound impact upon the administration and institutions of the Island. Indeed as all three chapters in this section illustrate, the development of Manx society and culture accurately reflects the shifting balance between the two opposed tendencies of isolation and integration. When external forces have dominated, there has been the introduction of outside elements, but during the intervening periods of calm when external forces have been at a minimum, the Island has developed its own characteristic insular variants of regional artifacts, structures and institutions. The accretion of these unique insular characteristics has produced an identity which is perceived both within the Island and off the Island as strongly independent. And it is this strong sense of independence and uniqueness which makes the Isle of Man such an interesting region for the geographer to study, for at the core of that discipline has always been the unique character of places.

PART THREE

THE ISLAND'S INFRASTRUCTURE

Introduction

Part Four of this book describes the main phases through which the Manx economy has progressed over the last two hundred years. Naturally the development of the Island's infrastructure has been closely tied to changes in the economy. The dominance of successive economic sectors has demanded specialist infrastructure, much of which can still be seen in the landscape today, albeit in relic form. A glance at Douglas or Foxdale proves the point. However, in addition, the quantity and type of infrastructure available on the Island has also reflected the Isle of Man's political status and its relationship to larger economic systems.

In the earliest stages, the Isle of Man was a very introverted, even parochial, society in which local needs were met from local resources. The Island was bound to the mainland, but was never a part of its economic system. Few goods were exported on a consistent basis and few were imported in bulk. Self-sufficiency therefore prompted the need for very limited and only local infrastructure. Mills were provided in each locality and goods were transported very short distances by horse-drawn sled. As Chapter 11 makes clear, the road network was rudimentary. Ships were built near to sources of timber and then dragged to the sea, and vessels were simply left on the beaches when not in use.

The nineteenth-century growth of tourism and mineral extraction heralded a new relationship with the mainland. The Isle of Man became almost an internal colony of the type described by Hechter (1975). Specialist infrastructure was installed extensively and rapidly but it was all orientated to external demands and developments. The harbours were constructed and extended so as to facilitate the export of minerals and the ingress and egress of day trippers and long-stay tourists. Roads were laid to open the interior to mainland tourists. Railways were built for the same purpose and also to speed the shipment of ores. Few indeed of these initiatives sprang from the internal dynamics of Manx society or economy. And even within those developments which pre-dated the 'externalisation' of the Isle of Man, it was the 'mainland-orientated' enterprises which thrived. Those flour mills which were able to take advantage of bulk imports of grain survived. The purely local did not. Those woollen mills which orientated themselves to external markets thrived, others did not.

The twentieth century has shown the costs and disadvantages of being dependent upon external stimuli and of being a peripheral part of a much larger system. Demands have changed and much of the Island's externally-orientated infrastructure is now redundant. The closure of the rail network provides one example as does the change in use of seafront boarding houses. The demise of the mining industry is yet another. In the post-war period the progressive re-assertion of Manx independence from mainland Britain has provided the Island with the confidence and resources to seek out a new and independent role in larger systems. Increasingly this has also been reflected in the Island's infrastructure. Developments now occur because of internal initiatives and internal demands. They are planned and controlled from within the Island. The expansion and internalisation of the telecommunications industry is a case in point. The creation of an indigenous airline is another. And the decision to modernise and expand the power generation industry independently of the Central Electricity Generating Board yet another. No longer is the Isle of Man simply a peripheral region supplying cheap materials, labour, and leisure to an advanced metropolitan core. It has instead usurped many of the higher-order functions which allow it to make its own decisions and chart its own progress.

CHAPTER 11

Social infrastructure
Richard Prentice

Introduction

The Isle of Man lacks statistics on the distribution of incomes. As such, the extent to which the recent increased prosperity of the financial sector has increased the income of the average household, and, say, the incomes of the poorest fifth of households, is unknown. Indeed, when *real* incomes are considered, that is, incomes after price inflation has been taken into account, the recent prosperity may have worsened the living standards of some groups. Foremost amongst these potentially disadvantaged groups are young households seeking to enter the rented or owner-occupied private property market, and finding rapid house price inflation during 1987 and 1988. For example, in the one year between 1986 and 1987 average house prices on the Island increased by over a quarter (Webster, 1988). However, writings on the assessment of the 'quality of life' go beyond money incomes and tend to disaggregate this broad term into a series of domains. These are designed to encompass both the basic and higher needs of mankind. They include nutrition, shelter and health at one end of the continuum through to the need for security and the desire for education and recreation at the other. Not all of these components of the level of living are relevant to a geographical study of the Isle of Man but three, in particular, are of some significance and interest: housing, social policy and leisure resources (and their usage). These, then, are the major concerns of this chapter.

Housing

The Island's housing stock is largely a mixture of pre-1919 and post-1960s stock: in 1984 houses of these ages each represented an estimated 37% of the stock (Department of Local Government and the Environment, 1987). Not only is this bimodal age distribution unusual when compared with Great Britain, it also tends to conceal some less than desirable attributes of the stock behind an appearance of estates of newly constructed dwellings. As much of the older stock is located in Douglas, the general impression outside of Douglas is of new housing, and expectations, therefore, of good amenities and good repair. In reality, these expectations are not fully met. In terms of amenity the housing stock is generally well provided, and at the 1981 census only 3% of households lacked the exclusive use of a fixed bath or shower, and only a like proportion lacked the exclusive use of an inside water closet. The 1984 House Condition Survey for the Island showed that just over half of the dwellings on the Island which lacked at least one amenity were owner-occupied (Department of Local Government and the Environment, 1987), although in comparison with their overall proportion of the stock, privately rented housing was found, as would be expected from experience in Great Britain, to be disproportionately lacking amenities. Likewise, eight out of ten dwellings which lacked at least one amenity were of pre-1919 construction.

Whereas the general picture is one of housing which is well provided with basic amenities, the same cannot be said about the basic repair of some Manx housing. Most housing on the Island is in good repair, but a substantial minority of the stock is not. The recent House Condition Survey has enforced a reappraisal not only of policies, but also of descriptions of the stock. The delays in publishing the survey were indicative of the political surprise caused by its findings. Two findings were of particular pertinence. First, just over one in seven houses on the Island were estimated from the survey to be statutorily unfit for human habitation, nearly nine out of ten of these houses and flats having been built before 1919. Secondly, a third of all homes on the Island were estimated to require expenditure in excess of £3,000 on basic repairs. These findings may be compared with those for neighbouring countries. In Northern Ireland just under one in five houses are officially estimated to be statutorily unfit for human habitation, in Wales one in eleven houses, and in Northern England, one in thirteen (Prentice, 1986a; Sizer and Kelly, 1987). Clearly, the extent of the problem of unfitness on the Island is worse than generally found in Great Britain. In part, this may reflect the small size of the Island and the age of its housing stock as well as policies, as in Great Britain and Ireland, favouring initial ownership of housing, rather than continued maintenance of owner-occupied property. On the Island, housing unfitness is distributed across all settlement types; in contrast, for example, in Wales, specific surveys have suggested much greater concentrations in certain districts, notably Cynon Valley and the neighbouring districts of comparable nineteenth century housing (Prentice, 1986a). Compared with the rate of unfitness of 50% found for Cynon Valley, the Island's problems seem less severe, but still undesirable. Rates of disrepair may also be compared with the Island's neighbours. Whereas a third of the Island's stock required estimated repairs in excess of £3,000, the same was true for a quarter of the stock in Northern Ireland and a fifth of the English stock. In Wales slightly different categories of costs have been used; for 1981 the Welsh Office estimated that nearly a quarter of the housing stock needed repairs of in excess of £2,500, and that one in six households in Wales were living in housing conditions which were officially unacceptable. Clearly, compared with its neighbours it would seem that much of the Manx housing stock is in particular need of repair.

The House Condition Survey showed that unfit houses on the Island were characterised by greater than average levels of deficiency in amenities and by high levels of disrepair. A fifth of unfit houses were estimated to lack at least one of the five basic amenities, and in excess of four out of ten were estimated as requiring repair expenditure in excess of £12,000. The average repair cost for unfit houses was estimated at £10,820 giving an estimated total repair bill for the unfit stock of £45.92 million. The most common recurrent defects in the unfit stock were disrepair, dampness and ventilation deficiencies, found in an estimated nine out of ten, three-quarters and two-thirds respectively of unfit houses. The latter deficiency is particularly unusual by British standards.

When Manx housing unfitness is expressed as a percentage of the total housing stock of the areas of the Island, the highest rate of unfitness is found for owner-occupied houses in the rural areas. This reflects a problem of single property-based action, rather than one of area-based actions as have been used in Great Britain, particularly in the 1970s. The Manx problem in this regard indicates a typical pattern of unfitness in rural housing. In contrast, in Douglas it is the privately rented stock in which unfitness was found to be concentrated. This suggests the need to develop a range of policies on the Island, rather than a single policy, to counter this problem.

In deciding future policies, the 1984 Survey has implied both descriptions and prescriptions. The essential problem is of housing disrepair, and not of housing unfitness, for both fit and unfit housing were found to be in disrepair. As noted above the problem is different in urban and rural areas. Further, the problem is not just associated with privately rented housing, as is often imagined. These descriptions lead to certain prescriptions. In the urban areas, and principally in Douglas, area-based rehabilitation of housing for continued renting is needed. In the rural areas, the rehabilitation and subsequent sale back to owner occupation of housing in disrepair is implied. Vacant housing, and housing put up for sale, if in disrepair, could first be improved by a publicly funded agency. In Great Britain housing associations have been active in both of these roles, but on the Island they have not developed extensively. The problem of making people aware of housing disrepair is also suggested. Loans may need to be made available, or substantial grants, to owner occupiers to finance these schemes. Under the Housing Improvement Act of 1975 grant aid and loans are available to bring residential property up to a prescribed standard of amenity and repair. However, the maximum grant is small by recent British comparison, £1,250: instead, the Act relies on loans of up to £10,000 to encourage improvements. For properties built before 1953 more recent schemes have provided up to a half of the costs of repair in grant, up to a maximum grant of £5,000. This, the Residential Property Modernisation Scheme of 1988, and antecedent schemes since 1981, have given greater grant aid, but still little when compared to the British initiatives of the early 1980s which resulted in the 'grants bonanza'. Indeed, the antecedent schemes of the 1980s on the Island only met up to a quarter of the costs of repair; proportionately much less than in the United Kingdom. As a sizeable proportion of the owners of houses in disrepair on the Island are elderly, interim action by a public agency to make their houses temporarily weatherproof may be needed. In short, the Island's problems, and their possible solutions, mirror the changes of policy being experienced in Great Britain. However, in one important respect the Island is showing a lead to the United Kingdom: much more recently built houses are eligible for grant aid than in the United Kingdom. In particular, inter-war houses are eligible for grant-aided works on the Island. Of fundamental long-term importance is the need to encourage owner occupiers to maintain their properties, and not just to encourage them to purchase them in the first place. Equally, the influx of New Residents cannot be assumed to have solved this problem as increased purchase prices effectively deter younger households from improving their first homes as mortgage repayments are so much higher.

Most of the Island's housing stock is owner-occupied, and this in itself emphasises the need to encourage the maintenance of this stock. At the 1981 Census just over six out of ten houses were owner-occupied, and notably higher rates in the Island's villages and parishes. The greatest concentration of owner-occupation was in Michael where over eight out of every ten houses were owner-occupied. Surprisingly, in comparison with Great Britain owner-occupation on the Island in the past has not been financed by building societies. Tax exiles and many immigrants have bought housing outright on the Island, but for many mortgage finance has traditionally been provided by the Island's banks or the Isle of Man Government (Prentice, 1984). The present Government scheme was set up in 1978 and in general applies to first time buyers who would otherwise be unable to purchase a property of their own. The scheme is only available to three kinds of residents; namely, Manx born, persons holding an indefinite work permit, and persons residentially qualified under the Control of Employment Act. In 1988 the scheme was further modified, in an attempt to counter the effects of rising house prices, by the introduction of low start interest schemes in which part of the interest repayments of a mortgage are deferred. For British readers the past absence of building societies may be surprising. Manx law was seen by British building societies as barring their activities on the Island. This was further complicated by the previous prohibition on building societies by British law of setting up subsidiaries or making payments to them. In 1984 both the Halifax and Leicester Building Societies were seeking to set up offices on the Island, not to finance Manx owner-occupation, but instead to take deposits from investors and to apply interest without deducting tax as is required in the United Kingdom. These initiatives were blocked by the British Chief Registrar of Friendly Societies. British building society law has since changed with the Building Societies Act 1986 and in 1986 Manx law was also amended to facilitate building society activity, and in 1988 the first British building society set up an office on the Island. By the end of 1988 four major United Kingdom building societies had opened offices on the Island. This change is part of the Island's Government's desire to reduce public spending by having building societies as well as banks finance owner-occupation, rather than the Government itself.

The Island's public sector stock of rented housing represented about a fifth of the stock in 1981. This was a smaller proportion of the stock than pertained in Great Britain at the time, although as the Island has no comparable provision to the British 'Right to Buy' this disparity has been reduced in the 1980s. However, on the Island new construction of publicly owned housing for rent has shown a marked decline in the 1980s (Table 11.1). Publicly rented housing on the Island is concentrated in the towns. For example, in 1981 twice the national proportion of households living in publicly rented housing were public tenants in Castletown: in Douglas, such tenants represented a third more households than found nationally. The one notable exception to this urban concentration is the parish of Jurby. Here, in proximity to the airfield, over half of the 198 enumerated households were found to be public sector tenants. More usually, in the Island's rural areas public tenants were enumerated as under one in seven resident households. In

Table 11.1 *New housing construction, 1972–85*

	Public sector	Private sector	Total
1972	144	567	711
1973	146	636	782
1974	132	465	597
1975	124	443	567
1976	94	360	454
1977	202	428	630
1978	81	438	519
1979	44	361	405
1980	56	288	344
1981	138	266	404
1982	79	261	340
1983	32	174	206
1984	1	167	168
1985	1	126	127

Source: Isle of Man Local Government Board.

the extreme case, that of Michael, no household was enumerated as a public sector tenant. Public housing on the Island is provided by Douglas Corporation, the Town Commissioners for Castletown, Peel and Ramsey, and various of the village and parish commissioners. Elsewhere on the Island public housing is provided by the Manx Government directly, now through the Department of Local Government and the Environment, but formerly through the Local Government Board. Local authorities and the Department are required by statute to consider housing conditions in their district, and the needs of their district with respect to the provision of further housing accommodation. This is a comparable duty to that placed on housing authorities in Great Britain. In practice, most public rented housing is either owned by the Department of Local Government and the Environment, or by Douglas Corporation; roughly a quarter and four-tenths respectively of the total stock of 1981. Traditionally, each housing authority on the Island has defined its own criteria for allocating public housing. However, the freedom of local authorities in this regard was questioned by the Local Government Board in 1984 (Isle of Man Local Government Board, 1984) and indeed since legislation in 1976, local authorities have been required by statute not normally to house persons who have been resident on the Island for under ten years. Enacted at a time when British local authorities have progressively relaxed residential requirements, the latter restriction implies a preference in terms of immigration for affluent migrants, in practice, owner-occupiers. The local authorities are in general, as in Great Britain, required by statute to give preference in rehousing to persons occupying insanitary or overcrowded housing, or having large families, or who are living under unsatisfactory housing conditions. These preferences clearly favour rehousing families from what were traditionally thought of as 'slums'. These preferences may still apply in terms of housing in multiple occupation, but, as is needed in Great Britain, a restatement may be desirable in terms of giving statutory preferences to other need groups, in particular the elderly or households living in housing in structural disrepair. Further, the Island has no comparable provision to British legislation on homeless persons. This need group is ignored in the statutory preferences which have applied since the mid 1950s.

Local authority housing on the Island is in comparatively good structural condition: only one in twenty of these lets was found to be unfit for human habitation by the 1984 Survey. An exception are the 38 dwellings forming Cronkbourne village which were built in the period 1830 to 1850 to house workers at the nearby flax mills (see Chapter 10). For their period of construction these houses represented a model village; and, arguably, are the Manx equivalent of Port Sunlight or Bournville in Great Britain. Owned for the past two decades by Braddan Parish Commissioners, a report of 1986 by the architect John Cryer showed many houses to be unsatisfactory in terms of internal plans, low standards of amenity, repair, rising damp and condensation. Mr Cryer's proposals included the rehabilitation of these dwellings to modern standards and their reduction in number from 38 to 25 by amalgamations and demolition. The estate demonstrates both changing standards and the difficulties of modernising properties while retaining the unique environment of the village. The estate is also indicative of the need to look beyond the facades of houses when deciding housing policies.

In 1984 the Local Government Board reported to Tynwald on the Island's housing policies (Isle of Man Local Government Board, 1984). Included in the Board's recommendations were proposals to increase the central supervision of the Island's public housing. In particular the Board recommended legislation to compel local authorities to adhere to the Board's policy directions in such matters as rents, rebates, allowances, criteria for acceptance on to housing waiting lists and forms of tenancy agreements. The Board also proposed a limited system of spot checks to verify that local authorities were achieving an adequate standard of public housing maintenance. It proposed that local authorities should retain their housing responsibilities only if the local authorities '. . . face[d] up to their responsibilities'. This implied centralisation was enabled by the Local Government Act of 1985, under which a Board inspector could provide technical assistance or undertake a local authority's functions. This power was applicable to all enactments concerning local government and public health. Further, the Act enabled the Board to undertake inquiries into local government matters. To date, no Department or Board inspector has undertaken the housing role of a local authority.

Public attitudes to housing on the Island are not generally at variance with the policies of the Department, or former Board. In particular, a survey of 1985 showed very strong support, even from public sector tenants, for the encouragement of owner-occupation, both generally and for persons of limited means (Prentice, 1986b). However, the survey did show general public support for greater public assistance towards housing repairs and improvement, and from owner-occupiers for the sale of public sector houses to their tenants. There was extensive support for the centralisation policies in terms of common rent setting, and, in particular, for common rules of application for public housing. In general, the survey found substantial support for continuing public sector intervention in the Island's housing markets. It is not that the public on the Island questions the value of public intervention in housing in *general*. Questioning centres on the *nature* of the intervention. The effects of the 1984 Housing Condition Survey have yet to be felt in terms of policy changes. It is this latter Survey and the former Board's commitment of 1984 to monitoring changes in housing standards which should prompt developments in the Island's housing policies. Similarly, the rapid rise in house prices on the Island in 1988 has brought public pressure for Government to act, to enable Manx first-time buyers to afford properties. The response of Government and local authorities in promoting starter home developments again indicates a preparedness to intervene in the private market, at least at the margins of supply.

Social policy

The National Health Service on the Island is similar to that in the United Kingdom and the National Health Service (Isle of Man) Act of 1948 closely follows the British Act of 1946. Administration of the Island's health service is carried out by the Island's Department of Health and Social Security, and formerly by the Isle of Man Health Services Board. It is the duty of the Department 'to promote the establishment in the Isle of Man of a comprehensive health service designed to secure improvement of the physical and mental health of the people in the Isle of Man and the prevention, diagnosis and treatment of illness'. The Department also administers the Island's social security schemes other than the health service; formerly this was the responsibility of the Isle of Man Board of Social Security. The Island's social security legislation is similar to that of the United Kingdom, although a number of differences exist, including the absence of a state redundancy pay scheme on the Island. Education policy on the Island is administered by the Island's Department of Education and the Isle of Man Board of Education; formerly educational policy was administered by an antecedent Board of Education. The present Board consists of fifteen elected members, who by statute cannot also be members of Tynwald.

Recent population changes have particular implications for the Island's social policies. Between the censuses of 1981 and 1986 the proportion of the Manx population aged under 14 years declined markedly. There were approximately 12,200 people age 14 and under in 1981, but this had dropped by about a thousand five years later (Gatrell, 1987). Since the population aged under four years has also declined over this period, this reduction may to some extent be due to a lower birth rate. However, there was also a decline in the number of persons aged 30–34 years, which suggests that a substantial proportion of the reduction in those aged under 15 years is due to emigration of families with growing children. Taken together, emigration and a lowered birth rate have, if they continue, an immediate implication for school provision on the Island. A further population change of the early 1980s is of importance to health care. Between 1981 and 1986 the proportion of the population aged 75 and over increased, from about 8% to 9%. It is unclear whether this is due to immigration or to a reduction in mortality. However, the likely need for increased medical services on the Island if this trend continues is clear. About a fifth of the Island's population was aged 65 years or over in 1986 (see Chapter 10). The spatial concentration of these persons in Ramsey, Lezayre and Maughold in the north-east and in Port Erin and Port St Mary in the south-west of the Island implies the need to target spatially community health care for the aged. The extent to which aged widows or widowers, who had emigrated to the Island with their spouses at retirement, return to their families in the United Kingdom when unable to manage by themselves is unknown; this outflow of frail elderly may reduce the impact of the aged on the Island's health care facilities, but clearly will not remove this impact.

The Island's welfare services range widely. Health services include general, psychiatric and maternity hospital accommodation, family practitioner services, personal care services by dentists, pharmacists and opticians, preventive medicine and community services. If a particular specialist treatment is unobtainable on the Island, Manx residents are then treated in the United Kingdom. Social security schemes include provisions for retirement pensions and benefits to the unemployed, low-paid, sick and others in need. A residential qualification may be necessary before benefit is paid. Certain other Government schemes also have a welfare objective, although run by other Departments. For example, in 1986 the Isle of Man Industry Board introduced a scheme for retraining unemployed persons, and to provide work experience to the unemployed. This scheme directly subsidised employers willing to provide employment training. Schemes of this kind, which have both welfare and other objectives, mean that any analysis of welfare expenditure on the Island is hampered by definitional problems. Such problems are by no means unique to the Isle of Man, and imply service-by-service analyses. As with health care generally in the western world, hospital costs account for the lion's share of Manx National Health Service spending; upwards of two-thirds of all public health service spending in 1982–83. Broadly defined community services, including general medical, pharmaceutical, dental, ophthalmic, health visiting and home nursing, accounted for up to 28% of expenditure in the same year; of which pharmaceutical services was by far the largest component. Specific non-hospital based geriatric services accounted for only just over a twentieth of the total expenditure; this is however a gross underestimate for it takes no account of general practitioners' time or community nursing (Webster and Dawson, 1984). The Island's hospital services go back to a charitable gift of the original Noble's Hospital in 1888, and the establishment by statute of a Trust for this hospital and dispensary in 1909. The original hospital site and building now house the Manx Museum. Several of the community-based professional health services offered on the Island have expanded during recent years, notably general practitioner, district nursing and health visitor services. In 1983 the Island had a third more general practitioners than a decade earlier, a quarter more full-time district nurses and twice as many health visitors (Webster and Dawson, 1984).

As in the United Kingdom, expansion in social security payments in the 1980s has in large part resulted from higher levels of unemployment than were experienced in the previous decade. In 1984 the Board of Social Security had just short of three times as many unemployment benefit claimants than were estimated for a decade earlier by the Board (Webster and Dawson, 1984). This compares, for example, with an increase of just less than a sixth in the number of retirement pensioner claimants in the same period. However, even with high unemployment, in 1984 the Island still had twelve times as many retirement pensioner claimants as it did unemployment claimants. Means-tested supplementary benefit claimants are also substantially retired persons; in 1984 this group accounted for just over half of all supplementary benefit claimants on the Island. However, the 1980s have brought a substantial increase in the proportion of young unemployed persons, who are often ineligible for unemployment benefit and who, instead, will often receive means-tested supplementary benefit because they have no other income or substantial savings.

In the three-year period 1981 to 1984 the number of persons required to register for work who were receiving supplementary benefit increased six-fold. However, 1988 saw a marked reduction generally in unemployment on the Island, with consequent reductions in social security payments.

Concurring with the population figures noted above, the 1980s have brought a decline in the number of primary school pupils attending the Island's schools: in the three years 1980 to 1983 the number of primary pupils decreased by nearly one in ten, to 5,193 in number. This number represents a decline from the recent high point of 1977 when 5,864 primary pupils were being taught (Webster and Dawson, 1984). In contrast, pupils attending secondary schools increased annually in number from 1973 to 1982, although this increase now seems to have stopped. As in the United Kingdom, the increase in the proportion of secondary pupils compared to primary pupils implies the need for greater expenditure in total, as secondary pupils generally cost more to teach. Whereas in 1983 the Island's primary schools were on the whole more favourably staffed in terms of pupil numbers compared with those in north-west England, the reverse was marginally the case for secondary schools: the ratio of primary school pupils per teacher was nearly 10% less than in northern England in 1983, but for secondary pupils per teacher it was nearly 4% greater. These staffing ratios reflect both the demographic changes occurring on the Island and the increasing tendency for secondary school pupils to remain at school subsequent to the statutory minimum leaving age. With recent youth unemployment, job scarcity and the present demand for *skilled* labour, the latter trend is likely to continue. The same is true for the numbers of Manx school leavers entering higher education. In 1982–83 twice as many students than in 1972–73 were in receipt of awards for degree or higher national diploma courses.

As planning of the Island's welfare services is in part dependent on the changes in United Kingdom policies which are frequently enacted also by Tynwald and on circumstances to various degrees outside of the Island's control, such as unemployment or population changes, service developments are not easy to predict. In 1984–85 the Isle of Man Health Service Advisory Council's attempts at health services forward planning were considered inappropriate by the then Health Services Board (Isle of Man Health Service Advisory Council, 1985). The Island has not, however, experienced the substantial swings in the ideologies of social policy comparable with those of Great Britain resulting from political change, and on the Island the post-war consensus on social policy largely remains and is fundamental to the Island's *Prosperous and Caring Society* policy introduced in 1987 as a means of raising the Islanders' standards of living.

Leisure provision on the Isle of Man

In the absence of a comprehensive survey of Manx leisure interests and activities, reference has to be made to British studies for guidance as to leisure preferences. A recent leisure study by the Countryside Commission for Scotland has emphasised the importance to all ages of television and radio, visits to friends, shopping, outings and walks; and of outdoor sport, clubs and dances to the young, and of do-it-yourself, gardening and voluntary activities to older persons (Thomas and Gershuny 1985). Likewise, Countryside Commission surveys for England and Wales of countryside recreation have emphasised the importance of drives and outings to the countryside, visits to historic buildings, to friends or to the sea coast, and of long walks (Central Statistical Office, 1987). Participation in leisure activities is clearly affected by opportunities to engage in activities, and it is likely that, on the Island, these patterns found for Great Britain are more, and not less, likely to pertain, simply because of the pro-

11.1 Bishops Court Glen originated as the garden for Bishops Court. It is a peaceful but little visited National Glen

11.2 This waterwheel was part of the illusion the Edwardians sought when promenading in the Island's glens

ximity of coastline and countryside to Manx residents.

As the countryside forms at least a background to much outdoor leisure activity, in a period of economic change the conservation of landscape is critical to the retention of this major recreational resource. The Island does not have areas in which development is forbidden as a matter of policy. However, a major landowner in the upland area of the Island is the Department of Agriculture, Fisheries and Forestry, with its areas of forest and other lands. The Forestry Act 1984 clearly restated that the promotion of forestry is the first duty of the former Board, and its successor, but that recreational use is a recognised secondary activity. The Board's successor is charged with promoting the interests of forestry, the development of afforestation and the production and supply of timber. The meaning of 'forestry' given in the Act is, however, wide and includes 'the growing of fruit trees and decorative shrubs, and the cultivation of trees in the interests of amenity' (Section 12). Recreational provision is one of eighteen specific duties outlined for forestry provision by this Act. However, effective provision is much wider as the Manx public have statutory access to the unplanted areas of the Department's forest lands for walking (Section 9). More specific provision has recently involved a visitor centre at Ayres, nature trail delimitation, overnight huts and recreational impact and development studies (Isle of Man Forestry, Mines and Lands Board, 1984, 1985, 1986). The current Manx Forestry Expansion and Hill Land Improvement Programme incorporates specific recog-

11.3 Emphasising the pleasure park origin of some of the Island's National Glens is this water-powered roundabout at Silverdale

nition of the need to integrate afforestation with other landscape interests, including the Manx Museum and National Trust.

The Department is responsible for the Island's seventeen National Glens, to which the public are freely admitted. Mostly developed as Victorian tourist attractions, these wooded glens are generally in need of substantial replanting, and in recent years a programme of replanting has been under way. Recent surveys of residents, both urban and rural, have shown that the glens are popular leisure attractions and are regarded as part both of the Manx heritage and beauty of the Island (Prentice, 1986; Prentice and Prentice, 1987). Residents are of much the same views regarding the glens, and disagreement primarily concerns their development. This is to be expected as people are often hesitant of change, particularly fearing the loss of the 'conserved naturalness' of the glens. The strengths of the rural residents' views were analysed for any general inter-relationships and these are shown in Figures 11.1 and 11.2. The importance of views about the Glens as an essential part of the Island's heritage as a predictor of other views is clear from Figure 11.1. In Figure 11.2 the importance of a national conservation plan is indicated. The public's use of the glens, particularly that of rural residents, showed some localisation of visits and of preferences. Some glens are, however, more generally popular throughout the Island, notably Glens Helen, Maye, Groudle and Silverdale. The surveys have shown that the glens are a popular leisure asset, to which educational and interpretative developments may sensitively be applied.

The Manx National Trust was formed by

Act of Tynwald as recently as 1951. Its development has favoured the protection of landscape rather than 'stately homes', and as such, mirrors the policies of the National Trust for Scotland, rather than the National Trust in England and Wales. The Trust has secured ramblage areas and coastline, and is now able to enter into voluntary agreements restricting the use of private land (Manx Museum and National Trust Acts 1959 to 1986). In particular, by ownership, the Manx National Trust controls the coastline at Spanish Head, the Calf of Man, coastline at Maughold and Eary Cushlin, land around Cregneash, at the Ayres and the Curraghs (as foundations of proposed national Nature Reserves) and St Michael's Island off Langness (Harrison, 1986). Further access to the countryside is promoted by the Manx Conservation Council, using public rights of way for short walks and hill walks designed for walkers arriving by car (Figure 11.3). Founded in 1970 as a contribution to European Conservation Year, the Council has also developed nature trails.

The Island's non-landscape heritage is also an important leisure and educational resource. The Island's folk culture figures highly in this resource. The principal agencies involved in the conservation of this heritage are the Manx Museum and National Trust, the Manx Heritage Foundation and the Isle of Man Arts Council. The Manx Museum is the oldest of these three agencies. The museum and Ancient Monuments Trustees were established by Act of Tynwald in 1886 with the aims of preserving the Island's cultural heritage and of establishing the Island's first national museum (e.g. Harrison, 1986). In 1922 the former Noble's Hospital building in Douglas became the museum. Expansion since has been both at the Douglas site and at secondary sites. Inspired by Scandinavian folk museums, in 1938 Harry Kelly's cottage was opened to the public at Cregneash as the nucleus of the Manx National Folk Museum. This was the first open air folk museum in Britain. In 1951 Tynwald created a single combined body of Trustees known as the Manx Museum and National Trust to give statutory protection also to the Manx countryside. The same era saw the Irish Folk-

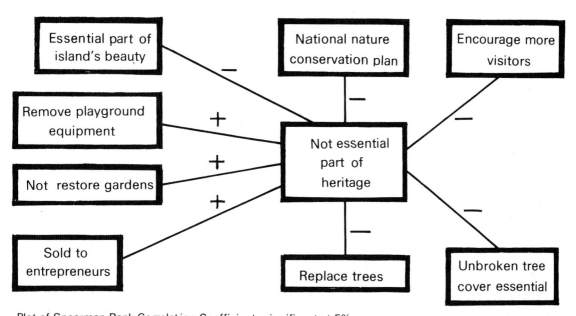

Plot of Spearman Rank Correlation Coefficients significant at 5%

Fig. 11.1 The importance of residents' opinions on the glens as an essential part of the Island's heritage

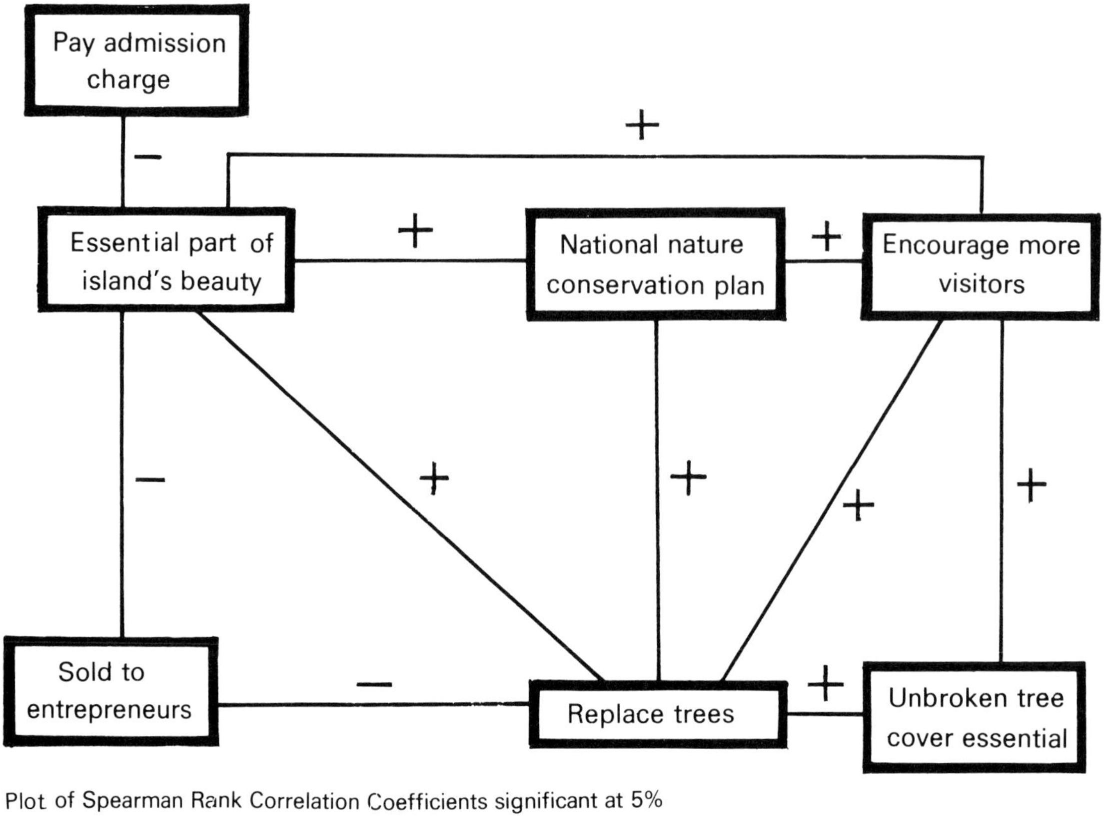

174 THE ISLE OF MAN

```
                    Pay admission
                       charge
                          │           +
                          │ ─  ┌──────────────────────────────────┐
    ┌─────────────────┐   │    │                                  │
    │  Essential part of │ + │  National nature  │ + │ Encourage more │
    │  island's beauty   │───│ conservation plan │───│   visitors     │
    └─────────────────┘        └──────────────┘      └──────────────┘
           │  ─        +              +               +        +
           │                                         
    ┌──────────────┐                          ┌──────────────┐
    │   Sold to    │ ─       Replace trees  + │ Unbroken tree │
    │ entrepreneurs│────────────────────────│ cover essential│
    └──────────────┘                          └──────────────┘
```

Plot of Spearman Rank Correlation Coefficients significant at 5%

Fig. 11.2 The importance of residents' opinions on a national nature conservation plan

lore Commission record Manx speech; a programme which gave the impetus for the Manx Folk Life Survey. Other museum sites are the Nautical Museum at Castletown and 'The Grove' Rural Life Museum at Ramsey. The Nautical Museum was developed around a late-eighteenth-century Manx yacht, Peggy, which was found in her walled up boat cellar in 1935. 'The Grove' depicts a Victorian shipping merchant's house, and is furnished in period styles with a costume gallery and toy room. Recent developments at the Museum in Douglas have involved both the redesign of galleries to include a sectional display showing how layers of archaeological history build up in the ground, and the expansion of the Museum's educational services. The Trustees intend to develop the Museum's displays with modern *interpretative* devices, of which the sectional display is one (Manx Museum and National Trust, 1986). Recorded visitor numbers (Table 11.2) show the popularity of the Museum and its branch museums to residents and tourists.

The other agencies noted above were the Arts Council and Heritage Foundation. The former was established in 1965 to increase public accessibility to the fine arts, and to encourage, foster, promote and assist in the publication of the arts. 'Fine arts' have effectively been defined four-fold: Manx culture, the visual arts, music and drama (Isle of Man Arts Council, 1985). As an example of the former the Council supports the festival, Yn Chruinnaght, which promotes Manx culture, including folk dancing. The Manx Heritage Foundation was created by Act of Tynwald in 1982. Its functions are essentially educational as well as recreational, principally to promote and to assist in the retention of the Island's

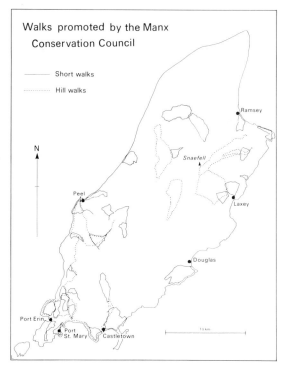

Fig. 11.3 Walks promoted by the Manx Conservation Council

heritage and to provide facilities for the public to acquire knowledge of the Manx heritage. The statutory definition of 'cultural' heritage is broad and ranges from crafts and language to ecology, law and industrial development. The Foundation is enabled to fund schemes by loans and grants in furtherance of its objectives.

The discussion of leisure provision on the Island has so far considered specifically landscape and heritage resources. As noted at the outset of this discussion, other popular leisure pursuits include gardening, sports, television and radio. As in Great Britain, gardening has historically been popular in the Isle of Man (e.g. Garrad, 1985), and awareness of older species has led to replanting of gardens at some heritage attractions. The year, 1985, was designated as the 'Year of Sport' on the Island, and sixty-six sports and recreational activities were actively promoted from the outset of this year, at the official opening ceremony. In 1987 the Departments of Local Government and the Environment and Tourism and Transport proposed the development of major new sports facilities in Douglas, for residents and tourists (Departments of Local Government and the Environment and Tourism and Transport, 1987). But perhaps of wider significance is Manx Radio. Originating in 1964, Manx Radio has operated under various official forms, most recently as a Manx Government-owned company. Manx Radio has survived only through substantial Government subsidy, amounting to between £150,000 and £280,000 per annum between 1978–79 and 1982–83 when the company's activities became politically contentious on the Island (Home Affairs Board, 1984). Dependent on the Manx market the company had sought to regionalise its programming by advertising and directing programmes to the north-west coast of England and the north coast of Wales. The result of this programming was to cause unpopularity on the Island, particularly by broadcasting British regional news and selecting programmes to suit the wider market. An investigation by Tynwald reported that not only was this expansion unpopular on the Island, it was also unprofitable, that the original local focus should be regained, and that deficit financing from the Manx Treasury should continue so to enable this (Home Affairs Board, 1984). The subsequent change in Manx Radio's programming was substantially in line with Tynwald's requirements, and the whole issue is illustrative of the difficulty of providing a distinctively Manx leisure product for a small population. Likewise, the Manx Post Office Authority is sometimes criticised on the Island for issuing too many special stamps, in order to attract foreign earnings. In this sense the Post Office Authority, like the previous policies of Manx Radio, is exporting leisure activities to earn revenue to help support its Manx activities.

Table 11.2 *Visitors recorded at the Manx Museum and branch museums*

Site	Visitors recorded			
	1982/83	1983/84	1984/85	1985/86
Manx Museum, Douglas	98,156	96,418	88,698	103,735
Cregneash Open Air Folk Museum	12,729	13,537	11,915	16,080
Nautical Museum, Castletown	8,604	9,807	7,720	9,116
'The Grove' Rural Life Museum, Ramsey	7,301	8,053	7,721	10,453

Source: Annual Reports of the Manx Museum and National Trust.

Conclusion

The Isle of Man has not developed public policies in housing comparable with those in Great Britain, although its social policies are in most respects comparable with those of the United Kingdom. In particular, the condition of some of the Manx housing stock should be a cause of concern; this does not, however, imply the necessity of public intervention nor of policies comparable with those of Great Britain. Matters of public policy are clearly in the realm of political choice. Where the Island's Government may, however, be criticised is in not having surveyed previously the condition of the Island's housing stock, unlike the governments of Britain and Ireland. However, on the Island the public mood is more of concern for rapidly inflating house prices resultant of in-migration than with housing conditions. The former results in the loss of housing opportunities for Manx residents, in particular first-time buyers, and the associated increase in rental charges in private rented housing in Douglas. As in Britain the existence of poor quality housing, albeit in the Manx case often well provided with standard amenities, can easily be ignored by a well housed majority. Increased property prices are however more generally pervasive and noticeable.

The social trends affecting Manx social policy and provision are much the same as in Great Britain. The need for increased health spending but reduced schooling expenditure, resultant of demographic changes, is common to both countries. Similarly, the propensity for Manx social policy-makers to follow British policy effectively means that social policy developments on the Island are likely to remain in large part consequential of changes in British social policies. Leisure expectations on the Island can be presumed to be much the same as in Britain. Of note are the seventeen National Glens, a conservation and leisure concept distinct from British practice. The National Glens concur with an emphasis on informal countryside recreation and heritage provisions. However, the provision of focused leisure facilities on the Island is again similar to recent British experience.

Our general conclusion must be that in social infrastructure the Island tends to follow British experience and policy, but with notable exceptions which invite comparison.

CHAPTER 12

Economic infrastructure

Vaughan Robinson, Richard Prentice and Gwyneth Davies

All islands, even those rich in raw materials and possessing a huge local market for finished produce, depend for their survival upon the transport links which exist between themselves and their major trading partners. This is even more true for those islands which are not well endowed with basic raw materials and which are forced to depend upon external markets for the sale of their finished produce. In these cases the quality of transport infrastructure may well be *the* single most important factor determining the economic health of the island and therefore of its people. This section therefore considers the transport infrastructure at the disposal of the Manx population, not unnaturally beginning with a discussion of maritime transport.

The movement of people and goods: sea and air transport

Passenger traffic by sea

The Island's Government established a regular sea packet for the transport of passengers and mails to and from the Island in 1767. Before that, communications depended upon the irregular sailing of vessels from Whitehaven. With favourable conditions the journey lasted six hours, but a final arrival from Douglas at Sunderland Point after a 53-hour journey is on record. By 1815 increased trade had led to regular packet services from Douglas to Whitehaven, Liverpool and Dublin. James Little's Greenock service was soon added, and in 1825 a weekly mail service was established between Douglas and Whitehaven using the 30-ton Steamer *Triton*. Three years later this mail service was transferred to Liverpool and sailings were increased to twice-weekly in summer, once a week in winter. This increasing activity led to the formation in Douglas in 1829 of the Monas Isle Company, 'to provide a more efficient and reliable shipping service to the Island', and in 1831 the name was changed to the Isle of Man Steam Packet Company Ltd. In that year alone, 20,000 passengers were carried at a time when the Island's population was only 40,000. The Steam Packet, as it became affectionately known, developed steadily during the nineteenth century and fought off attempts by both Ramsey and Castletown companies to establish additional passenger services to Liverpool. An increasing tourist trade with the Island led to the construction in 1857 of the Princes Landing Stage at Liverpool to avoid the row out at low tide to the Steam Packet. However, on arrival at Douglas, passengers were disembarked into small boats and only reached the Red Pier 'by a scramble across the slippery stones' (Moore 1900). Few escaped without a soaking by either sea or rain. The Red Pier, constructed in 1793, continued to serve until 1873 when the Victoria Pier was opened being used in its first year by 90,000 passengers. The early 1890s saw two more harbour developments in response to the increasing tourist trade (see Table 16.1), these being the extension of the Victoria Pier and the opening of the Queens Pier at

12.1 Mona's Isle III at Douglas in 1890. Vessels such as this were the lifeline of the Island's visiting industry

Ramsey. The Victoria Pier and the later King Edward VIII Pier have continued to handle all the passenger traffic.

By 1914 the Steam Packet owned 16 ships with a total of 23,000 gross tonnage and operations were proving highly profitable. Though the vessels were requisitioned during the war, the 1920s saw continued growth and the Company demonstrated its ability to cope with the greatly expanded number of passengers especially at the summer weekends. The Company in 1928 took over the Heysham to Douglas service previously run by the London Midland and Scottish Railway.

The Second World War saw vessels again requisitioned, the Port of Liverpool was closed to Manx ships and the mainland link was transferred to Fleetwood. From April 1946, when the Liverpool link resumed, heavy tourist traffic encouraged a re-equipment programme and the building of new vessels, but this heyday period was short lived. Passenger numbers were now falling. In the period between 1 May and 30 September 1949, 611,286 were carried whilst 10 years later this had fallen to 495,804, due in part to air travel but more particularly to changes in holiday concepts. The Isle of Man, unlike the Mediterranean beaches, could not guarantee sun. The reduced numbers of passengers emphasised the seasonal, even Saturday only, nature of the tourist traffic. For nine months of the year most of the fleet was laid up either at Birkenhead or Barrow. The season was now concentrated between Whit weekend and late August. As early as 1961 the Fleetwood to Douglas summer service was closed; a service which had at times carried more than 100,000 passengers on one season. During the 1970s the Heysham service was closed and services to Belfast and Ardrossan were reduced. Furthermore the Company's vessels, built in the late 1940s, were becoming due for replacement and the general economic climate, including escalating oil costs, was not conducive to the necessary investment. The only encouraging aspect was the newly introduced (1974) car ferry service between Dublin and Douglas.

The late 1970s also saw disputes over the

establishment of roll-on roll-off (Ro-ro) services, the construction of a link-span at Douglas harbour and in February 1978 the appearance of a new rival company, Manx Line. A multi-purpose Ro-ro vessel of 2,800 tons was to be used on a service to the mainland port of Heysham. A shorter crossing, better motorway accessibility and cheaper harbour dues confirmed Heysham's advantages over Liverpool. The *Manx Viking* made her first official sailing into Douglas on 26 August 1978. The company, however, was plagued by disasters both man-made and natural. So by November 1979 the assets of the Manx Line were transferred to a locally registered company, James Fisher and Sons. Part of the agreement involved Sea Link taking over later, and in October 1980, they increased their holding to 86% of the Manx Line shares and included the Heysham to Douglas route in their advertising. Their share of traffic increased markedly.

Although the 1980s marked the 150th Anniversary of the 'Steam Packet' they were not auspicious times for the Company. They had already lost both passenger and freight traffic to Sea Link, but worse was to come. On 9 July 1984 the three largest Island freight agents transferred their allegiance to Sea Link. In the same month Sea Link itself was sold to Sea Containers Ltd., together with the Douglas and Heysham service. The position of the Steam Packet Company was therefore deteriorating at a time when their major competitor was benefiting from new business acumen brought in by the Sea Container takeover. Difficult and protracted negotiations started between the two shipping lines and early in 1985 agreement was reached to transfer all passenger services from Liverpool to Heysham. The last sailing by the *Monas Queen* left Liverpool for Douglas on 30 March and so ended a connection of more than 150 years. The 1985 season continued to

12.2 Panorama of Douglas taken from the Head. It shows the Ro-Ro vessel Peveril in the foreground and the sweep of Douglas promenade behind the characteristic ferry terminal building

12.3 Ramsey harbour in 1902. A photo taken from the same place today would reveal that little has changed in the intervening years

be a troubled one, the only vessel carrying out schedules properly being the Sea Link's *Manx Viking*. The problems of *Monas Isle* continued and she was sold at the end of the season and the *Manx Viking* suffered the same fate in September 1986.

The first half-year's financial results for 1986 showed pre-tax losses reduced from £1.8 to £1.4 million and this reduction was maintained in 1987. The new company is thus on target to meet its prediction of profitable operation within five years. The regular Douglas to Heysham service continued in 1988 with additional sailings during the summer season to Belfast, Dublin, Fleetwood, Liverpool and Stranraer. These services are maintained by the car/passenger ferries *Lady of Man* and *Monas Queen* whilst the *Tynwald* serves as a multi-purpose vessel and the *Peveril* for freight only. The lack of stability of services since the merger has renewed calls for Government intervention and, although the name Isle of Man Steam Packet appears on the side of the vessels, the future has none of the certainty enjoyed in the past.

Freight traffic by sea

Though passenger and commodity transport was initially combined when trading to and from the Island, as the nineteenth century progressed the two elements separated. Even so, for one-and-a-half centuries the Isle of Man Steam Packet Co. predominated in freight services as well as in passenger. By the 1960s the Steam Packet cargo vessels provided the Liverpool to Douglas service and made occasional landings at Ramsey. They transported most of the Island's merchandise and general goods, but were subject to increasing complaints about the high cost and slow movement. Bulky cargoes such as coal, grain and cement, were carried to the Island by boats belonging to the Ramsey Steamship Company Ltd., a company formed in 1913 and providing an efficient coastal service often to the smaller ports of the Island.

During the 1960s, the introduction of containerisation was one of the most significant changes taking place in sea trade. In 1967 a long established Preston firm—with interests in Ronaldsway Air Freight—formed a group known as Ronagency Shipping Ltd. They introduced a container service from Glasson Dock to Castletown and were soon attracting trade away from the Steam Packet. In 1975 they transferred their mainland operations to Preston and in 1978 became absorbed in Manx Line and its subsequent history. During the 1970s the Steam Packet not only suffered this competition, but also much industrial strife and closure of its traditional dock base in Liverpool. The Steam Packet elected to have the *Peveril* converted to a container vessel, but the introduction by the Manx Line of a Ro-ro service and its continuation by Sea Link proved more attractive to the freight shippers. By 1985 continuing delays and high tariffs encouraged Island Carriers Ltd.—known as Island Express—and Edmundson-Ronagency Ltd. to begin their own cargo service using Ro-ro facilities (Roll-on, Roll-off/Lift-on, Lift-off vessels). Mezeron Ltd. of Ramsey also announced a scheduled cargo and container service from Ramsey to Glasson Dock. This company, which began operation using a single fishing boat, had two vessels on this route by April 1985. The continuing increase in trade enabled them to purchase another larger vessel in 1987. By the end of 1986 the Steam Packet was operating one multipurpose Ro-ro ship into Douglas and achieving a modest profit; this was maintained in 1987 and by early 1988 two vessels were involved, the *Peveril* and the *Tynwald*.

The role of the other ports has also changed during the 1980s. Ramsey's recent growth is

12.4 Many people's first introduction to the Isle of Man, the unusual ferry terminal. Douglas Head to the right

reflected in its present programme of harbour works. Between 1981 and 1986 the number of vessels handled annually increased from 63 to 155 and an extra 20% of tonnage was recorded, largely under the heading of general cargo. Like Port St Mary and Peel, Ramsey's salt herring export trade (6,081 tonnes 1981) ceased in that year, whilst its grain trade has been growing in importance during the 1980s. In 1984, 2,333 tonnes of barley were exported, though this amount was reduced to 1,429 tonnes in 1985. Grain imports, largely of wheat, continued to rise and reached 3,158 tonnes in 1985. Impetus for development at Ramsey has come from the town itself and from dissatisfaction with Douglas harbour. Of the smaller ports of the Island, Peel has seen an increase of 30% in general cargo handled over the period 1980–87; here increased fertiliser imports from 147 tonnes (1981) to 789 tonnes (1986) contrast with the rapidly diminishing imports of coal, gas and oil. The last consignment of coal was received in February 1985 and gas and oil imports were down 25% between 1981 and 1986. The chief shipper, Glenlight, is favourably regarded and is willing to move any available general cargo around the Irish Sea.

Port St Mary lost its salt herring exports in 1981 and ceased to import coal after June 1982. Only scrap metal continues to be exported and this is less than 1,000 tones annually. At Castletown, after the activity of the 1970s there is now little trade. In 1986 only small quantities of coal (951 tonnes) and timber (603 tonnes) were imported and 165 tonnes of scrap metal exported. The southern ports, as working units, have shown marked decline. Locally-based fishing continues and leisure sailing has become the growth development here. Little change can be expected in the near future from the present pattern of freight traffic, though the individual carriers themselves may change.

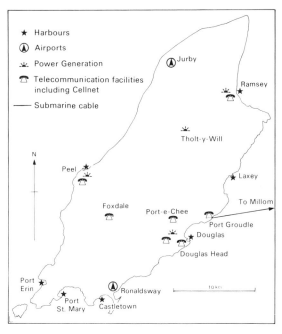

Fig. 12.1 The location of key economic infrastructure, 1988

Air transport

Although sea transport is the oldest and most conventional means of communication between islands and their economic partners, it is by no means the only one. The twentieth century has seen the development and exponential growth of air transport as a major competitor in the movement of people and high-value goods. The Isle of Man has had a long and distinguished association with air travel dating as far back as November 1902, when the first manned balloon took off from the Island engaged in naval survey work. From that time until the early 1950s the Island played a prominent role both in civil aviation and flying as a sport. Indeed, as late as 1965 Ronaldsway, which is the Island's main airfield, was the third busiest civil airport in the United Kingdom after Northolt and London (see Figure 12.1 for the airport's location).

The development of air transport in the Isle of Man has been inextricably linked with two factors, neither of which result from the country's geographical status as an island. These two factors are the tourist industry and its continuing search for new novelties, and the importance of the Island's location for the British during the Second World War. This last factor emphasises that the Island's strategic location in relation to the trade of the western seaboard of the United Kingdom has not really diminished since Viking times, despite advances in modern warfare and the geographical expansion of its impact.

The beginnings of air transport in the Island are closely associated with tourism since the novelty of aircraft acted as a powerful attraction for visitors either as by-standers or as passengers. The earliest recorded flight from Manx soil took place in 1911 when the organisers of events to celebrate Douglas' Jubilee as a municipality invited two leading aviators to the Island to give a display of flying. Only one aircraft actually participated in the display on 4 July, but it established a pattern which was to last until 1969 when the Island hosted the final Manx Air Derby as a tourist attraction. Immediately after the First World War various companies offered tourists joyrides around Douglas Bay and this was the start of commercial flying on the Island (see Kniveton, 1986 for details of the early history). The war itself had also seen the use of flying boats to get essential supplies through to the Island without fear of attack by German U boats, and this mode of goods transport continued after the war when newspapers and magazines were flown in early in the morning. The TT motor-cycle races were periods of peak demand for this service, and during the appropriate week, *The Motor Cycle* magazine chartered three-engined airliners to get its wares to eager spectators. These larger planes tended to land in the unobstructed field belonging to Ronaldsway Farm near Castletown. When scheduled seaplane flights for passengers began between Blackpool and Douglas in 1932—incidentally the first scheduled internal airline service in the UK—they used the same field at Ronaldsway when conditions prevented a landing on the sea. Indeed by 1933 Ronaldsway was hosting services from Liverpool and Blackpool and possessed a wind-sock and hut in which crew and passengers could shelter. In the following year the first hangar was built there and a passenger terminal was planned. However, Ronaldsway was not the only airfield in use on the Island in the 1930s since a second existed at Close Lake near Ramsey. Despite its ability to handle larger aircraft, rationalisation among operators led to its closure in 1939. By that time Ronaldsway had been considerably improved by its private lessee who had levelled and extended the runway and built workshops and a booking office.

The Second World War was to have a profound influence upon the facilities available for aviation on the Isle of Man. In 1938 the Royal Air Force announced its plans to build an airfield at Jurby in the north of the Island where it intended to create a Training School for Bombing and Gunnery. This was completed in 1939 and provided facilities for up to 1,000 trainees at any one time. The Ministry proceeded to build a second facility at Andreas which was fully operational as a fighter station by March 1942. Aircraft from Andreas were used to patrol the Irish Sea and escort and protect merchant shipping. Finally in 1943 the Admiralty took over and rebuilt Ronaldsway as a training school for torpedo bomber crews. Four permanent runways were laid, land was drained, generators were installed, and the site was provided with the full range of support facilities such as a control tower, radar, and a maintenance department.

The end of the Second World War saw the Isle of Man with three high-grade airfields and it was immediately clear that military and commercial demand could not support such a high level of provision. Andreas closed at the end of 1946 and has all but disappeared, whilst Jurby eventually closed in 1963 after a protracted period of rundown. This left a revitalised Ronaldsway as the Island's main civil airport when it was purchased from the Admiralty by the Manx Government in March 1948. In 1949 the airport handled over 100,000 passenger arrivals and departures for the first time, with peak weekends seeing more than 100 aircraft landing. In 1951 the Airports Board invested further money in a new terminal building which was completed by 1953. The years through to 1966 saw a continual increase in traffic associated with the post-war resurgence of the Island's tourist industry. A variety of independent and state-owned companies flew passengers to the Isle of Man over twenty-two different routes, and at peak times operators had scheduled flights shuttling in every half hour from, for example, Blackpool. Runways were lengthened and widened, radar facilities were continually upgraded and the terminal was extended to handle the new traffic, which amounted to some 408,000 passengers in 1966.

The late 1960s saw Manx aviation in the doldrums, despite the expansion of facilities at Ronaldsway and the first appearance of jet airliners on the Island. Passenger numbers reflected the general decline in tourist trade as they fell back again below the 400,000 mark (see Chapter 16 for a fuller discussion of the tourist industry). Concerted efforts by the four main airlines then servicing the Island helped to reverse the trend temporarily in 1972 and 1973, and in the latter of these two years over 480,000 people used the airport. This is a figure that has not been exceeded since that time, as the oil crisis of 1973 and the continuing decline of tourism on the Island lead to a steady, if fluctuating, fall in number of passengers throughout the remainder of the 1970s and the early 1980s. By 1982, only 283,000 passengers used Ronaldsway.

Although the early 1980s saw air traffic shrink back to the levels of the early 1960s, they also witnessed a significant reorganisation of airline services to the Island. The Airports Board had discussed the idea that the Island should have its own airlines as early as 1967, but at the time it had come to nothing. However in 1982, with the progressive deterioration of profits on the routes to the Isle of Man and the fragmentation of those that still remained viable between different competing operators, the circumstances were much more favourable to a renewed discussion of reorganisation. Air UK, British Midland and the Government entered into discussions about the formation of a new merged airline and in November of that year Manx Airlines came into existence. The greater flexibility of this new 'local' airline has allowed it to respond more rapidly to market demands, and this has been reflected in the number of passengers it now carries and the profits it has started to record. Since Manx Airlines is now the major operator flying out of Ronaldsway, its success has also been reflected in overall traffic levels through the airport. In 1987, over 425,000 passengers used the airport, a 50% increase in only five years.

In the late 1980s Ronaldsway is still the second largest airport in the north-west after Manchester. It handles some 38,000 aircraft movements per year of which less than half are scheduled flights. Reflected within this total is the changing role of the Island. There are now charter flights from the Island to tourist resorts such as Majorca, and nearly 700 aircraft movements are accounted for by the business and executive sector. As financial services continue to expand on the Island, this figure is likely to rise significantly, and indeed a greater proportion of all air traffic is likely to be business—rather than tourist—related. Signs of this are already apparent in the relative success of different air routes: whereas Blackpool was still the dominant source of passengers as late as 1978, by 1983 it had fallen to fifth position. In contrast, even within the tourist season, Heathrow has climbed as an origin from fifth in 1979 to first in 1981. Although business traffic is without doubt of growing significance, it is important not to underestimate the contribution of air transport to the tourist industry of the Island, particularly given the attempt to move the Island away from cheap 'bucket and spade' holidays. Research in 1982 indicated, for example, that tourists arriving by air stayed longer on the Island and spent 55% more money in the local economy. As a result, whilst 20% of all tourists travelling to the Island did so by air, they accounted for 27% of total on-Island expenditure by tourists. Clearly, air travel will remain important not only for business but for the continued health of the tourist industry.

The movement of ideas and knowledge: telecommunications

The discussion of transport infrastructure outlined above depends upon an implicit assumption that the only items which move into or out of islands are human beings or tangible commodities. In the latter part of the twentieth century this is rarely the case, since post-industrial societies depend more for their existence upon their ability to provide services than they do upon their ability to manufacture tangible commodities. These services are of two types: consumer services which exist for the direct benefit of the individual consumer; and producer services which cater for the needs of corporate customers. The latter would include the financial sector, the research and development function and professional services such as marketing, distribution and advertising. It is this sphere that has seen such rapid growth in the Isle of Man in the last ten years (see Chapter 15 for fuller details). Central to the post-industrial society is information and knowledge, and the rapid communication of these between individuals, between different parts of the same organisation, and finally between different organisations. Clearly the telecommunications industry is central to the success or failure of such societies.

The first electric communication link between the Isle of Man and the mainland came into existence in 1864 when a telegraphic cable was laid across the bed of the Irish Sea from Port Cornaa to St Bees. This link was to remain in use for the best part of ninety years, albeit constantly improved. One of the staff employed to maintain this cable

12.5 Part of the modern fleet operated by Manx Airlines. This is a Shorts SD360 on the tarmac at Ronaldsway. These commuter-liners are manufactured in Belfast and used by Manx Airlines on lightly-loaded routes. Shorts have a design office on the Isle of Man

was George Gilmore who, early in 1889, was granted a licence to operate the first telephone service on the Island. The early success of the venture prompted Gilmore to open an exchange in Atholl Street, Douglas from where he serviced his first customers, numbered amongst which were the Steam Packet Company and the Isle of Man Railway Company. Gilmore was not blind to the potential of such a service, but felt that the full realisation of this depended upon being associated with a larger company which possessed the necessary resources. This would be crucial to establishing a permanent telephone link with the mainland. As a result, Gilmore sold his licence to the National Telephone Company which was eventually to become part of the British General Post Office. 28 June 1929 saw the realisation of part of Gilmore's vision when the first call was made between the Island (Douglas) and the mainland (Liverpool). From 1942 onwards islanders had an alternative means of communicating with the UK, since the Isle of Man was the site of the first application of frequency modulation to a commercial radio voice link. In the post-war period the Island's telecommunications network continued to be run under the auspices of the General Post Office and more recently British Telecom. It was closely allied to the Liverpool telephone district, and developments on the Island tended to match those taking place within this district. Thus, for example, when telex facilities were provided on the Isle of Man all calls were directed through a main switch in Liverpool rather than one sited on the Island itself. Regardless of this dependence upon a service which was essentially run from outside the Island, telecommunications provision had kept pace with the technology employed in the UK. By 1978 there were some 24,600 subscribers on the Island making 64,000 calls per day either through the 2,000 miles of copper cable on the Island or through the 372 channels across the Irish Sea: these channels included both microwave and submarine cable channels with approximately two of the former for every one of the latter. There were eleven exchanges located upon the Island and two radio towers at Foxdale and Douglas.

The pace of technological change within the telecommunications industry and the changing nature of the Manx economy have progressively made the industry more central to the future of the Island in the 1980s than ever before. As a result, when in 1985, the Manx Government asked for tenders for the licence to operate the Island's telecommunications service for the next twenty years, competition was very keen and the two major protagonists elected to highlight the quality and range of the services they proposed to install, rather than simply emphasise the revenue they intended to offer the Government.

Although Tynwald had said that it would decide upon the licence issue early in 1986, the first developments in the contest between Cable and Wireless and British Telecom had taken place in 1983. No doubt spurred by the forthcoming competition for renewal, British Telecom instigated a major investment programme in 1983 that ran over a three-year period. In that time they invested £7.5 million in the network, installing a new electronic exchange at Douglas, and converting the microwave link from analogue technology to digital technology. Work also began on the process of replacing old copper cables on the Island with new fibre optic channels thereby improving the quality, reliability and capacity of the service. In 1984, British Telecom also overhauled its management of the Island's telecommunications by registering a new company, Manx Telecom, which is based on the Island. Control and management of the service thus returned to the Island and the new operation is semi-autonomous from its mainland partner. This move not only created a more politically acceptable image for British Telecom on the Island, but it also created 32 new jobs, mainly at the new headquarters in Asgard House. Subsequently, Manx Telecom has also moved the billing function to the Isle of Man and invoices are now distributed internally by the Island's own post office.

Manx Telecom has a policy of buying goods and services from local suppliers wherever possible and they also seek to recruit local staff.

Manx Telecom's competitor for the licence, Cable and Wireless, seemed somewhat unimpressed by all this activity when they submitted their own proposals in October 1985. Their central argument was that by being tied to British Telecom's technology and purchasing policies, the Island had not benefited from the major advances which had taken place in telecommunications technology, particularly in the vital business sector. They described how five of the eleven exchanges were of a type dating back to 1938 whilst a further five relied upon 1960s semi-electro mechanical technology. All of the equipment relied upon analogue technology which is inherently less efficient than the newer digital technology; as an example of this, they point to transmission speeds of c.9600 bits of information per second for the former as against up to eight million bits per second for the latter. Clearly, digital technology offers enormous time and cost savings for business customers (such as banks) who regularly transmit large banks of data, for example, between different computer installations linked by telephone. Not surprisingly, Cable and Wireless, through their locally registered subsidiary, offered a completely digital system for the Island, to be installed within two years of taking over the licence. They also suggested a new satellite teleport for international traffic (to other offshore banking economies such as the Channel Islands and the Cayman Islands) and for transit traffic, which would earn a royalty for the local company. A new submarine cable was to be laid to the mainland, and Cable and Wireless offered new and unusual tariff structures both for business and domestic users. Finally, they agreed to maintain British Telecom's new policy of being a significant direct contributor to the local economy through employment of existing staff and local purchasing. Indeed they offered to improve upon their predecessor's record of training local people for internal promotion.

Spurred by the new threat of competition, Manx Telecom responded with the offer of a thorough-going overhaul of its facilities and services. It promised the replacement of all exchanges within four years as part of a £34 million investment programme for the licence period, and the new exchanges were to be of the latest type employing digital technology. A new satellite teleport was to be constructed in order to stimulate new traffic rather than simply catering for existing demand. It proposed new local tariff structures and a renegotiation of the fees earned from calls to and from the mainland. This, it was thought, would add £1.8 million to the revenue of Manx Telecom each year. And the company was at pains to stress its commitment to the Island and its future by pointing to the £2 million per annum which it contributed directly or indirectly to the local economy. Their proposals concluded with the sentiment that Manx Telecom would be happy to be judged on its past service to the Island and that, in particular, the financial, legal and commercial communities already had available to them 'a range of speech, text data and value-added services equal to the best in use anywhere in the world.'

In the event, both the consultant called in to advise the Manx Government and the Government itself opted to grant the licence to Manx Telecom. There seem to have been several reasons for this decision: first, Manx Telecom was thought to have offered a good and reliable service in the past; second, although Cable and Wireless were offering all-digital technology two years earlier than Manx Telecom, they were proposing to do so at a much higher price; third, Cable and Wireless were intending to fund the satellite earth station from within the Island; and finally, Manx Telecom were offering to be exceedingly generous in their financial dealings with the Government. Not only were they proposing a flexible annual income from a licence fee, but they were also offering a

12.6 The new Manx Telecom satellite station at Port-e-Chee ooutside Douglas. The Island's service sector is increasingly becoming part of a world financial community that depends upon telecommunications for the rapid transmission of information

multi-million pound one-off payment in exchange for being granted the licence. This last element may well have been crucial, given the Island's recent and future expenditure on major capital projects and its limited tax base.

The Island's telecommunications infrastructure is thus poised for rapid redevelopment over the next ten years, particularly with regard to the provision of services to the business community. The satellite station has already been commissioned on a site at Port-e-Chee, north of Douglas, and a new submarine cable has also been laid to the mainland (see Figure 12.1). Manx Telecom envisage steady growth in the demand for their services over the remainder of the century, with international calls at the busiest part of the day increasing by as much as 540%. Local calls will experience less rapid growth but should, nevertheless, double over the same period, whilst there is still some potential for greater densities of residential connections. At present, Ramsey and Peel have much lower densities than Marown and Kirk Andreas.

The generation and movement of power

Modern economies do not just depend upon information and knowledge for their success, they also require the ubiquitous provision of power. Increasingly, the locational link between sources of power and forms of economic activity has broken down as power has become more transportable and therefore more readily available across the economic landscape. This is a point which is discussed more fully in Chapter 14, which looks at the past and present manufacturing sector on the Island.

An inventory of indigenous sources of power reveals the Island to have been reasonably well provided for prior to the onset of industrialisation. When the demand for energy was largely limited to domestic heating and cooking, the prime source of power was local peat, cut by farmer-fishermen for their own use. Key areas of extraction were Beinn y Phott—west of Snaefell—the Curraghs of Ballaugh and Lezayre, and the east side of South Barrule (Corran, 1977). Evidence of these turbaries can still be seen in the landscape, in areas such as those near the Ballure plantation, south of Ramsey. Peat also provided the source of power for some of the earliest industries on the Island. Birch (1964), for example, describes how peat was used in the lime kilns which burnt carboniferous limestone in order to produce agricultural lime. Water power was also available to the earliest economic enterprises. Accounts of the milling industry of the Sulby area stress the importance of water power (Quilleash, 1964), whilst there has also been more general research on the Island's water mills (Jones, 1966). Lady Isabella at Laxey is also a prime example of man harnessing the power of running water for industrial use. However, despite repeated attempts to locate coal deposits on the Island, none has been found in commercially viable quantities. The search has concentrated on the Point of Ayre region, with the earliest explorations taking place in the seventeenth century. Subsequent borings by a Liverpool company were responsible for the discovery of brine in the area in 1892, but again coal deposits proved illusive (Lamplugh, 1903). More recently Rio Tinto Zinc has also drilled in the vicinity on a one-year licence. Their explorations, which took place during 1985 and 1986, were no more successful than earlier attempts. Activity has not been solely restricted to the Northern lowland, however, with borings having taken place near Peel, Derbyhaven and Ballasalla.

The lack of any commercially exploitable sources of fossil fuel on the Island has meant that the provision of power has become an exercise in importation and reprocessing. As on the mainland, coal was the prime source of domestic and industrial power from the end of the nineteenth century until the late 1950s. Much was burned to provide power and heat directly, but an increasing proportion was also processed to provide alternative forms of energy. Five gas undertakings, for example, used coal to produce town gas for the areas of Douglas, Castletown, Peel, Ramsey and Port St Mary. Bawden et al. (1972) describe the facilities which existed in Castletown for such a purpose, and they also note how coal was transported to the plants in small coastal vessels. In contrast the earliest electricity generating station on the Island used oil as its raw material. It was opened in 1893 in Douglas to provide power to the new Manx Electric Railway then being built to Groudle. As the line was subsequently extended, new plants were also constructed at Laxey, Snaefell and Ballaglas. As well as providing motive power to the railway, these facilities also supplied direct current electricity to domestic consumers who were fortunate enough to live near the tracks. However as demand for electricity expanded, it became clear that new undertakings would be required to oversee the generation and distribution of electricity throughout the Island. The first of these was established by Douglas Corporation, and in 1923 it opened a new generating station on the North Quay in Douglas. This was equipped with four diesel generators capable of producing a total output of 615 kw. Six years later, the Corporation added a second power station at Pulrose, north of Douglas. Powered by coal supplied through Douglas harbour, this station served not only the Corporation's own customers, but also those of the newly-formed Isle of Man Electricity Board (IOMEB), which had come into being to serve the remainder of the Island. Rising demand from both sets of consumers in the immediate post-war period stimulated successive additions to

12.7 Two of the Island's generating stations found on the same site at Pulrose, near Douglas. On the right is the original station, opened in 1929 as a coal-powered installation. On the left is Pulrose 'B' opened in 1989, and diesel-powered. Together they form the core of the Island's generating capacity

generating capacity. Pulrose was extended in 1948 in anticipation of the closure of the North Quay installation, and the IOMEB became the owner of its own generating capacity in 1950 when it built a new station at Peel. It subsequently erected additional plant at Ramsey in 1960, both being diesel-powered.

Since that time, both the gas and electricity undertakings have seen a progressive change in the raw materials that they employ. The gas industry experienced a technological upheaval in the early 1970s when it became clear that the production of gas from coal was an outmoded process and one which was becoming increasingly out of line with practice on the mainland. There, the gas boards were beginning to switch to natural North Sea gas. From 1971 onwards, Douglas also began to abandon the production of coal gas, substituting in its place a new product called Butane-Air which is a mixture of Liquid Petroleum Gas (LPG) and Air. Within five years this was the sole form of gas 'manufactured' on the Island, although in reality the manufacturing process involved only the fully automatic mixing of imported LPG with air. LPG is now brought into the Island from mainland refineries through Douglas harbour where there are bulk storage tanks on the Battery Pier. From there it is distributed either in vapour form or as a bulk/cylinder liquid. The electricity generating industry has also abandoned coal as its main raw material. From the 1950s onwards the relative cheapness of diesel oil has encouraged a transition to this source of power. The last station to convert was Pulrose, which had moved over to oil by 1972.

Perhaps of greater significance than these changes in technology have been the parallel changes in organisational structure. Again these have affected both the gas and electricity undertakings. In the case of gas, restructuring has essentially involved denationalisation and rationalisation. The local authority gas utilities, most of which had become part of the Isle of Man Water and Gas Authority, have been privatised, as has the Douglas Gas Light Company. From the end of 1983 onwards both were taken over by the Calor Gas company which operates on the Island in two guises: as a supplier of piped gas to the Douglas area (still called the Douglas Gas Light Co.) and as Kosangas, a division that sells bottled and bulk liquified gas throughout the Island and also distributes gas in vapour form through the old piped networks in Ramsey, Peel and the south of the Island. The reorganisation of the electricity supply industry has been a good deal more protracted and complex. Although the IOMEB began life as a retailer of electricity produced by Douglas Corporation, it soon developed into a parallel and independent organisation. In 1954 the two undertakings came to an agreement about power swapping at times of peak demand and in 1966 they were both represented on a new Joint Electricity Authority. Despite the existence of such linkages, the notion of complete merger always foundered on the two companies generating and distribution costs. With the largest generating station and 12,000 spatially concentrated customers, Douglas Corporation was always able to produce and distribute electricity at a lower unit cost. The IOMEB, in contrast had two small generating sites and 15,000 customers spread over some 200 square miles. It relied upon state financial subsidies to survive. Not unnaturally the residents and politicians of Douglas feared that merger of the two organisations would lead to a levelling up of the price of electricity in Douglas. Thus it was not until a Committee of Enquiry in 1982 lent its full weight to the idea of merger that rationalisation developed any momentum, and it then took until April 1984 for a unitary Manx Electricity Authority to come into being.

In the late 1980s the Isle of Man thus has a rationalised organisational structure for the provision of power. The gas industry continued to make considerable headway with 2,500 consumers now choosing this form of domestic central heating. Increased demand has been reflected in new investment in harbourside storage facilities and a fleet of road tankers for bulk distribution. Growing demand has also forced change in the electricity industry. The Energy Committee of Tynwald was asked in 1985 to consider the various options which existed for new capacity. They reported in February 1986, and their recommendations were then accepted by the Government in full. They rejected the very popular option of laying a submarine connection to the mainland's National Grid and using power generated by the Central Electricity Generating Board. It was felt that this would have made the Island too dependent and therefore vulnerable. They also rejected the immediate use of alternative sources of power such as that generated by wind and waves. In this case the technology was felt to be insufficiently developed at present, although this would not preclude future deployment especially were the UK Government to select the Island as a site for an experimental station. Lastly, the committee also refused to recommend the extension of the HEP programme beyond the current sites at Block Eary and Sulby (see Figure 12.1). Instead, the committee plumped for a policy which relied upon proven technology and with which the Electricity Authority were already familiar. As a result, £25 million is now being spent on installing a new diesel-powered generating station next to the current building at Pulrose. The opportunity has also been taken to rationalise capacity, and the site of Peel is to be progressively run down whilst the Ramsey station will become an emergency facility only. Second-hand and experimental generators have been installed at Peel to secure supplies until Pulrose B is fully operational.

Although the various reorganisations and infrastructural improvements have ensured that the Island will have a secure and adequate supply of power in the foreseeable future, the key issue for most Manx consumers is one of cost. Power in the Island is still expensive relative to the mainland and to other islands such as Jersey and Guernsey. In September 1988, for example, a unit of electricity cost 8.79 pence for Manx consumers outside Douglas, 5.79 pence on the mainland (Norweb) and 5.5 pence in Jersey. Higher energy costs are not a new feature of life on the Isle of Man, as evidenced by the report of the 1959 Coal Enquiry Commission which noted that transport costs of coal from Garston dock in Liverpool to Douglas added over 20% to the cost of each ton. Slightly later Birch (1964) commented that even diesel—with its lower handling costs—was up to 14% more expensive on the Island than on the mainland. Even so, higher energy costs are an important factor not only for the industrial development of the Island but also for hoteliers and domestic consumers. The fact that all raw materials for power generation have to be imported also places a burden upon the Island's 'balance of payments'. In 1984, for example, the Island imported 64,000 tonnes of gas oil, 9,000 tonnes of butane and 33,000 tonnes of coal (Webster and Dawson, 1984).

The movement of goods and people within the Isle of Man

Although the Isle of Man is critically dependent upon the importation of people, goods, knowledge and power from the mainland—and therefore upon the infrastructure which exists for the movement of these—facilities must also exist for their distribution within the Island. These are of two kinds: rail transport and road transport, but in both cases their development was directly stimulated by the growth of the tourist industry.

Rail transport

The inter-urban railways of the Island reflect the rise and decline of narrow gauge railways throughout the British Isles. A preliminary comment is apposite:

> The narrow gauge railway was a product of its own era. It was conceived when the railway knew no serious rivals, and economical construction and the ability to penetrate awkward terrain made it the only effective form of transport possible in many places. The birth of road motor transport altered life almost beyond recognition, and in these new conditions, the traditional narrow gauge railway had little chance of survival. That a few lines have survived is due to tenacity, effort and an ability to move with the times to meet a new role. (Hendry, 1979, p. 2).

The railways of the Island were developed in the second era of railway construction in the British Isles; that of the opening up of remoter areas. The Festiniog Railway in Wales had in 1863 demonstrated publicly and very successfully the potential of narrow gauge railways using steam locomotives. In the 1870s there was a rapid increase in such railways. Using a three-foot gauge, rather than the narrower gauge used in Wales, the technical and loading restrictions of the earlier lines were avoided. By the end of the century not only had the Island a narrow gauge railway system, comparable systems could be found, for example, in Counties Donegal and Clare. Narrow gauge systems relied upon a mixture of private and public funding, and demonstrate both the *competitive* nature of railway development and an early attention of governments towards economic development of marginal areas by *public policy*.

The Isle of Man Railway Co. Ltd. was registered in December 1870 with projected lines linking Douglas, Peel, Ramsey and Castletown. The Company expected further to extend the line from Castletown to the village of Port Erin as the latter was to be developed as a steamer port for Holyhead (Boyd, 1977). Funds were short, and the difficulty of raising capital on the Island caused the Company to abandon the projected Ramsey line in January 1872. For want of local funds, the Company was forced to raise three quarters of its capital in Great Britain, and in July 1873 the Douglas to Peel line was opened, followed in August 1874 by the line to Port Erin.

These lines left Ramsey without the benefit of a railway. Two general routes were favoured: the east coast route which was direct from Douglas but expensive and heavily graded, or routes along the west coast using the Peel line to cross the Island from Douglas (Figure 12.2). The latter routes involved two via Glen Helen, one via Jurby, and the route actually built. Seen as a potentially unprofitable venture, Lieutenant Governor Loch encouraged Tynwald to guarantee effectively a dividend on certain preference shares for twenty-five years. Loch's objective was to develop the north of the Island; similar

ECONOMIC INFRASTRUCTURE 193

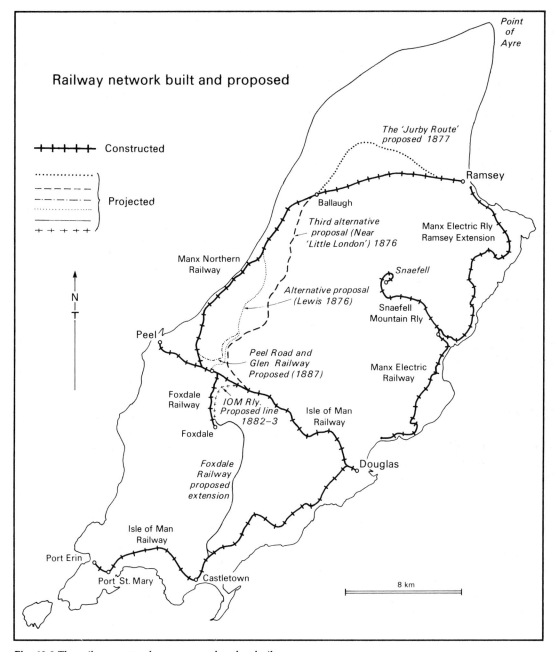

Fig. 12.2 The railway network: as proposed and as built

developmental objectives were later to result in like guarantees by local authorities in Ireland in the subsequent decade (Hendry, 1979). All of the guaranteed shares were sold to investors off the Island. Authorised by the Act of Tynwald in 1878 the line from St Johns to Ramsey opened in September 1879. The economic viability of this line, the Manx Northern Railway, would have been more marginal had the Laxey and Ramsey Railway Company ever built its rival line from Douglas to Ramsey, as authorised by Tynwald in 1882.

As it was, the northern line was by the end of the century to have a rival for both passenger and freight traffic in the form of the electric tramway which finally connected Douglas with Ramsey, via Laxey, in July 1899. Representing new technology in the 1890s the electric tramway was capable of economical construction and operation on steep gradients and sharp curves. It was progressively extended to Ramsey from Douglas, first to Groudle, then to Laxey, and following the construction of the Snaefell line, to Ramsey from Laxey (Pearson, 1970; Hendry and Hendry, 1978). Douglas was also provided with a further tourist tramway, the Douglas Southern Electric Tramway, which linked Douglas to glen developments in the south, offering some competition to the southern railway line.

The Island also developed a railway line with a more transient purpose. In the early 1880s the Foxdale mines were very productive: coal was needed for pumping and had to be imported from Cumberland, and ore needed to be transported across land to the coast for export. In response to this commercial opportunity the Foxdale Railway Co. Ltd. was registered in November 1882; this line was to be worked by the Manx Northern Railway, whose Board saw valuable traffic to and from its harbour connection at Ramsey. Indeed, so determined were the Manx Northern to gain this traffic, they negotiated an arrangement which meant that if the ore and coal traffic declined, the loss would be *not* to the Foxdale Railway but to themselves. The Foxdale line opened in the summer of 1886, but by 1898 after the collapse of the coal and ore traffic, the Manx Northern were paying the Foxdale Railway twice as much for operating the line than they were receiving in receipts from it!

The original railway company, the Isle of Man Railway, was financially sound. Less could be said of both the Manx Northern, the Foxdale Railway and the electric tramways. As early as 1887 there were merger talks between the steam railway companies (Hendry and Hendry, 1980). In 1900 the electric tramway went into liquidation, and in 1902 a new, and largely British-owned company, the Manx Electric Railway Company Ltd, bought the line. This new company set out explicitly to develop the tramway as a tourist line, and built hotels and developed the nearby glens as attractions (Hendry and Hendry, 1978). Finally, in 1903, after Tynwald refused to help the financially ailing Manx Northern, the Isle of Man Railways Act enabled the Isle of Man Railway to absorb the other two steam operated inter-urban railways, and unify the system.

One other line needs a mention. The Island participated in a third stage of railway building, that of the pleasure line. In 1896 the Groudle Glen Railway opened. It was of two-foot gauge, unlike the other steam operated lines which were of three-foot gauge. As an early example of a pleasure line, it used conventional steam locomotives, rather than miniatures which have often become associated with such lines in Great Britain. The line operated until the 1950s, was resuscitated for 1961–62; then closed a second time, and was recently again resuscitated in 1985. In 1987, one of the original steam locomotives returned to the Railway after extensive reconditioning in Cumbria.

As in the British Isles generally, the motor bus proved difficult competition for the narrow gauge railways, both steam and electric powered. In this general perspective two steam operated narrow gauge railway systems survived longer than most, the County Donegal Railways and the Isle of Man Railway. Both survived into the post-Second World War era by adaptation; the County Donegal Railways developed railcars and the Isle of Man Railway developed bus services integrated with its railway services. On the Island two lines suffered from motor bus competition, in particular the Electric Railway between Laxey and Douglas and the Northern Line. The latter was clearly the long way around to Ramsey; the former lacked a central Douglas terminus.

A major economy was first made on the Foxdale line, unprofitable since the collapse of the lead industry (see Chapter 14). In July 1940 the passenger service was withdrawn, but for four years a substitute bus service was run. As early as 1949 Tynwald was advised by a railway consultant that the Ramsey section of the Electric Railway should be closed; and in 1956 a committee of Tynwald recommended its full closure (Hendry and Hendry, 1978). However, the tramway was saved by a decision of Tynwald to nationalise the line and operate it as a tourist asset, and by the Manx Electric Railway Act 1957 it was vested in a new Board of Tynwald. Yet again in 1966 the Ramsey section of the tramway was proposed for closure; a recommendation which was repeated in October 1973 by consultants. In December 1975 Tynwald approved the closure of the Ramsey extension of the tramway, but in 1977 the line was reopened.

The main steam operated line fared little better than the Electric Railway in the post-war era. In January 1965 shareholders of the Isle of Man Railway were advised that train operations, already suspended for winter maintenance, were to cease as uneconomic. In response, the House of Keys set up a transport commission, which reported in 1966. No trains ran on the railway in the summer of 1966, but in 1967 the system was leased to the Marquis of Ailsa and associates for 21 years, for operation largely as a tourist railway.

The lines to Peel, Ramsey and Castletown were operated for the summer of 1967, but in 1968 the Ramsey line was not operated. The financial results were disappointing and in 1969 Tynwald approved a Tourist Board proposal to subsidise the lessees to operate the Port Erin line. The Marquis changed his company's name to the Isle of Man Victorian Steam Railway Company Ltd. and tried to emphasise a Victorian image generally in staff uniforms and the like. However, in the spring of 1972 the Marquis announced his withdrawal. Prompted by this Tynwald approached the Isle of Man Railway once again to operate their own Port Erin line on a subsidised basis. The closed sections of the system remained intact until March 1973 when they were offered for sale for scrap. With track lifting completed in 1976, only the Port Erin line now remains of the original steam operated system. For the years 1972 to 1975 only the Port Erin to Castletown section was open, but in 1976 operations extended to Ballasala, and in 1977 back in to Douglas. Reduced both in extent and dependent on public subsidy, in 1978 the line was finally sold to the Manx Government by the Isle of Man Railway Company.

By the late 1970s both the electric tramway and the Port Erin steam operated line were owned by the State, and reliant on subsidy as tourist assets; the common carrier role of the lines having been transformed into products to differentiate the Manx tourist offering from that of competing destinations. Finally, in 1980 Tynwald took responsibility for most public transport on the Island, including the bus services within and outside of Douglas.

Road transport

The road network and its use by private cars

The earliest known map of the Manx main road system is Fannin's map of 1789. The road pattern was quite basic at this time (Figure 12.3) and ignored the extreme north and south of the Island. Roads linked Castletown directly with Douglas and with the north side of the Island, via Kirk Michael. Douglas was directly linked by road with Ramsey, Peel and Castletown. Other roads, not shown on Fannin's map, were largely local in function, for example, enabling farms access to the sea for fishing and seaweed collection. Two hundred years later the contemporary main road system of the Island is more extensive, intensive and focused (Figure 12.4). In particular, the contemporary Island capital, Douglas, is

12.8 The Island's tramways continue to provide attractions for tourists, and form part of the Island's 'heritage tourism' resource

clearly the focus of the present road system. However, the local road classification recognises the antecedent route pattern, in that the routes of two hundred years previously are the first numbered A-class Island roads, A1 to A5. The present network extends over the whole Island, and is more extensive particularly in the extreme north, extreme south and middle south of the Island. Similarly, Snaefell is no longer avoided as a route between Douglas and Ramsey.

Vehicles licensed on the Island have substantially increased in the past two decades, although in the early 1980s this increase slowed markedly (Table 12.1). Presently there are nearly twice as many private cars licensed on the Island than there were in 1970, and it is amongst this type of vehicle that the recent increase in vehicle numbers has occurred. Whereas in 1970 there was approximately one private car licensed for every three persons resident on the Island, in 1987 the proportion had fallen to two residents for every car registered. This is despite an increase of a fifth in the Island's population in the same period.

Bus services

As a new technology, motor buses became a strong challenger to the narrow gauge railways of the Island on inter-urban routes in the late 1920s. In May 1927 Manxland Bus Services was registered, a company promoted by Cumberland Motor Services Ltd. from England. Realising the threat to its trade the Isle of Man Railway, along with other interested parties, formed Manx Motors Ltd., as a rival company. However, in June 1928 the Railway sold its interest in Manx Motors and instead began its own bus operations. The

ECONOMIC INFRASTRUCTURE   197

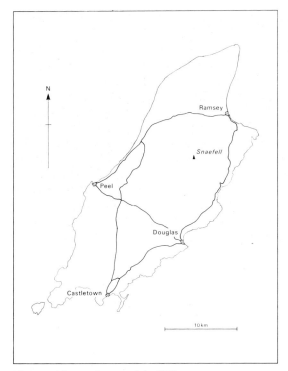

Fig. 12.3 The road network in 1789

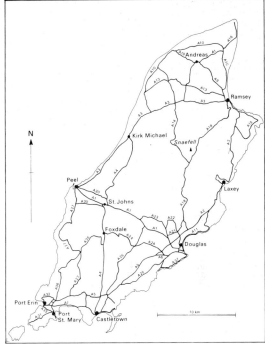

Fig. 12.4 The road network in 1989

ensuing cut-throat competition demonstrated the over-supply of bus services and in February 1929 both Manxland Bus Services and Manx Motors sold out to the Isle of Man Railway (Boyd, 1977; PSV Circle, 1980). The three bus operations were then combined by the Railway into Isle of Man Road Services which was formed in June 1930. From this date the Railway sought to integrate its railway and motor bus operations, and from the 1930s the limited substitution of buses for trains was effected for some of the out of holiday season services. Although the inter-urban bus services of the Island were born out of competition, this was substantially between companies and not, as in some parts of the British Isles, between former carriers using buses or individuals operating only single routes (e.g. Lacey, 1985). The inter-urban stage bus services were also integrated into a spatial monopoly at a comparatively early date, and competition with the railways minimised. This is not to imply that private coach operators failed to exist on the Island, to cater for the tourist sightseers. The contrary is true. Figure 12.5 shows, as an example, Blair's motor coach routes of around 1928.

Until recently the Island's bus and coach services were operated by the Isle of Man

Table 12.1 Numbers of motor vehicles licensed on the Isle of Man

	Private vehicles (excluding motorcycles, scooters, tricycles)	Total vehicles
1970–71	16,902	23,473
1975–76	22,219	29,862
1980–81	26,940	35,837
1981–82	27,496	36,156
1982–83	27,755	36,227
1983–84	28,098	36,422
1984–85	28,350	36,450
1985–86	28,942	36,915
1986–87	32,695	41,377
1987–88	31,594	39,372

Source: Webster (1988).

198 THE ISLE OF MAN

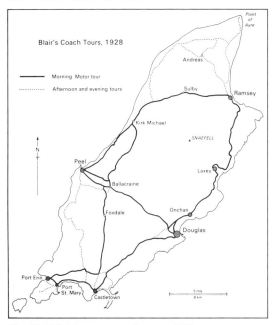

Fig. 12.5 Blair's motor coach tours, 1928

Road Services Ltd. (outside of Douglas), Douglas Corporation (within Douglas) and private coach tour operators. It is the latter who have shown the greatest contraction of services; their number had reduced from 56 listed by Lambden (1964) to 11 in 1984. The two large bus operations were amalgamated into the Isle of Man National Transport Ltd. in 1976, and in 1983 railway services were fully integrated with the creation of the Isle of Man Passenger Transport Board, the predecessor to the present department. Although the pattern of bus routes on the Island has shown little change since the 1930s, fleet sizes have been reduced. Whereas in 1936 the two main companies operated 127 buses, by 1984 the Board operated only 75 vehicles, although their average seating capacity was of course greater. Figure 12.6 shows the stage bus routes operated in 1983 outside of Douglas. Whereas some thinning of the network is apparent when compared with routes advertised for 1934 (Figure 12.7), this is not extensive. The spatial pattern of services shown for 1983 can be found in that for 1970, and represents recent stability in routes despite a continued decline in patronage. However, a comparison of the scheduled services operated on the routes between towns shows a general and substantial decline over the past half-century, particularly on the routes from Peel to the other towns on the Island (Table 12.2). The figures shown in the table take no account of relief buses, and it is known that these were a common feature of past bus operation on the Island (e.g. Lambden, 1953a). Further, traffic abstraction in the post-war years by private cars from public transport on the Island is more extensive than from the buses alone as until the 1960s the Island's railways were also carriers.

Traffic censuses on the Peel to Ramsey and Ramsey to Jurby and Bride bus routes undertaken in late 1983 revealed very light loadings over most of these routes on certain services (Prentice, 1985a). Income from fares in 1982–83 and 1983–84 covered little more than the direct operating costs of the buses, excluding servicing, repairs and maintenance. Clearly, in the 1980s the buses have become a subsidised service. This is despite early attempts at one-person operation on urban routes (e.g. Lambden, 1953b, 1958) and in the past twenty years, sizeable and repeated purchases of second-hand buses from Great Britain. The latter have usually been bought prior to the end of their service life, either in the early 1980s as the result of premature restocking to gain subsidy by British operators or, more recently, as a result of fleet reductions to cope with the deregulation of bus services as occurred in Great Britain in 1986. For example, in 1984 only a third of the Board's bus fleet had been purchased new by itself or constituent operators. Buses second-hand from Liverpool, Manchester, Sheffield, Newcastle-upon-Tyne and Preston have in the 1980s been common sights on Manx roads, as in the 1970s were buses from Stratford-upon-Avon and Bournemouth (Prentice, 1985b). Recently, however, this policy of substantial second-hand purchases has been reversed and in 1988 new vehicles began to arrive for stage service operation.

ECONOMIC INFRASTRUCTURE 199

12.9 The Island's former transport companies are here represented among preserved vehicles at Douglas

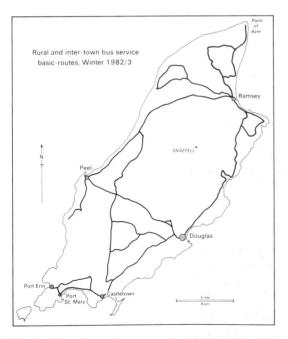

Fig. 12.6 Stage bus routes operated outside Douglas, 1983

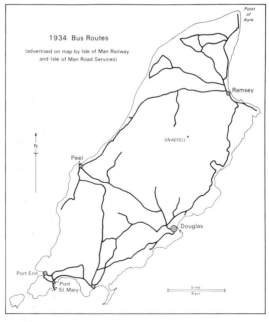

Fig. 12.7 Stage bus routes operated outside Douglas, 1934

Table 12.2 *Numbers of direct scheduled bus services between selected places, 1929 and 1986 (summer weekday services)*

Origin	Douglas	Castletown	Destination Port Erin	Peel	Ramsey	Port St Mary
Douglas						
1929	—	31	31	30	19	31
1986	—	21	21	19	21	21
Castletown						
1929	31	—	43	13	0	31
1986	21	—	23	3	0	23
Port Erin						
1929	31	43	—	12	0	31
1986	20	23	—	3	0	24
Peel						
1929	30	13	12	—	12	0
1986	22	3	2	—	4	2
Ramsey						
1929	19	0	0	12	—	0
1986	19	0	0	5	—	0
Port St Mary						
1929	31	31	31	0	0	—
1986	18	20	20	2	0	—

Source: Timetables 1929 and 1986.

In view of the losses incurred on the rural routes outside of the tourist season, it is pertinent to ask how the Manx population regards its service. Such a survey was undertaken in villages with bus services in April 1983. It involved the traffic habits of 467 rural residents, and not only confirmed the low usage of the buses but suggested that many needs could perhaps be met with less than daily frequencies on some services. Most persons did not use the bus service, and most had the effective use of a private vehicle for, in their own judgement, all of the time they needed one (Prentice, 1985a). Including as passengers school children making journeys to school, nearly eight out of ten residents did not use the bus services at all. Only one in six persons interviewed used the bus over three times per week, including school children making journeys to school. Other than for school journeys, most users were females, and many were retired, both making largely shopping trips. However, the effects of *totally* withdrawing the bus service were seen as inconvenient or seriously impairing mobility by a third of the retired population. For the non-retired, mild inconvenience or no effect predominated, even among users of the service. The conclusion that, other than for school services, less than daily frequencies may suffice for the rural population in the less populated parts of the Island is hard to avoid; but, equally, total withdrawal of services would be unpopular. In the absence of route-by-route subsidisation as now pertains in Great Britain on socially desirable services, further analysis is impossible. However, it is likely that in the future many rural routes will show further decline unless subsidies are maintained.

Conclusion

Not surprisingly, the economic infrastructure of the Isle of Man strongly reflects its economic past. Each economic sector that has gained a position of prominence has left its legacy in the landscape. Perhaps more interesting, though, is the way that infrastructure is also now pointing to the future, and being used to create opportunities rather than simply responding to them. In the case of telecommunications and air travel, private enterprise is installing new infrastructure and providing new services, often in advance of the market. Equally, older facilities are being turned to new uses where this is profitable: a clear example of this is the way that the steam and electric railways now form the focus of special interest holiday breaks. This more adventurous and innovative provision of infrastructure not only reflects the current buoyancy of the Manx economy but also bodes well for its future.

PART THREE: THE ISLAND'S INFRASTRUCTURE

Conclusion

The location of the Isle of Man upon major trade routes has historically been the key factor which has fashioned the development of Manx society and economy. As late as the Second World War, the strategic location of the Island was instrumental in it being chosen as the site of three military airfields, one of which provided the basis for the current growth in air transport to the Island. Similarly, the location of the Isle of Man has also played a significant part in the development of its telecommunications industry, since it acts as an ideal staging post both for Irish Sea traffic and for satellite communication with the eastern seaboard of the United States.

However, whilst the location of the Isle of Man has obviously been beneficial in some respects, the Island possesses three distinct disadvantages for those responsible for providing services and infrastructure. First, and most obviously it is an Island. All consumer goods and industrial raw materials have to be imported by sea (or possibly air). This not only increases costs, but also introduces an element of uncertainty, since transport is dependent upon the weather and upon industrial relations. In addition all infrastructure must be self-contained. Power, for example, cannot simply be cabled in from adjoining areas which have excess generating capacity. Nor can the telecommunications sector easily rely upon extant facilities on the mainland. Being an Island therefore enforces the duplication of facilities. This need not be a problem in itself, but when it is combined with the second disadvantage—namely the small size of the local population—it does become one. Infrastructure has to be provided, even though the local population numbers only 64,282. Inevitably unit costs are inflated, and these must be passed onto the consumer either directly through charges or indirectly through taxation. The electricity generating industry provides a case in point. It has not only consumed some £25 million of public funds in recent years, but also continues to charge prices above those of the mainland. Neither of these alternative methods of funding services provision is desirable. The Manx population is still relatively poor by European standards and enjoys a per capita income which is significantly below that of the mainland. In addition, the importance of a low tax regime for attracting new employment and new residents is such that the Treasury has limited scope for raising direct Government income. And the third disadvantage which the Island possesses is that the use of infrastructure is both spatially and temporally uneven. Excess capacity has to be incorporated in many elements of service provision in order to meet demand in peak periods. The holiday season imposes considerable demands on the electricity and water industries whilst also requiring extra facilities in the transport and leisure sectors. The uneven distribution of local people across the Island provides similar contrasts to those planning infrastructure. Over half the Island's population now resides in Douglas and Onchan but the remainder is often sparsely distributed, particularly in the upland regions. Despite this level of urban primacy, planners still face the task of catering for those who commute into the capital. Again resources have to be committed to meet the day-time parking needs of commuters whilst those same facilities are under-utilised at other times.

The three disadvantages described above

are by no means unique to the Isle of Man, nor to the late 1980s. What does make these disadvantages more of a potential problem is the Island's uniquely coherent attempt to move up-market. As the Isle of Man Government (1987) itself noted in a recent strategy document:

> 'The Island is developing a broadly-based consistent image reliant on quality and high reputation. Beginning with the Island's natural attributes this has been extended into banking, insurance, manufacturing, ship management and is increasingly the marketing message of tourism. These complementary elements work together to project an attractive and wholesome image . . .'.

Whilst this concerted pursuit of excellence is already conferring benefits, it will inevitably also incur costs. In certain cases, the private sector and Government are successfully meeting the challenge, but in others they appear not to be. Per capita spending by tourists is still low indicating an absence of sufficient facilities and services. Businessmen still appear dissatisfied with the cost of energy and the quality and cost of travel (MORI, 1986). And New Residents are critical of travel both within and to the Island, and the quality of retail facilities. It is clear, therefore, that the Isle of Man Government will not only have to continue to invest its own resources in infrastructural improvements, but it will have to be rather more successful in its attempts to stimulate private sector involvement.

PART FOUR

THE ECONOMY

Introduction

Writing in 1964, Birch described how the Manx economy had developed through a series of distinct phases. Such a chronology provides a valuable backcloth against which the evolution of particular economic sectors can be set. This brief introduction therefore outlines Birch's chronology and brings it up to date by discussing recent changes in the Manx economy. Concise details of the current form of that economy are also provided as an overview, prior to separate chapters on agriculture, forestry and fishing, manufacturing and extractive industry, producer services and consumer services. These chapters detail the historical development of their respective sectors and they also contain discussions of the contemporary problems of each, as well as the potential which they offer for the future.

Birch (1964) identified three phases of Manx economic development and hinted at a further two. The first covered the period up to 1830 and was characterised by insularity and a gradual recovery from the debilitating effects of the eighteenth-century smuggling trade. Until the Revestment of 1765, smuggling had been by far the most profitable economic enterprise, and both farming and fishing had suffered as a consequence. Neither the economy nor its labour was specialised and the Island was therefore little involved in external trade. Cured herring, linen and paper were sent from the island, but the bulk of economic effort was designed only to satisfy the insular market which was neither large nor particularly sophisticated. This introversion was further accentuated by the terms of the Revestment which restricted and regulated those goods from the Island which were most likely to compete with domestic English products. Denied export opportunities, the Manx economy adjusted to the end of smuggling by absorbing dislocated labour into the two staple enterprises: agriculture and fishing. It is doubtful whether either of these benefited greatly, although it is possible that the temporary shortages of labour brought about by the coincidence of harvest time and the herring season might have eased somewhat. Tithes on agricultural income were a further brake on efficiency.

The period 1830–95 witnessed a dramatic transformation of the economy, with major developments in all sectors. Improved external and internal communications stimulated the explosive growth of the tourist industry with its demands for infrastructure and consumables. The Tithe Commutation Act of 1839 freed the tenant farmer from the worst of his feudal obligations and encouraged both specialisation and professionalisation of agriculture and fishing. The extractive industries were meeting with sustained success for the first time, and progressively the English were dismantling tariff and non-tariff barriers which had restricted Manx exports. In 1866, fiscal reforms completed this task and triggered not only a growth in those manufacturing sectors associated with farming, fishing and mining, but also a parallel period of investment in public infrastructure such as harbours and private infrastructure such as hotels. By the 1880s the Manx economy was more robust than previously and also more broadly based. Agricultural acreage was at its peak, the mining industry was producing record output, fishing had become an efficient industry, and 300,000 tourists were visiting the Island each summer.

The period 1896–1935 was marked by recession and increasing specialisation in only one economic sector, tourism. The collapse of the mining industry in the early years of the century and the growing dominance of non-Manx boats in the fishing of Manx waters would, themselves, have had a profound impact upon the Island economy. Given the dependence of much manufacturing industry upon these two sectors, the impact was considerably magnified. Tourism, however, continued to enjoy relative advantages and, in 1913, over 634,000 visitors came to the Island. After the First World War even this industry began to experience recession and by the mid 1930s visitor levels had dropped by 15% with a consequent loss of market for farmers and manufacturers as well as those directly employed in tourism.

Birch regards the years 1935–49 as something of an interlude before the return of recession. He points to the post-war revival of the tourist industry, the fillip presented to farming by the presence of internees and military personnel on the Island during the war, and the recovery of the fishing industry at Peel. Construction work both during and after the war reduced the problem of male unemployment and provided work for associated local manufacturers and suppliers. Whether these short-lived improvements really constituted a distinct phase in any chronology of economic development or merely a temporary reprieve is a moot point since, by 1949, the Isle of Man was again plunged into recession.

Birch's account effectively ends with the return to recession, but with the benefit of greater historical hindsight, it is possible to extend his chronology through to the late 1980s. Two further phases can be identified, the earlier running from 1949 to 1961, and the later extending from 1961 to the present.

Although the Manx economy continued to decline between 1949 and 1961, this period deserves to be regarded separately from the previous era of decline between 1896 and 1935. There are two reasons for this. First, the Manx Government began to re-assert its traditional independence during the 1950s and with this came far more frequent state intervention in economic affairs. Second, this intervention sought unique solutions to problems of existing industries even if these diverged from accepted mainland policy. However, despite the new context and content of economic policy, its primary objectives remained unaltered. State intervention sought only to react to the failing fortunes of its existing enterprises by maintaining them in as good health as possible. The focus of policy was thus essentially to underpin the existing economic system rather than radically reshape it. To this end, measures were put in hand to provide a stable market for herring, assistance was provided to existing industrial concerns and to similar ones which sought to relocate to the Island, agriculture received a more generous level of State support and the Government re-assessed the viability of an island mining industry in the light of new market trends.

By 1961 it was clear that reactive policy in support of the traditional economic structure was inadequate to meet the needs of the day. The continued decline of the tourist industry was making the costs of over-specialisation abundantly clear. The progressive loss of custom to alternative destinations and the consequent reduction in profits and therefore investment, had become permanent features of the Manx economy. Winter unemployment, public works schemes and emigration signalled the failure of reactive policy and pointed the way to a pro-active restructuring of the economy. The initial phase of this sought to diversify the economic base through the expansion of policies designed to attract light industries and through the overhauling of the tax regime to attract affluent New Residents. In parallel, the Government also sought to provide continuing support to the tourist and agricultural industries, but this was progressively more dependent upon their willingness to restructure and address new market opportunities. The emphasis of the tourist industry has increasingly moved away from the volume end of bucket-and-spade family holidays towards a more limited number of activity holidays, short breaks and the attraction of a high-spending and more discerning clientele. The 1970s witnessed a continuation of each of these policy strands, but increasingly the development of new service sector enterprises became the dominant theme of pro-active policy. Commercial banking, finance and insurance became the new generators of wealth, if not jobs, and Government has worked hard to ensure that an attractive legislative climate exists for such ventures. However, progress towards economic restructuring received a setback in the early 1980s. The abolition of exchange controls in 1979, the world recession of the early 1980s, and the failure of the Savings and Investment Bank in 1982 coincided to produce a Manx recession which lasted from 1982 to 1986. By 1985 the Government had responded with tighter financial controls, a new New Residents policy and the focusing of manufacturing growth on the Freeport at Ronaldsway. February 1986 is now commonly regarded as the time when the Manx economy regained its strength and moved out of recession.

CHAPTER 13

Agriculture, forestry and fishing
Gwyneth Davies

Both Chapter 10 and the Introduction to Part Four make it clear that agriculture and fishing have long been key parts of the Manx economy. Although both have tended to decline in economic importance during the twentieth century they still retain considerable significance. Moreover, agriculture and forestry are important not only for their ability to generate employment and income, but also as key determinants of landscape aesthetics. Farming, too, is inextricably bound up with the maintenance of rural life and customs, whilst the fishing industry has also traditionally enriched Manx culture. Each of the sectors will be discussed in turn.

Agriculture

The Isle of Man currently has 47,100 hectares being farmed. This represents 86% of the Island's total area. The exact use to which this land is put is strongly influenced by four factors: relief, soils, climate and the fact that the region is an Island.

The relief consists of two upland areas cut by the Douglas-to-Peel trough and surrounded both by the gentle slopes of the dissected plateau and the north and south lowlands. The soils reflect this conformation, the northern drift lowlands having mainly sandy areas with small clay enclaves (e.g. the Ballaugh Curraghs), the uplands having thin poor soils and the southern lowlands having a good medium loam. This type of loam is also found on the western coastal fringes, but a heavier loam soil—due to the presence of clay—developed in the central trough and the eastern coastlands. The bare climatic statistics suggest a benign environment for Manx farmers. Ronaldsway records mild winters, which average 5.6 degrees centigrade, and cool summers (average 14.6 degrees centigrade). Rainfall varies between 675 mm at Port St Mary and over 1,500 mm in the central uplands and there is only less than one degree centigrade temperature range between the north and south of the Island. However, in reality the Manx farmer does have to contend with difficulties. These include strong winds. Adaptation to the prevailing south-west component of the wind can be seen in the siting of farm houses, the high hedges and windbreaks and the characteristics of many growing trees and bushes. In early spring, cold north-east winds can persist which not only affect conditions for early lambing but also delay the beginning of the growing season. In addition, although temperatures vary little across the peripheral lowlands marked differences do occur as a reflection of altitude. At sea level only February records a temperature below that at which grass will grow: 5.6 degrees centigrade, and even then some days reach 7.5 degrees centigrade; grass growth is little restricted in these areas. By contrast, at the upper limit of cultivation, 275m, a mean temperature in excess of 5.6 degrees centigrade is often not reached until late April. These physical conditions suit grassland farming and so grass production associated with cattle and sheep forms the most important section of most Manx farms. To this physical background should be added historical and social factors.

208 THE ISLE OF MAN

As early as 1577, Bishop John Merrick was writing that the Island 'is rich in flocks, fish and corn and not only produces sufficient for its own use but annually exports a great deal'. During the 18th century reports by Basil and Thomas Quayle found that Manx practices were behind the mainland, in that they were slow to introduce the rotation of crops or to use turnips as winter feed. Yet before the end of that century Robertson (1794) was able to comment that 'marshy land is now drained, wasteland enclosed and nourished with marl and seaweed. Eggs, butter, poultry are plentiful and beef sells at two pence per pound; the mutton is very good and wild pig is a delicacy'.

During the nineteenth century the clear division on the Island between the land of the Duke of Atholl and other large farmers and the traditional family farms, often associated with a fishing-farming economy, began to change. Thomas Quayle could say in 1812 that the greater part of the Island's farmland is worked by its owners, but returns in 1887 showed that by then only 32% was owner-occupied. This change was connected with the growth and development of other sectors of the Island economy. It was about 1840, that with the increasing importance of the herring fleet, men were making either fishing or farming their full occupation. Later in the century the same process occurred with the development of mining in the central trough and peasant farming was only continued on the poorer marginal areas. By the end of the century, the expanding economy, changing developments in cereal production and an increased demand for livestock products—both on the Island from tourists and in the mainland industrial markets of north-west England—led to the development of a more mixed farming system based on livestock and fodder crops. This has provided the basis for the developments and refinements of this century.

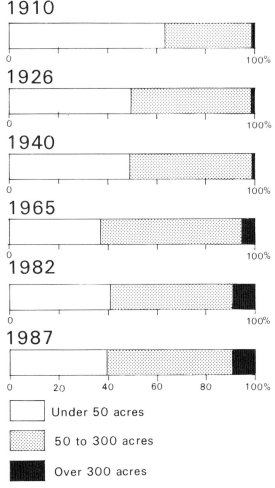

Fig. 13.1 Changes in farm size, 1910–87

Agricultural structure 1980

The 'typical' Manx farm consists of units of animals and related crops with at least 75% of the land under grass. The average size of farms has been increasing during the 1980s, the largest gain being in the 8ha–20ha category with an increase of +16.7% between 1986–87 but the total number of farms remains around 800 (798 in 1987, see Table 13.1 and Figure 13.1). Of the 74 holdings over 120 ha, about 20 are hill sheep farms. This utilisation of the uplands along with a variety of crops and animals carried on the lowlands has led to the development of spatial patterns of specia-

Table 13.1 *Farm size*

Size	Percentages 1910	1926	1940	Numerical (percentages in brackets) 1965	1982	1987
Under 8 ha	63%	49%	339 (29.5%)	251 (24.7%)	217 (26.9%)	206 (25.8%)
8–20 ha			222 (19.3%)	124 (12.2%)	111 (13.7%)	112 (14.0%)
20–40 ha		28%	320 (27.8%)	249 (24.5%)	144 (17.8%)	145 (18.1%)
40–60 ha	36.9%	13%	161 (14.0%)	162 (15.9%)	86 (10.6%)	96 (12.0%)
60–80 ha		9%	97 (8.4%)	171 (16.8%)	97 (12.0%)	91 (11.4%)
80–120 ha					76 (9.4%)	74 (9.2%)
120 ha+	0.7%	1%	10 (0.9%)	56 (5.5%)	74 (9.1%)	74 (9.2%)
TOTAL No. of farms	—	—	1149	1013	805	798

Source: Agricultural Returns, Ministry of Agriculture and Fisheries and Forestry.

Table 13.2 *Land under cereals, 1907–87, in hectares*

	1907	1914	1939	1945	1965	1985	1987
Oats	7,480	7,800	5,800	5,856	3,012	556	540
Wheat	176	112	48	272	188	240	256
Barley	1,040	736	156	213	1,488	3,884	3,864
Mixed corn	no record	2	76	760	188	24	16
TOTAL	8,696	8,650	6,080	7,102	4,876	4,704	4,676

Note: (i) effects of war time cultivation 1939–45; (ii) the marked decline of oats but growth of barley.
Source: Agricultural Returns, Department of Agriculture and Fisheries.

lisation. The recent developments in improved grassland management with new techniques, increased use of nitrogenous fertilisers and better winter feed conservation have encouraged this development.

Arable crops

Among the cereal crops barley has shown a marked increase from 212.6 ha in 1945 to a figure for 1985–87 of over 3,800 ha (see Table 13.2). This increase has been positively encouraged by the granting of a subsidy and by a new recognition of the value of barley as a feed grain by many farmers. The lowland areas of the north and south have the largest concentration, but most farmers grow a small acreage. Some surplus has resulted and this has led to an export trade from Ramsey.

Similarly, oats which are grown for home consumption occupy 12% of cereal acreage whilst wheat occupies only 5%. The latter is used by the Island's bakeries which mix it with imported hard wheat. Mixed corn, so prominent during the Second World War at 10.7% of the acreage, had been reduced to 0.2% in 1987 (see Figure 13.2 and Table 13.2).

210 THE ISLE OF MAN

Interestingly the marked decline in area for fodder crops—turnips, swedes, kale, mangolds, fodder beet and rape—has been arrested. Returns for 1986–87 show gains as farmers maximise their own production and reduce the cost of imported feed. Potatoes, which long featured as a staple part of the Island diet, are showing a slow decline and the impact of earlies from mainland sources still continues. The seed potato industry, benefiting from the Island location, continues to export to England and Wales.

Horticulture

Green vegetable production has fluctuated during the 1980s; the Island is nearly self-sufficient in cauliflower and brussels sprouts,

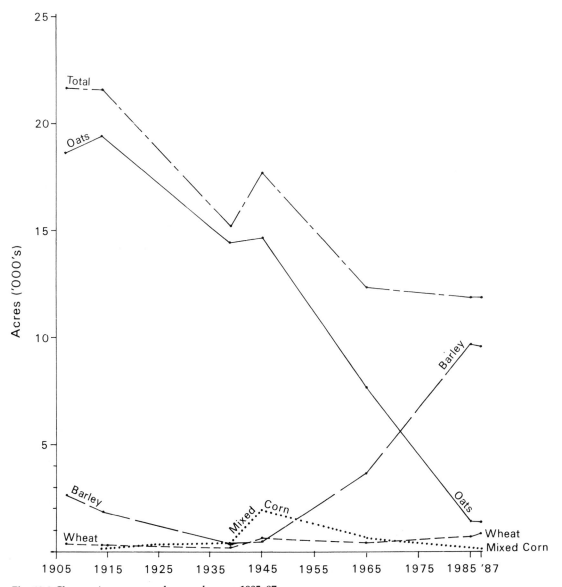

Fig. 13.2 Changes in acreage under cereal crops, 1905–87

Table 13.3 *Hay and grassland, 1945–87, in hectares*

	1945	1983	1985	1987	% change 1986–87
Temporary Leys:					
for mowing this season, hay	} 15,044	2,180	2,188	1,736	−12.9
for silage		2,024	2,116	2,536	+7.2
for grazing		8,888	7,324	7,164	−4.5
Permanent Grassland:					
for mowing this season, hay	} 4,264	1,480	1,472	1,428	+18.8
for silage		664	816	1,048	+4.5
for grazing		10,092	11,532	12,400	+3.3
TOTAL	19,308	25,328	25,448	26,312	+0.9

Source: Agriculture Statistics, Department of Agriculture, Fisheries and Forestry.

but continues to import other vegetables and additional supplies during peak periods. The recent introduction of daffodil production starting at Knockaloe Experimental Farm in 1983 is a new venture, revenue being obtained both from the cut flowers and from the sale of bulbs. Four hectares are now under this crop on the Island. Greenhouses are used to produce small quantities of tomatoes, lettuces and asparagus. The majority is marketed locally, but Ballakinnish Nurseries south of Douglas have a fully-integrated rose nursery which is geared to the export market. Two pick-your-own fruit farms, Croit-e-Caley, predominantly strawberries, and Cooil Cam are being developed near Port Erin. In addition to horticultural crops, field vegetables and free range eggs are produced for sale both on the site and in local shops. With costs calculated at 10% higher than on the mainland and returns 10% less, these become marginal ventures where diversification is essential for success.

Grassland

The total area under temporary leys and permanent grass is increasing slowly (1987, 26,312 ha, 1% growth over 1966). Within this figure a decline in leys for hay is balanced by a marked growth in permanent grasses for silage and grazing (see Table 13.3). Upland improvement is also reflected in a reduction of 3.5% between 1986 and 1987 in the area of rough grazing. The farmers are increasingly utilising newer methods of cattle feed involving silage production and out-grazing the more marginal areas.

Dairy farming

Because of its maritime climate and areas of free draining lowland, the Island is well suited to milk production. The dairy herd stood at 5,711 in 1987, an increase of 4.2% over 1986 (see Table 13.4 for details). The numbers have shown a small rise throughout the 1980s. Whereas current European Community Agricultural Policy, with its quota system, has had profound effects on the mainland, the Island has aimed at self-sufficiency with expansion based on the smaller herds (Manx average 30 cows) and their more intensive grass use and better silage production of the last few years. This improved technology resulted in all the Island's dairy farmers in 1987 having refrigerated tanks for milk storage prior to collections by the Isle of Man Milk Marketing Association IOMMMA. The milk collection has been the direct responsibility of the IOMMMA since 1960 and in 1987 it was using six tankers of 9,000 litre capacity. This Association started cheese production in 1962, moving to a new creamery at Tromode in 1974, where at peak capacity 60,000 litres of milk are processed into six tons of cheese daily. Of an annual production of approximately 1,300 tons, less than 200 tons (15.4%) are sold on the Island, the rest going for export. The quality of the product was recognised in 1987 by the award for the best British Cheddar Cheese. The value from cheese production in that year reached £3 million out of a total Creamery sale of £6.5 million. The Creamery output in 1988 included eight varieties of cheese, full and clotted cream, and whey butter. As the quantity produced annually is now 15 tons for full cream and 26 tons of whey butter, it is insufficient to meet local demands. Increased production is dependent on the quantity of milk required for cheese-making. A gap in the local market has been identified by a Baldrine dairy farmer who early in 1986 started the production of 300 litres of yoghurt per week, an item not produced by the Creamery.

The dairy farmers of the Island view their dairy enterprise as providing a 'stable and regular return', but the Association shows concern at further development and has instead backed milk quotas during the peak production periods of May/June. However, penalties had not yet been imposed by June 1988.

Beef

The dairy herd has a secondary role in that it supplies nearly 50% of the Island's beef requirements. Cross-bred stores from these herds are a cheap basic source of beef production, provided quality is maintained. It is a traditional method, well suited to the upland grass areas, requiring low labour inputs and has been encouraged by the Government. As Tynwald itself said, in 1986, 'it cannot be explained too strongly that the hill livestock farmer is an important and integral part of fatstock production'. On the mainland by 1987, the effects of dairy herd reduction due to the European Community milk quota system, had produced a shortage of calves. This may provide the Island with a temporary export opportunity. Beef cattle also derive from the pedigree herds of the Island. Knockaloe Experimental Farm has been conducting trials on fattening Friesian bull calves. Three methods have been employed: (i) on summer grass, in yards during the winter and selling at 18 months; (ii) a group fed as 'barley beef' on an intensive indoor system and sold at 12 months; (iii) indoor all the year round based on a silage system and selling at 18 months. These are methods which could be adopted elsewhere on the Island, but by the late 1980s policies of low input in both feed and labour involving lower conversion rates in beef pro-

Table 13.4 *Stock*

	1914	1939	1945	1965	1985	1987
Cattle and calves	21,519	20,072	23,885	31,307	34,558	32,488
Sheep and lambs	77,446	98,704	74,522	118,332	138,017	147,315
Pigs	3,954	4,693	2,946	4,586	6,204	8,425
Horses	5,851	3,707	2,665	499	1,015	885
Poultry	—	—	107,641	—	73,412	71,098

Source: Agricultural Returns, Department of Agriculture, Fisheries and Forestry.

duction appear to be more acceptable to individual farmers.

Economically the local and export markets have been controlled by the Fatstock Association since 1966. Although there was a 59% increase in home-killed cattle in the 10 years 1967 to 1977—reaching 9,946 in the latter year—this figure has changed little since then. Export shipments of live cattle finished in 1975 and since then only carcase meat has been exported. The return to the Island from beef exports in 1984 was £3.5 million but overall numbers of stock continue to decline, between 1986 and 1987 by 3%. So producers have not responded to the Manx Government proposals giving a minimum price, which guarantees farmers a price equal to, if not above, that of the UK. However, due to the European Community 'beef mountain', the UK price has itself been depressed and the future for beef exports from the Island does not appear to be strong, unless markets can be found beyond the European Community.

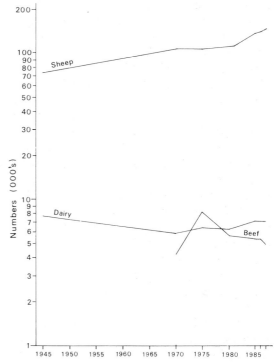

Fig. 13.3 Changes in cattle and sheep numbers, 1945–87

Sheep

The Island's physical regions have been well integrated into sheep production. A long historic tradition has led to high standards of flock management and the production of lambs of good quality. The system involves hardy ewe production on the hills, from which stock the upland flocks are able to produce high quality ewes who themselves provide the lowlands with lambs for fattening. In addition, both lowland and upland farms maintain a balance between cattle and sheep populations to their mutual benefit. Sheep and lamb numbers have been increasing steadily and reached 147,315 in 1987, a gain of 4.7% over 1986 (see Figure 13.3). The Report of the Committee of Inquiry into Agriculture and Horticulture (1975) commented on the lack of integration between mountain and hill grazings and lowland farms. The hill enterprises often suffered from a lack of nearby enclosed lower land. This trend is being reversed and land improvements and more integration have taken place since 1975, these leading to higher stocking rates and increased sheep numbers.

Pigs

Though the total number of pigs has increased by 25% between 1986 and 1987 (to 8,425), fewer than 20 farmers on the Island keep pigs as their main enterprise. The scale of production is generally small compared with the mainland. It is a volatile market with increasingly rigorous standards demanded and high cost inputs. By tradition the producers breed and fatten their own pigs using locally grown barley and, in summer months, swill. The output of this sector, of which 75% is pigmeat, covers local demand but there is a shortfall of bacon pigs and over 50% of the Island's bacon is imported. It has been suggested that a separation between weaner producer and fattening unit may be advantageous for the Island's conditions. There is little evidence as yet of this taking place.

Poultry

A steady decline has been seen in this sector during the 1980s such that by 1987 the number of poultry on the Island was 71,098. This figure is itself only two-thirds of the 1945 number. Few farmers have poultry as their main enterprise, but many 'keep hens'. The small size of the egg units, the seasonal nature of demand and the high cost of imported feed open the Island market to eggs from the mainland. The Department of Agriculture did provide a measure of protection through import controls, but this is now precluded under the Island's special relationship with the European Community. There are only two broiler units and these suffer similar disadvantages to the egg producers. Most of the Island's requirements in eggs and chickens are imported.

Government assistance

Demand for agricultural products is at its highest during the tourist season and yet local surpluses occur at other times. A pattern emerges of major imports in any one year being bacon, eggs, poultry, wheat, early potatoes, vegetables and salad crops, set against major exports of fat cattle and sheep carcases, wool, barley, cheese, seed potatoes and flowers. Because of the variability in local demand and the economic disadvantages faced by many of the Island's agricultural enterprises, the Government has chosen to intervene to assist with marketing and price support. The official producer organisations market nearly 80% of the total farm output, whilst the rest moves through private trade arrangements. The three producers' organisations for milk—using the trade name AGRIMARK—work under the Agricultural Marketing Society founded in 1934 and re-organised in 1954. The overall effect of the Society has been to stabilise price fluctuations and similar producer price structures exist for milk and fatstock as in the UK. Potatoes and eggs fall under a more direct statutory price control.

The Government is committed to an efficient, if not necessarily expanding, agricultural industry. In a policy statement of 1986, Tynwald commented: 'the Island's only raw material is grass and the backbone of our industry is milk, beef and lamb produced from the grain'. To this end grants, loans and subsidies are available under the Farm and Horticultural Improvement Scheme, including milk and sheep subsidies for upland farms. Advisory services centre on Knockaloe Experimental Farm founded in 1924. One of its major functions is livestock breeding, and the Board's Artificial Insemination centre is here. The early experimental work with cereals has been scaled down and use is now made of mainland research centres when necessary. Both seed potato inspection and marketing, and milk and beef recording continue, with the addition in September 1983 of field trials in daffodils, as both flowers and bulbs. The lecture room at the farm is available for use by agricultural organisations, but is primarily for the day release course for young people working in agriculture.

Overall the agricultural picture on the Island is a mirror of similar situations in European peripheral areas. Expansion is now limited to reducing costly imports and maximising all available advantages, including financial ones.

Land values, for example, have remained relatively stable during the last five years, reflecting the more conservative approach of the Island's farmers who have borrowed less capital than their mainland counterparts. The current interest rate of 8% (1988) for Manx born farmers is not available to United King-

13.1 Sail-powered fishing smacks in Ramsey harbour, 1931

dom farmers who move to the island. Since 1987 the buoyant residential market, which has boosted house prices by 50%, has introduced a new element into land values. It is possible that the demand for rural properties will cause some fragmentation and it may also introduce a new social element into rural communities.

Fishing

Historically the traditional Island pattern of a farmer-fisher economy allowed the summer months to be spent at sea. The boats fished local waters especially in the south and south-west of the Island. In 1881 Peel was recorded as the home port for 80% of these boats and herring provided the major catch. This century has seen increasing competition from the mainland and the dominance of Scottish and Irish fishing boats and by 1937 there were only 47 Manx boats active; by 1960 this had been further reduced by a half. Government encouragement, including financial assistance, has resulted in this number increasing recently. In September 1987, 86 working vessels were recorded (Laxey 36, Ramsey 27, Douglas 16, Port Erin 4, Castletown/Derbyhaven 3). These are inshore boats whose average size is only 12 metres.

The Isle of Man has sole jurisdiction of the fishing grounds within the three-mile limit, whilst between three and 12 miles it is shared with the UK Government. The Island's special relationship with the European Community results in the Community's Common Fisheries Policy not applying to the waters within the 12-mile limit.

216 THE ISLE OF MAN

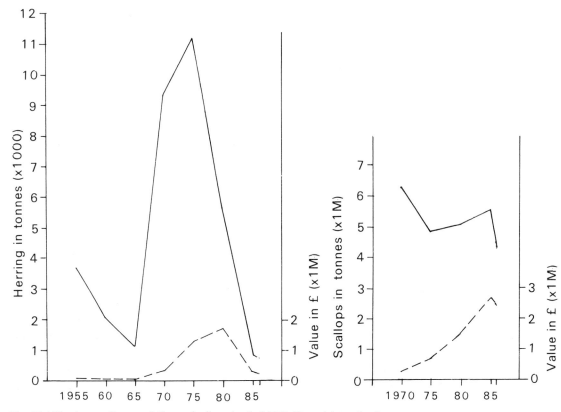

Fig. 13.4 Herring, scallops and Queen Scallops landed 1955–85; weight and value

Herring

The over-exploitation of the north Irish Sea fish stocks, especially for herring, during the 1970s and early 1980s has led to a number of control measures (see Table 13.5). To the fishery limits previously mentioned were added a herring catch quota. This quota was a share of the internationally agreed total allowable catch known as TACs. Even these measures have not prevented over-fishing as the declining catches show (Figure 13.4), and so the TACs have been annually reduced. In 1980 the TAC for north Irish Sea herring was 10,000 tonnes, of which 9,000 tonnes was claimed by the UK and the Isle of Man. By 1984 the European Community-approved TAC for herring had dropped to 4,000 tonnes and the UK and Isle of Man allocation was only 2,550 tonnes. In that year, as the catch quota for the summer period was reached early, herring fishing only occurred between 4 June and 31 August. This accelerating reduction in landings (see Table 13.5) affected related on-shore activities including fish processing units. The disposal of landed herring involved meal and oil production up to 1973, and pickled, cured, and frozen herring up to 1982. Although the Manx herring stock is showing slow recovery, careful management is needed before both stock and catches can be considered to have reached a more economic level for the future.

Shellfish

A new marine resource, scallops (*Pecten maximus*), has produced lucrative growth to parallel the decline in the herring trade. Scallops and their smaller relative, queenies (*Chlamys opercularis*), were first dredged commercially around Port Erin in 1937. They thrive in the sandy inshore areas extending now to Peel and most recently to Douglas. The growth in value of these shellfish landings (see Figure 13.4) by the 1980s meant that Manx fishermen were relying heavily upon them as a source of revenue. To conserve stocks there is both a four-month summer closed season and regulations regarding the minimum size to be removed. It was estimated in 1983 that nearly 50% of all scallops and 25% of all queenies currently caught in the Irish Sea were landed in the Isle of Man, and that of the north Irish Sea stocks of scallops and queenies, 65% and 40% respectively were located within the Manx 12-mile fishing limit. The overseas marketing of this high value scallop meat in frozen form has been largely through Ronaldsway Airport, with the USA taking 75% and the European Continent 25%. France, and more recently Spain, are the chief continental markets.

In 1980, 193 males and four females were employed on the fishing fleet, whilst 49 males and 146 females were involved in fish processing units, this total being 1.5% of the Island's labour force. By 1985 some increase in the workforce had occurred, and it was estimated that on the basis of a full-time all-the-year equivalent employment calculation, 200 were at sea and 250 employed in fish processing.

Government encouragement had also resulted in four small starter units for fish processing being built at Peel.

Forestry

It appears from early evidence that the Isle of Man had a woodland cover largely of oak and ash and this was found on the lower slopes of the uplands and in the glens (see Chapter 7 for further details). However, by the seventeenth century a contemporary writer commented: 'I could not observe one tree to be in any place but what grew in gardens', and this lack of woodland also attracted comment from Thomas Quayle in 1812.

By the middle of the nineteenth century some small-scale planting had been undertaken by private individuals; and in 1884 a policy of afforestation on Crown Land was started with plantations at Archallagan, Greeba and South Barrule. Progress was slow until 1931 when the Forestry Board was created.

The maritime nature of the Island's climate produces severe conditions for tree growth: aspect, exposure to salt spray and the condition known as physiological drought have all to be contended with. Most of the plantations are found between 150 and 300 metres and Birch (1963) in his land use survey of the Island estimated that woodland covered 3,500

Table 13.5 *Landings in tonnes—live weight and value*

Year	Scallops and Queen Scallops tonnes live weight	Value £	Herring tonnes live weight	Value £
1955			3,815	60,465
1960			2,093	41,743
1965			1,135	26,212
1970	6,412,575	309,058	9,740	319,004
1975	4,882,295	633,623	11,113	1,313,367
1980	5,019,460	1,657,832	5,862	1,720,655
1985	5,664,459	2,706,194	857	159,842
1986	4,552,871	2,521,961	798	125,366

Source: amended from Isle of Man Government, Department of Agriculture, Fisheries and Forestry, Fisheries Statistics, 1986.

acres, of which 759 acres (21%) were in coniferous plantations.

In 1987 woodland was estimated to cover 5,500 acres, of which 80% are coniferous plantation. The newly-formed 1987 Department of Agriculture, Fisheries and Forestry now either owns or manages about 17% of the Island, and has instigated the early stages of a vigorous plan extending to the end of the century. Under this, the total acreage of forest on Government land would be doubled, and based on past experience a planting plan of 80% Sitka spruce, Scots and Corsican pine with at least 10% hardwood, broad-leaved trees, would be adopted.

It is recognised that, along with this new and additional planting, many of the older plantations need management and re-stocking. The new schemes are based on the need for viable economic units to be left for the tenant and for the woodland to integrate into the current agricultural use of the area. A process of partnership is involved and, with present resources, the Department is only able to stock and manage an additional 250 acres of woodland annually. Even so, private contractors are being used for such work as ploughing and planting, leaving fencing and maintenance to the Department's own staff of 67 employees.

The Department's centre at St Johns includes administration, a nursery section and a modern saw mill. The latter, besides producing timber, fence posts and gates, is now adding picnic tables and garden play equipment to its range of products. These will be sold direct to the public.

The nearby National Park of 25 acres, to commemorate Millennium Year, and the Island's National Glens are also managed by the Department. The Glens are being subjected to schemes involving clearing and restocking, as well as improvements both to access and paths. The Department is also re-evaluating their recreational and amenity use (see Chapter 11 for further details).

Conclusion

Agriculture, Fishing and Forestry on the Island in the 1980s reflect current mainland and European trends without their extremes. Self-sufficiency, conservation and environmental awareness are now key attributes.

CHAPTER 14

Extractive and manufacturing industries

Vaughan Robinson

The introduction to this part of the book described the various phases through which the Manx economy has passed. Such a chronology highlights the change which has taken place over the last forty years in how the Island earns its living. Although these changes are associated more accurately with the period since 1961, their roots can be traced back to the early 1950s, and there is consequently some justification for regarding the Second World War as a watershed in economic affairs. This chapter therefore contains two sections: the first describes the growth and decline of traditional extractive and manufacturing industry in the Isle of Man prior to 1945; and the second looks at the period 1945 to date, during which there has been a significant change in the shape and orientation of Island manufacturing.

The extractive and manufacturing sectors prior to the two world wars

The Isle of Man possessed advantages and disadvantages for the development of traditional forms of manufacturing and resource extraction. Although it lacked indigenous coal supplies (see Chapter 12), the Island was not bereft of sources of power. Peat provided fuel for some industries (e.g. lime-burning), as did wind power. Although only one windmill is still visible in the present day landscape —at Castletown (see Qualtrough, 1969)—this represents a remnant of what was a considerable number of such mills in the nineteenth century. Probably of greater significance, though, was the availability of water power from the main streams which flowed off the central upland. Jones (1966) has calculated, for example, that over 150 industrial plants were water-powered in the late 1860s, including mining plants, saw mills, a brickworks and fifty-one corn mills. He also notes that the equipment employed in these plants to harness water power was the equal of any in use in contemporary England and Scotland. In conjunction with these sources of power, the Isle of Man was also relatively well endowed with exploitable resources. Mineral reserves, including lead, iron, copper, silver, and zinc (see Chapter 3) were the most obvious, but to these must be added limestone, black 'marble', pure soft water, timber, fish and shellfish, and the agricultural produce of the fertile lowlands. Given the problems of transport inherent in the area being an island, such resources were likely to stimulate local industries concerned with their extraction and processing. In turn, these industries would necessitate local support industries to supply their needs. In this instance, then, insularity was a positive benefit since it stimulated a parallel industrial superstructure to that of the mainland, but protected by the inadequacy of contemporary transport technology. This was also true of those enterprises supplying the local consumer, rather than industrial, market since again they were shielded from the full force of mainland competitors. Lastly, in terms of advantages, the Island had frequently enjoyed direct intervention by its rulers to stimulate economic development. This no doubt stemmed from the Island's status as a possession and as a source of income for its 'owners'. Mathieson (1957a) describes the Dukes of Atholl's attempts to develop an eighteenth-century mining industry, whilst Harrison (1968) notes the even earlier efforts of the Derbys to 'modernise' their estate through the introduction of linen manufacture and brick-making at the end of the 1600s.

Set against these potential strengths, the Isle of Man also possessed demerits. Its insularity was not only a commercial defence, but also a commercial burden for those exporting to a mainland market or importing raw materials. Mathieson (1957a), for instance, notes how a cotton printer on the Island suffered financial hardship when he tried to export but could not collect his debts. Less adventurous producers relied solely upon the home market for their sales, but this denied them the economies of scale which ultimately determine profit. The decline of the Island's flour-milling industry is largely attributable to this cause. Coupled to the small size of the home market was its volatility. The inter-relatedness of the Manx economy ensured a high degree of mutual interdependence. Thus, when the mining industry was booming, other manufacturers benefited, as indeed did agriculture from the success of the tourist sector. Conversely, when a major industry declined, others such as net-making or ship-building, followed suit. And finally the Manx market was not only small and volatile but also segmented. Until the latter part of the nineteenth century internal communications were poor (see Chapter 12) and many enterprises served only a very restricted local market. Developments in the road and rail network inevitably led to rationalisation and the loss of many firms providing a solely local service. The unique status of the Isle of Man as a feudal estate for much of its industrial history imposed an additional burden on many enterprises. At various points in time the potential entrepreneurs had to apply for licences to establish a business. Michael McDaniel had to wait ten years before being granted a licence to open the Island's first paper mill and the onerous financial conditions of this licence ensured that it remained open for less than twelve months. Similarly, tenant farmers were assigned to particular mills for their grinding and could be fined if they contravened this monopolistic imposition. In association with the economically repressive aspects of feudal ownership, the unique status of the Island also ensured that it was not accorded equal treatment by the mainland. After the Revestment of 1765, any goods which could be successfully exported to England were either prohibited or burdened with excessive taxes. Restrictions and regulations added non-tariff disincentives. The last economic disability faced by Island entrepreneurs was the labour force. Difficulties here varied through time: when farmer-fishermen formed the majority of the labour force the lack of specialisation prevented the same degree of professionalisation enjoyed in many parts of England; when the tourist industry had come to dominate the economy, the ease with which labour could find well-paid summer jobs discouraged those entrepreneurs seeking a stable, year-round workforce. And lastly, many of the newly-developed nineteenth-century industries depended upon imported skills, management and capital for their futures. The pool of local skilled labour was limited.

Clearly then, in the pre-war period the Isle of Man offered industrialists considerable potential but in order to realise this, disadvantages had to be overcome. The extractive and manufacturing base that eventually developed from this mix of opportunities and threats can be divided into three: those companies that sought to extract Manx raw materials either for a domestic or mainland market; those that manufactured or processed products for use by other enterprises; and those that supplied a finished product direct to the public. Each will be discussed in turn.

Agriculture and fishing represent forms of extractive industry but these have been covered in Chapter 13, leaving the exploitation of mineral resources as the other major member of this group. Mining in particular was central to the economic history of the Island for most of the nineteenth century. However, records show that the industry is much older than that. Indeed the first mention of the industry related to 1246 when Harold Olaveson of Norway granted a Charter to the Monks of Furness and Rushen Abbeys which allowed them to mine, transport and sell minerals from the Island. Clearly mining pre-dated this Charter. Shortly afterwards, in 1292, the right to extract lead from an existing mine on the Calf of Man was granted to John Comyn the Earl of Boghan (Ralfe, 1924). A document from Henry IV in 1406 mentions both lead and iron mines and in 1656 the range of metals broadened even further with the discovery, at Bradda Head, of lead ores with a high silver content (Mackay and Schnellmann, 1963). By 1699 total output was 164 tons, and a letter of 1703 confirms that copper was also being commercially mined. By the beginning of the Atholl period in 1736 mining had also spread geographically and was known to be taking place at Bradda, Glenchass (near Port St Mary), Foxdale, and Maughold (Radcliffe, 1971). The Atholl period marked the first coordinated attempt to develop the industry but without great success. The Atholl papers reveal the Dukes' initiatives to expand mining: they offered money to any individual discovering workable lead or copper lodes; they entered into a succession of business partnerships with mainland entrepreneurs; they even worked the mines on their own account; and when this failed, the rights were leased out (Mathieson, 1957a). Disappointed by his return, the fourth Duke eventually sold all mining rights back to the Crown in 1828, some 63 years after the Lordship of Man had gone the same way. Although none of the Dukes became rich from mineral extraction, this period did lay the foundations for the subsequent development of the industry. Seven sites had been either worked or explored, including both Laxey and Foxdale, the two locations which

14.1 A period view of Lady Isabella at Laxey. Built in 1854, the wheel was designed to drain the mines higher up the valley. It is now one of the largest remaining wheels in Europe, and is unusual because it is a 'backshot' wheel

were to prove the most significant in the nineteenth century. Steam pumps had been introduced and skilled English labour had been imported since, as a contemporary account notes, none of the local workforce 'had any notion of mining at all'.

The nineteenth century saw both the growth and decline of Manx mining. Commentators seem somewhat undecided about dating these two phases or specifying the period of maximum success. Mackay and Schnellmann (1963) suggest a chronology as follows: the 1820s saw the first substantial growth in the industry; the 1830s and 1840s was a period of capital investment and steady development; 1850–90 witnessed the peak of exploration and exploitation; and 1880–1919 was a phase of rapid recession and closure. Birch (1958) and Kinvig (1975) are rather more precise about the dating of peak production. They both mention the 1870s and 1880s, but Birch suggests that the 1840s and 1850s were the key period for profit generation. Regardless, what is clear is that the Manx mining industry assumed considerable significance in the latter part of the nineteenth century. In the 1870s and 1880s, Foxdale was the most important area for lead ore extraction in the whole of the UK. The same can also be said for Laxey in relation to zinc extraction. Indeed Mackay and Schnellmann (1963) estimate that, between them, these two areas were responsible for 20% of the entire zinc output ever to have been mined in the UK and some 5% of historic lead output. Ten mines accounted for 99% of Manx production, these being grouped into three areas, Laxey (three), Foxdale (five) and the South (two at Ballacorkish and Bradda) (see Figure 14.1). Together they were responsible for some 256,000 tons of zinc, 268,000 tons of lead, 14,000 tons of copper, 25,000 tons of iron and substantial quantities of silver. In the two decades of the 1870s and 1880s the industry provided direct employment for over 1,000 men and boys, with more in work at the Foxdale group than at Laxey. There were other distinctions between the areas too. Foxdale majored on lead and silver production whilst Laxey was more important for zinc and lead. The former, being sited at the head of the valley, required the use of steam-powered pumps to drain the mines whilst Laxey could rely upon water power for energy. And finally Laxey had its own port for direct exports whilst Foxdale had to rely upon rail transport to move ore to the harbour. Both, however, were very profitable in their heyday and investors made a healthy return. Stenning (1958) notes, for example, that £80 shares in the Great Laxey mine were worth £1,100 in 1852. Not all Manx mining was so successful and indeed not all of it was found in the two mining areas on which the literature tends to concentrate. Radcliffe (1971) describes the modest and short-lived

14.2 This 1900 view of Laxey lead mines shows the industry in its twilight years. It also demonstrates how the industry had a devastating impact upon local landscapes

14.3 Snaefell lead mines in 1898. Timber for pit props can be seen in the foreground whilst the line of the Snaefell Mountain Railway can be seen on the other side of the valley

success of the mines in the Maughold area, whilst Bawden (1976) gives an account of the repeated financial failures in the Glen Maye area between 1826 and 1870.

Although £91,000 worth of ore was shipped from Laxey in 1876 and nearly 200,000 ounces of silver were recovered in the following year, success was short-lived. Some authors blame a fall in lead and zinc prices and foreign competition for the demise of the industry from the 1880s onwards. Mackay and Schnellmann (1963), as consulting geologists, argue the reasons somewhat differently. Whilst acknowledging economic factors, they lay more emphasis upon the progressive geological exhaustion of the mines. Despite this difference of opinion, by 1900 only three mines remained operational with Snaefell closing in 1905, Old Foxdale in 1911 and Great Laxey in 1919. Attempts were made to rejuvenate the industry in the inter-war era. Birch (1964) notes how the Government commissioned a study in 1928 to assess the feasibility of re-opening the Laxey mines, but how, despite acute local unemployment, the idea was not proceeded with because of the need for substantial capital outlay. The Coltness Iron Company also investigated the Andreas area for haematite in the underlying carboniferous limestone. And finally, during the last war, explorations were undertaken for manganese in the Manx slates.

Mining was not the only extractive industry to exploit Manx geological resources. Quarries extracted four types of material either for building or agricultural use. Although Manx slate is not an ideal substance for building, the cost of importing alternatives forced its exploitation in the nineteenth century. Quarries opened at Douglas Head, Glen Mooar, South Barrule and Glen Rushen and their produce was not only employed for roofing but also for the construction of entire buildings. The last of these quarries was sufficiently large to employ some 120 men in 1863 (Corran, 1977). Peel sandstone was a very soft building medium too, but again this did not prevent its use for decorative and structural purposes. The Creg Mallin quarry on the north side of Peel Bay was the principal point of extraction and in its late nineteenth century heyday output exceeded 2,000 tons per annum. Further south, the limestones of the Castletown area were also extensively worked and Bawden et al. (1972) list five quarries, one of which is still in operation. Material from there was used not only for building, but also to provide agricultural lime. Birch (1964) describes how it was taken by cart and boat to all parts of the Island in the last century. And lastly, the same area also yielded the Poyllvaaish dark limestone, often termed 'marble'. This ornamental stone was first used by the monks of Rushen Abbey in the fourteenth century and was continually extracted at surface level throughout the sixteenth, seventeenth, eighteenth and nineteenth centuries. In the earlier centuries it was employed for paving, tombstones and mantelpieces, whereas in later years it became associated with ecclesiastical work. Extraction ceased in 1887.

Elsewhere in the Island one other notable user of earth resources was the Ramsey salt works. A substantial brine lake had been accidentally discovered at the Point of Ayre in 1892 during exploratory borings for coal. Early suggestions that Ramsey should become a spa town on the strength of this fell through, and instead the Manx Salt and Alkali Co Ltd. was established in 1902 to extract the material for industrial purposes (Kneale, 1987). They took land adjoining Ramsey harbour and laid a 6.5-mile long pipeline from the two 7–900' deep bore-holes to a new panning plant. Their initial output was scheduled to be 18,000 tons per annum, and the company bought a series of vessels to export their produce and import Scottish coal. The enterprise was always under pressure from the environmental lobby of those associated with the tourist industry, and with the post-war loss of most of its market it faded away. The site was cleared in 1957.

In addition to the extractive sector, a second category of industry also existed within pre-war Isle of Man. These were businesses which manufactured products which were, in effect, raw materials to be further processed or combined by other companies. The bulk of their business was thus not direct to the public. This sector tended to be associated with mining, agriculture or fishing, either for its raw materials or as its customers.

Mining and quarrying stimulated three other subsidiary enterprises: lime-burning, smelting and brick-making. Many farmers opted to buy limestone rock and transport it themselves to their own locality. There they had access to twenty-one small kilns where the rock could be burned over peat (Birch, 1964). Alternatively, between April and Autumn, four commercial kilns were in operation at Derbyhaven, Port St Mary, Billown and Ballasalla (Corran, 1977), and these bagged agricultural lime which could be bought direct from them. Brick-making, too, occurred on a number of sites throughout the Island and in the late nineteenth century there were ten works in production. The oldest of these, at Castletown, had opened for business in 1694 (Bawden et al., 1972), but prior to the mid nineteenth century there was very limited local demand for brick products. The plants which opened to meet demand later in the century were often associated with the warp clays from glacial drift. Six were located in the north of the Island (at Ballacoarey, Regaby Beg, St Judes, West Craig, Mooragh and Maughold) with the remainder in the central trough at Ballanard, Ballawillin, Glenfaba, and Douglas) (see Figure 14.1). The Glenfaba and Mooragh plants differed from the others in the raw materials they used: the

EXTRACTIVE AND MANUFACTURING INDUSTRIES 225

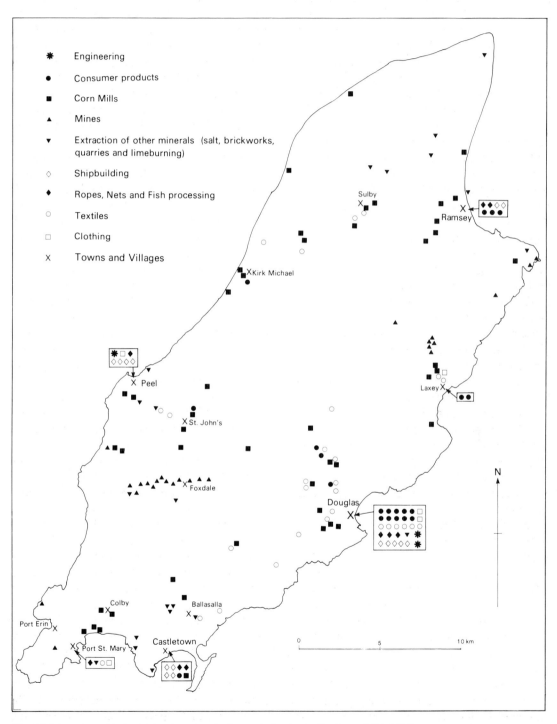

Fig. 14.1 Location of industrial enterprises in the period prior to the Second World War

former employed crushed Peel shale whilst the latter produced calcium silicate bricks from a combination of Bride sand and lime. A more typical plant was that at Ballacoarey near Andreas which employed a 40–50-foot thick deposit of boulder clay found in a 10-acre field belonging to the Christian family. The clay was excavated by hand and hauled up rope inclines to the top of the building. It was then ground, masticated, rolled, squeezed and eventually fired in a coal-powered kiln which had a potential output of 8,000 bricks a day. Although the plant was entirely re-equipped in 1925, production only lasted another twelve months (Caine, 1927).

Milling was the predominant processing industry associated with agriculture, although tanneries also existed. Grain and feed mills were perhaps the most common of these because of the feudal restrictions which the Lords of Man were able to impose upon their tenants. Qualtrough (1969) describes how each tenant farmer had an obligation to grind his corn at the manorial mill and how penalties were incurred by those who attempted to contravene this regulation. He also points out that the Lords of Man received a rental from each mill, and that it was therefore in their interests to encourage the erection of as many mills as possible. As early as 1511, there were already 32 recorded corn mills, and by 1870 the Ordnance Survey map shows a total of 51 such plants. Most were water-powered and were therefore sited on the streams flowing off the slate massif. Qualtrough considers that two types of mill existed: large 'town' mills which possessed threshing mills and bulk storage, and which could therefore accommodate a range of grains; and small 'country' mills which handled a limited range of grains on a day-to-day basis. Most were of the latter type, and Quilleash (1964) describes a typical example at Sulby. Several factors account for the decline and concentration of grain milling from the late nineteenth century onwards: first, technological change towards large diesel-powered mills which could reap the economies of scale; second, the withdrawal from local wheat production; third, the fall in the number of tied tenants; and fourth, the rise in demand for good baker's flour for an increasingly urban population. Three mills responded to these challenges by re-equipping for flour production at the close of the century. Only the Laxey Glen Mill was successful in competing with the larger mainland mills. Built in 1860 by the then Captain of the Laxey mines to supply the booming local market created by the success of the mining industry, the mill was modernised prior to the First World War to make it the first roller mill of the Isle of Man. Despite being gutted by fire in 1921 and bankrupt shortly afterwards, the mill is still in operation today (*Isle of Man Weekly Times*, 1964).

Textile mills were also numerous on the Island, but their presence dates from a more recent period. The production of woollen goods was a traditional craft industry on the Isle of Man which formed part of the economic activities of most farmers, and farmer-fishermen. Wool would be spun at home and then taken to a local hand-weaver who would undertake this task on a part-time basis to supplement the income from his own small-holding. The nineteenth century witnessed a radical change in this historic pattern, with the localisation and specialisation of the industry into purpose-built water-driven mills equipped with power looms. Changes in the allowable trade with the mainland also stimulated the industry. Cowley (1959) notes how, early in the century, restrictions on the export of woollen goods from the Isle of Man were lifted and how in 1844 duties and licences for the same products were relaxed prior to abolition in 1853. The coincidence of these commercial changes with the advent of steam power brought about a period of marked expansion in output in the last half of the nineteenth century, from a sharply reduced number of mills. Birch (1964) states that there have been 16 woollen mills on the Island at different times, although this number had fallen to only three by the out-

14.4 Corlett's flour mill at Laxey. One of the few large industrial enterprises on the Island, it was built by the Captain of the Laxey mines to supply his workforce. It has been thoroughly modernised during the twentieth century and is now state-owned

break of the Second World War (see Figure 14.1). Quilleash (1964) describes the history of one of the earlier and smaller water-powered mills at Sulby which manufactured flannel, shirting, and some ready-made goods for local farmers. In contrast, Birch (1964) and Cowley (1959) provide accounts of two of the newer and larger integrated mills which have survived to the present day by concentrating their production on high-value goods for export. The Tynwald Mill at St Johns is electrically-powered and dates from the inter-war period. St George's Mill at Laxey was opened in 1881 and resulted from an initiative of John Ruskin's Guild of St George. It, too, is still in production for export.

Woollens were not the only textile enterprises to be found on the Isle of Man. Harrison (1968) draws attention to the correspondence of the Earls of Derby who, at the close of the seventeenth century, were anxious to establish a linen industry on the Island using locally-grown flax. It is not clear how successful they were, but it is known that English weavers were introduced to train the local workers. By the eighteenth century Cowley (1959) describes flax mills as being common. Most were 'scutch mills' where the fibres were separated from the waste and prepared for spinning. By the nineteenth century the industry was far less common, and became concentrated upon the mill at Tromode owned by the Moore family. Prior to 1850 this specialised in linen sheeting. Another mill concentrated upon a different textile specialism, namely cotton. De la Pryme's Cotton Mill at Ballasalla was the first to use a Spinning Jenny on the Island, but survived for only a limited period at the end of the eighteenth century.

Fishing as an extractive industry also required a range of processing and support functions to provide it with its industrial equipment. Not least amongst these was the shipbuilding industry. Traditional practice in the Isle of Man was to build boats at the source of the raw material, i.e. wood. Completed craft would then be transported overland to the nearest creek. However, from the 1820s onwards shipbuilding became both more specialised and more localised in the major harbours. It also benefited from a discrepancy in customs duties which allowed Manx builders to import Norwegian timber duty free between 1853 and 1866 whilst their mainland counterparts were liable for tax. The golden years of the industry were short, spanning the period 1840–90. During that time there were five yards in Douglas (Rothwell, 1919), two at Ramsey (anonymous, 1937), four at Peel and four at Castletown. Most of their work was associated with the fishing industry and merchant marine, but pleasure craft and private yachts were also completed. Wood and iron were both employed, although steamships were usually equipped at English yards. At its peak the industry was launching anything up to 42 vessels per year and the Douglas shipwrights were an industrial élite earning 25% more than other manual workers. Rope-making also thrived temporarily on the Island. Garrad and Scatchard (1973) describe how what had been a successful industry was decimated during the Atholl period by a doubling of the duty on imported hemp. Nevertheless they note four more modern yards, two in Douglas and two at Castletown. Raw materials were brought in from Russia or Manila and, in 1871, 200 were employed in the combined rope/net-making sector. By the First World War the industry had suffered severe recession with only 70 recorded employees, but the Quiggins yard at Douglas had remained competitive by re-equipping in 1905. This yard did not close until the late 1950s.

Lastly among the maritime support sector was the manufacture of sailcloth. This was largely associated with the Tromode Mill of the Moore family which, from the 1850s onwards, concentrated solely upon linen sailcloth. At the time the plant was very modern and largely employed immigrant Irish workers. Finally, the Island also possessed an engineering and foundry sector which produced components for agriculture, mining and fishing. Jones (1966), in discussing Manx Water Mills, comments on the very standardised componentry of the Island's mills which had been almost entirely supplied by local foundries. On a larger scale, Lady Isabella at Laxey—a mine pump—also had a significant local engineering input.

The third and final sector of Manx manufacturing in the pre-war period contained those industries orientated to a consumer market and therefore supplying a finished product direct to the public. Naturally this sector was a very varied one, but a broad distinction can be made between industries catering for tourists and those supplying local residents. The former included both the manufacture of rock and mineral water for the Douglas holidaymakers and, to a certain extent, the brewing industry centred on Douglas and Castletown. The latter included an even wider range of enterprises such as the production of paint, wallpaper, starch, bacon, cheese, and clothing. Clothing production is of interest because it developed only in the inter-war period and had its roots in the hand-knitting of garments during the First World War. In the 1920s, four knitwear factories and two cloth-garment factories were established on the Island at Douglas, Peel, Laxey, and Port St Mary. They largely employed female labour and were attracted by the ready-trained workforce and the relatively low rates of pay. Three other components of this consumer sector have been studied in detail and therefore merit a more extended description. These are the cotton-printing, paper and herring curing/kippering industries. Mathieson (1957b) provides a detailed account of the cotton-printing plant at Port-e-Chee near Douglas which he considers to be one of the first Manx manufacturing ventures. The factory remained in production throughout the period between the 1760s and 1789. During that time it imported both raw cotton and cotton cloth from the north-west of England. The former was spun and woven under contract at various locations and was then sent along with the imported cloth to Port-e-Chee for dyeing and making-up. The Douglas plant specialised in one-, two- or three-colour dyeing and produced handkerchiefs and socks. The predominant range of colours was red, chocolate, purple and indigo, dyes being made from chemicals and natural substances such as sour beer and cow dung. The plant closed in 1789 because of the weight of bad debt.

The paper industry was rather more extensive and long-lived, and has been studied by Bawden (1971) and Williamson (1938–39). It first began in 1769 but did not really succeed until 1789. Indeed in the latter of these two years the Isle of Man exported more than 2,000 reams of paper to the mainland. Paper production continued on a combination of six sites into the present century, initially concentrating upon high quality papers but latterly specialising in machine-made, brown wrapping paper. If anything, the industry was condemned by its slowness to mechanise, for it was unable to remain competitive with the large Irish and English mills. Only two mills, the Woodside Mill at Braddan and the Baldwin Vale Mill survived into the 1870s.

Kinvig (1975) provides some historic detail of the kippering/curing industry which also peaked early in the nineteenth century. He notes that the curing of herring was already very common in Douglas by 1754, and that salted and pickled herring were exported in bulk to England, Ireland and the West Indies. They formed an integral part of the diet of many working people on the mainland, and one of the few staple foodstuffs of black slaves in the Caribbean. The industry enjoyed considerable success as a result, between the end of the eighteenth century and the 1830s, but by 1840 the export trade had almost ceased. The inter-war period saw a return to curing and kippering herring rather than exporting them as low-value fresh fish. Indeed in the 1930s the curing industry employed almost 250 men and women either at Peel or Port St Mary.

As the introduction to this part of the book makes clear, the first forty years of this century were far from successful for the majority of enterprises in the Manx economy. With the possible exceptions of the tourist and agricultural industries the underlying trend was one of recession and contraction. Thus by the Second World War the Isle of Man possessed a relatively limited manufacturing base. Indicative of this is the contents of Table 14.1 which lists those enterprises named in a Manx Trade Directory of 1900. Admittedly many companies would not appear in such a directory as a matter of choice, but even allowing for this, the contents make an unflattering comparison with the plethora of nineteenth century manufacturers mentioned above. What is also clear is that by the Second World War Manx manufacturing had become very introverted. On the whole it either employed Manx raw materials or looked to a Manx market. Very few manufacturers had a role in any World economy and many were independent even of the mainland market.

Manufacturing in the post-war period

The post-war period found Manx manufacturing in a different economic context. The Island was losing its historic localised advantages for tourism and was consequently experiencing a decline in this industry (visitor numbers, for example, fell by 19%, 1948–57). The multiplier effect from lost tourist business was such that most sectors of the economy began to experience the costs of economic over-specialisation. For manufacturing, these costs took the form of a shrunken home

Table 14.1 *Industry in the Isle of Man, 1900*

Douglas:	Rope-making	Peel:	Fishing Nets
	Ice-making		Timber
	Timber		Brickmaking
	Builders		Builder
	Engineering		Machinist
	Ironfounding		Boat builder
	Coal importers		Cabinet makers 3
	Corn milling		
	Seed merchant	Ramsey:	Timber
	Coach builder		Builder
	Harness maker		Grain Merchant
	Organ builder		Oil trade
	Boat builder		Mason
	Brewing		
	Wines and Spirits		
	Mineral waters	Laxey:	Woollens
	Coopering		Corn milling
	Nurseryman		
	Fruit importer	Castletown:	Timber
	Sanitary ware		Limeburning
	Trunk maker		Mason
	Baby carriages		Implement maker
	Mason		
St Johns:	Corn milling	Port St Mary:	Fishing nets
West Craig:	Brickmaking	Tromode:	Sailcloth
Ballanard:	Brickmaking		

market at a time when improved (air) transport was allowing mainland manufacturers to penetrate the insular market. In parallel, homogenisation in key consumer products was allowing British and European manufacturers to reap the competitive benefits of the economies of scale. Moreover, in the aftermath of the war the Isle of Man no longer seemed such an attractive proposition for UK investors. As a result, the 1950s were essentially a period of recession for the Manx economy. Winter unemployment rose to between 2,000 and 2,500 people, and despite the Government's work schemes, 200 or more residents were engaging in seasonal migration to the mainland each year (see Chapter 10 for fuller discussion).

To their credit the Manx Government were quick to intervene, albeit in a role designed to support and stabilise the existing economic structure. Agricultural support was raised, the fishing industry benefited both from a new fish meal plant at Peel and ten new kippering houses, and attempts were made to resuscitate the mining industry. In 1949, the Manx Government had taken back all mining rights from the Crown and in 1954 a Government-funded survey of mineral resources was instigated. The latter led to further exploration through until 1959 and also the reworking of waste tips at Snaefell by Metalliferous Holdings Ltd. Their operations lasted four years (1954–58) and during that period they shipped some 2,500 tons of mixed concentrates. Despite a further major report published in 1963 (Mackay and Schnellmann,

1963), the industry was never to recover its former position or importance.

All of this highlighted the likely future importance of manufacturing industry to the Island's economy and employment structure. Steps had already been taken in 1949 to create an environment more conducive to industrialists who might be considering relocating on the Island. The Development of Industry Act 1949 gave power to the Lieutenant Governor to grant monies to those engaged in establishing, reconstructing or developing industry on the Island up to a maximum of £100,000 per annum. Later in the same year a Commission was established to oversee the Development of Industry for a period of six years. The interim report of that commission, published in 1949, made it clear that the prime objectives of State intervention were to attract new industrial concerns and therefore ensure year-round employment. Even in 1949, however, they were sceptical as to whether the Government had committed sufficient resources to the task, given the competition from other development areas. In the event, the 1950s were a period of some success for policy designed to attract manufacturing enterprises. In 1951 the Ronaldsway Aircraft Company came to the Island and is, to this day, the Island's largest employer. In 1956, Aristoc opened a plant in Ramsey for the manufacture of stockings, with an initial workforce of 119. Indeed, between 1952 and 1963 twelve new factories opened employing 650 people (Birch, 1964). Qualitatively, however, the policies of the 1950s were less successful. Most of the new enterprises simply added to existing industries rather than diversified the economy into new products and new markets. Most were textiles, clothing or engineering firms.

At the close of the 1950s State intervention in industrial policy had been both introduced and proven, albeit directed at the rather modest goal of maintaining the existing economic fabric. The response of individual entrepreneurs to changing economic circumstances had also begun to crystallise. Some sought monopolistic positions within the insular market by re-equipping and using the costs of inward transportation as a defence against competition from mainland enterprises. The Laxey Glen Flour Mill, for example, modernised their production shortly after the war and continued to hold 80% of the domestic market (*Isle of Man Weekly Times*, 1964). Other industries pursued an alternative strategy, namely re-orientation towards high-value products for export. The St George's Woollen Mill at Laxey exemplified this trend after its 1954 modernisation. And finally State intervention had succeeded in introducing companies which followed a third strategy. This entailed the importation of low-weight raw materials to the Island which were then processed in a labour-intensive manner prior to re-exportation to a mainland customer. The Ronaldsway Aircraft Company was one such enterprise, manufacturing ejector seats for aircraft, more than half the cost of which is accounted for by labour (Kinvig, 1975).

Whilst policy had been developed and improved, the 1950s were a period of modest achievements and few of the central problems of Manx manufacturing had been addressed directly. Williams (1960), in his commissioned survey of Manx industrial development was very critical. He noted how newly-introduced industry accounted for only 666 people, barely 5% of the employed workforce. The administration of financial incentives was still on an ad hoc basis and was controlled by the Lieutenant Governor who operated without any policy guidelines. The Island still did not possess an industrial officer, and publicity material was very unprofessional. Those companies which had been attracted to the Island were not wholly satisfied, and indeed over half regretted their relocation decision. Williams (1960) was not, however, only critical of Government. He also lambasted the Manx labour force for its inadequate skills, its lack of industrial discipline and its willingness to abandon factory jobs during the summer for more lucrative work in the tourist industry.

Finally, he took local management to task for what he regarded as their complacent and apathetic attitude to exporting, marketing and growth.

In a more positive vein, Williams did contribute to the development of policy. He argued the case both for a permanent industrial officer on the Island and a Standing Advisory Council, and he suggested guidelines for the type of industry he felt ought to be the target for Government incentives. These would be small enterprises producing lightweight finished goods for which a market existed on the island (possibly in Summer) and which could also be exported to the mainland (in Winter). The production process would have to be environmentally sensitive, and new companies would ideally draw their labour from the ranks of the unemployed rather than poach them from existing employers.

The failure of Government policy radically to reshape the economy and prospects of the Isle of Man led in 1961 to a series of new policy initiatives. The attraction of New Residents, the low tax regime, and the stimulation of the service sector are all covered in other chapters, but they formed part of a package which marked an economic turning point for the Island. Policy towards manufacturing industry did not escape radical rethinking either, partly as a result of Williams' (1960) critical analysis. An Industrial Advisory Council was established in April 1961 and in the following month an Industrial Officer was appointed for the first time. Together, they formulated more specific industrial objectives, guidelines and policy. The Advisory Council's first report (1961) recommended a range of loans and grants which could be awarded in different circumstances, and established the objective of full employment, not only for those out of work on the Island but also for the disabled, and for those who had left the Island for higher education. Subsequent commentators have been somewhat unkind towards these early attempts at policy: they argue that policy, in so far as it existed, relied too heavily upon the attraction of low personal income tax and that beyond this there was really only an ad hoc collection of incentives—some advance factories, some loans, some grants and some dedicated housing. Such criticism is unfair. By the time the Industrial Advisory Council published its seventh bi-annual report in 1964 the problem was one of over-industrialisation and shortage of suitable labour: 'Although a limited amount of unskilled labour remains available, it is impossible to recruit skilled personnel in most trades, and this factor is seriously hampering the expansion of development programmes of many firms'. Indeed the Council went so far as to recommend that work permits should be more readily available for immigrants and that the emphasis should be shifted towards the quality of new employment opportunities, not their quantity. Writing in the same year, Birch (1964) also describes a thriving industrial sector that was short of labour. Clearly, policy had successfully achieved one of its objectives.

Trends did not change substantially during the balance of the 1960s. New industry continued to relocate to the Island. In 1965, Dowty opened a factory in Onchan to assemble aircraft components. This had an initial workforce of 187 people and immediately became the third largest employer on the Island. Indeed by 1971, January unemployment had fallen to its lowest level since 1960 with only 533 registering with the Employment Exchange. In the same year, manufacturing concerns employed 2,500 workers (Corran, 1977), and the manufacturing sector contributed 17.59% of the Island's National Income (Webster, 1988).

The period 1969–75 was, if anything, even more prosperous for the Isle of Man and, in particular, the introduction in 1973 of a coherent package of incentives served to stimulate further growth in the manufacturing sector. New grants covered first year expenses, training, plant, buildings and machinery although all incentives remained discretionary. The type of enterprise that the Government wished to attract remained essentially the same: revised guidelines stressed the attractiveness of small enterprises that would not prejudice the environment and which were managed by staff with a proven track record in business. More importantly, the guidelines added two further requirements: a high return on capital and high added value, the latter referring simply to the amount of value added to a product during its processing on the Island. Ideally, cheap raw materials would be imported and an expensive finished product exported. Increased spending on industrial assistance clearly produced results. In 1971–72 the Government had spent £135,000 on assistance, but by 1975–76 this had almost quadrupled to £502,000. In the period 1970–75, 23 new companies were attracted to the Island, creating some 180 jobs. Expansion elsewhere in manufacturing generated a further 220 full-time jobs. As a result, the sector accounted for 2,932 workers who produced 15.6% of the national income in 1975.

When PA Consultants (1975) completed their second review of the Manx economy in mid-decade, their report was far more complimentary than Williams' earlier survey. They described the success of Government policy (albeit somewhat overshadowed by the explosive growth in the service sector), not only in terms of the number of jobs created but also their diversity. They pointed to new companies in the fields of optical equipment, furniture-making and jewellery. However, they also remarked upon a number of potential weaknesses which they felt might become more significant in the later years of the decade. They described what they considered to be an unhealthy specialisation in engineering and textiles, and also the dominance, in employment terms, of only nine relatively large firms. In addition they also noted how population expansion would necessitate the creation of some 800 new jobs in manufacturing alone over the period 1975–80.

The recession which afflicted most industrial nations in the late 1970s did little to deflect manufacturing from its charted course. By 1978, the share that this sector contributed to National Income had fallen further to 12.3%, but increased Government assistance to industry from that year on helped stabilise the situation. Grants and loans, which had totalled £531,000 in 1977–78 rose through £746,000 in 1978–79 and £1.1 million in 1979–80 to a new high of £1.25 million in 1980–81. Despite the full force of recession, which was felt on the Island between 1982 and 1986, the manufacturing sector has only contributed less than 14% of the National Income in one year between 1980 and 1986 (the last year for which figures were available at the time of writing). And, at the last full census, manufacturing employed 3,467 people or 13.4% of the workforce.

However, the growth of unemployment on the Island has provided a considerable stimulus to attract further new enterprises and expand existing ones. January unemployment climbed dramatically from 646 people in 1980, to 1,020 in 1981, 1,617 in 1982, 2,061 in 1983 and ultimately to 2,505 in 1986. Against this background it became politically and socially expedient to redouble efforts, and the Government did not shirk its duty. In 1981 it created a new permanent Industry Board with expanded powers and responsibilities, and in 1982 this body enhanced financial incentives by extending them to cover marketing costs and also enterprises in the service sector (e.g. computer software houses). Expenditure on assistance has risen accordingly, and in 1986–87 the Department of Industry (as it had then become) pledged some £1.9 million of assistance (Webster, 1988). In 1983, two further initiatives were announced, the first of which entailed a more professional and aggressive advertising campaign to get over the advantages of relocating to the Isle of Man. In the years between 1980 and 1983–84 advertising expenditure rose from £5,000 to £250,000. The second initiative involved the announcement of a Freeport on a twenty-acre site adjacent to Ronaldsway airport. This was to have been the first development of its type in Britain, but

14.5 During 1989 the first buildings were erected in the much-delayed Freeport, sited next to Ronaldsway airport. De Beer was one of the first tenants

problems with successive developers have meant that the first tenant only took up occupancy in 1989. De Beers, the diamond merchants, have invested £30 million in a plant that will process imported industrial diamonds and that will employ some 150 workers when fully operational. Such an operation is ideally suited to the concept of a Freeport since goods can be imported, processed and exported without being subject to taxes or customs duties. It is hoped that over forty other tenants will join De Beers in due course, although progress, to date, has been slow.

1986 saw the increasing professionalisation of the activities of the Industry Board. They commissioned a MORI Survey (1986) of the business community and, in particular, new businesses. This was designed to explore the behavioural decision to relocate on the Isle of Man, the reasons why such a strategy had eventually been pursued, and attitudes to the Island as a business environment. Such information will be vital in honing future advertising campaigns and targetting assistance. Despite some reservations about the labour force, the cost of energy and the burden of transport charges, 75% of businessmen would still relocate to the Isle of Man if allowed the choice again.

The most recent statement on the manufacturing sector, and its future in the 1990s, was contained in the Government's strategy plan *The Development of a Prosperous and Caring Society* (Isle of Man Government, 1987). It is a reflection on the success of Manx industrial policy and, indeed, on the buoyant state of the Island's economy in general, that the document is concerned more with the quality and distribution of industrial employment than its sheer volume. It describes how the Industry Department will be increasingly selective about new applications and how preference will be given to those new enterprises which offer high levels of remuneration to their Manx staff. In addition, it proposes that, in future, a regional policy will operate and that this will discriminate in favour of those parts of the Island that have failed to benefit from the strong economic growth of the late 1980s. This last initiative is particularly interesting giving the wholesale disman-

EXTRACTIVE AND MANUFACTURING INDUSTRIES 235

tling of regional policy in the mainland during the same time period.

Manx manufacturing into the 90s

Detailed facts about contemporary Manx manufacturing and extractive industries are by no means plentiful. Recent estimates suggest that between 3,500 and 4,000 workers now find employment in these sectors, in perhaps 270 different firms (Faragher, 1986, 1988; Bawden, 1988). Together, these enterprises contribute 14% of the Island's employment and nearly 15% of its National Income: the latter represents a total in excess of £33 million. The largest single industrial category is engineering (43% of all manufacturing jobs); followed by miscellaneous manufacturing (24%); food, drink and tobacco products (21%) and textiles, clothing and footwear (11%). Not surprisingly, the different industrial categories are not equally important to both sexes. 79% of all engineering jobs are taken by men, for instance, whilst women hold nearly 60% of all jobs in the clothing and footwear category. As well as being specialised by industrial category, Manx manufacturing is also relatively unique in that it is dominated by small enterprises. The Ronaldsway Aircraft Company is the largest employer with a workforce of 560 and there are only one or two other enterprises which employ more than 100 people. Typically, a Manx plant would provide work for only 15–20. Indeed around 70% of all manufacturing plants on the Island employ less than 19 workers with a further 20% having between 20 and 49 workers (MORI, 1986; PA Consultants, 1975).

Of some concern to geographers is the spatial distribution of employment opportunities. Figure 14.2 thus maps those enterprises listed in the 1985 Government publication *Manx Industries*. Enterprises are divided into broad categories. Figure 14.2 makes an interesting comparison with Figure 14.1 and reveals how twentieth-century industry is considerably more agglomerated than its nineteenth-century counterpart. Gone are the scattered industries dependent upon indigenous agricultural produce—which was uneconomic to transport any distance because of the inadequate transport

14.6 A new addition to the Island's industrial base is Airship Industries' manufacturing plant at Jurby. This is a high-risk venture, but otherwise fits in well with the Island's aim of attracting enterprises that add value and employ leading-edge technologies

236 THE ISLE OF MAN

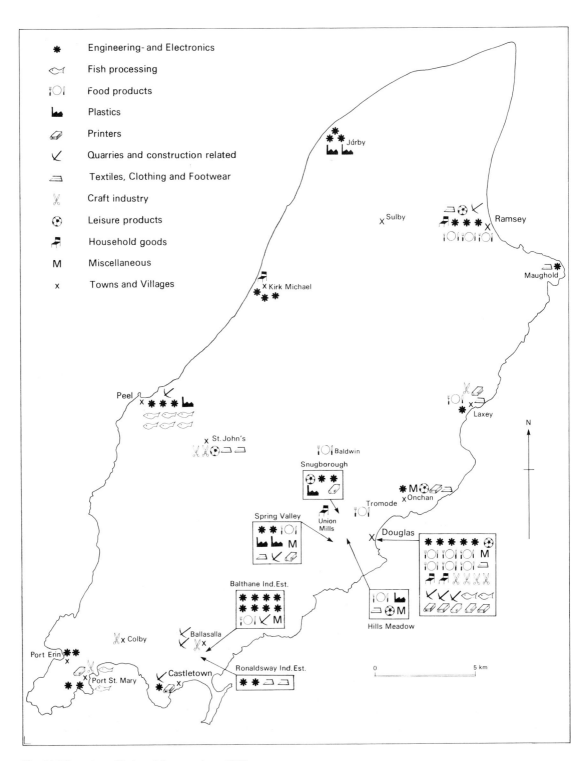

Fig. 14.2 Location of industrial enterprises, 1985

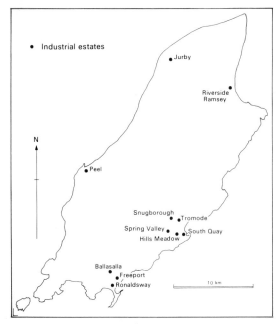

Fig. 14.3 Location of main industrial estates

network. Gone are the industries that were distributed to take advantage of water power from the streams flowing off the central massif. And gone, of course, are the localised sites of mineral extraction in the outlying Foxdale and Laxey areas. Instead footloose industry has become market-orientated or orientated towards the key transport nodes through which goods pass on their way into or out of the Island. Compounding these changes has been the role of property developers and planners. Both have sought to contain light industry within estates put aside for that purpose. Figure 14.3 shows the main industrial estates on the Island, most of which are privately owned, although two are owned by Government. The coincidence of these estates and the present-day distribution of industry (Figure 14.2) is striking.

The MORI report (1986) allows us to take a more behavioural and less neo-classical approach to the location of industry on the Isle of Man. As previous writings on the subject would lead us to expect, relocation to the Isle of Man was not simply a matter of rationally attempting to minimise transport or production costs. 70% of businesses felt that the taxation advantages were the major incentive, whilst 30% mentioned the lifestyle, environment and security available on the Island. Labour was also considered to be important; not only its availability but the absence of industrial disputes. And, finally, political stability was seen as an added attraction. Nearly three-quarters of those interviewed had prior knowledge of the Island either through a holiday, friend, colleague or relative. The converse of these findings is that, after arrival, many of the entrepreneurs had become more aware of the factors stressed by neo-classical theory. A significant number (15%) had been disappointed by the cost of transport to the Island, 8% commented on the difficulties of travel to the Island and a further 7% specifically mentioned the high cost of moving freight by sea. Additionally, 12% had suffered from the higher cost of energy.

Finally, a limited survey of industrialists in 1988 showed that manufacturing is facing the 1990s with great optimism. Of the fifteen enterprises interviewed, only four had shed labour in the previous six months. More importantly, 87% expected to recruit additional workers in the last half of 1988 because of full order books. They anticipated taking on an average of 3.4 more employees each, despite the relatively small size of the businesses concerned. Indeed, these additional staff would increase the workforce of companies in the sample by an average of one-third, although one company was proposing to add ten new jobs to its existing fifteen. Furthermore, 67% of companies also proposed to make substantial investments in new plant over the following twelve months. In fact, the salient problems for Manx manufacturing in the late 80s appeared to be a shortage of local labour, the absence of industrial skills amongst the indigenous workforce and the limited availability of larger commercial premises. Many regions on the mainland would be delighted to be faced with only these problems.

CHAPTER 15
Producer services
W. Dawson

The financial sector is the largest sector of the Isle of Man's economy in terms of contribution to the gross national income. It also has considerable potential for future expansion. Yet only 25 years ago when Birch wrote his book on the Isle of Man he made no mention of the sector. In those days, the banks and the insurance companies provided the services required by the community in their day-to-day activities and there was virtually no demand for services by non-residents.

Following the recession in the economy during the 1950s the Island Government took action in 1961 to stimulate the economy by reducing income tax to 21.25% and abolishing surtax and offering financial incentives to investment in manufacturing industry. The reduction in tax was marketed as an incentive to attract retired persons to become resident in the Island; it was of course a benefit received by all taxpayers. Decline in the population was arrested and reversed and during the 1960s the population was expanding (see Chapter 10), consequently the business for the financial sector also grew to meet the demands of the resident community together with the needs of the visiting tourists. By making the 1961 tax reductions the Government had accidentally taken the first step to encourage the development of the financial sector. In retrospect it is surprising that the expansion in world trade, international banking and low tax areas which was developing at that time, and the potential opportunity for the Isle of Man, was not apparently appreciated within the Island. Round about the mid 1960s a small section of the financial community started to provide services primarily for residents, which could be used also by non-residents. This manifested itself in the formation of a number of unit trusts which attracted a relatively small amount of non-resident funds and represented the first attempt by the financial sector to attract offshore business. Leaders of this trust development urged the adoption of controls to regulate the movement, and the Isle of Man Government decided to introduce legislation which would control the operation of unit trusts which were being offered to the public. These safeguards were included in the Prevention of Fraud (Investments) Act 1968.

Producer services encompass a wide range of activities within the financial/commercial field. However, within this chapter the emphasis has been placed on banking, insurance, non-resident companies and marine administration. Nevertheless, the role of the legal and accountancy professions should not be overlooked; in an off-shore area they are an essential factor in the provision of services to the non-resident and the members of these professions often hold directorships in many companies. The accountancy profession has experienced phenomenal growth and many of the large United Kingdom accountancy firms established a presence in the Island in the 1980s. Chartered accountants could muster about 40 members in 1970; by 1987 they numbered over 200. Advocates, who carry out the function of solicitor and barrister, are less numerous; whereas there had been only 24 in 1970, by 1987 the number had grown to 41, and had been joined by a small number of English solicitors.

Other changes in the rate of taxation were made during the mid 1970s and the late 1980s, and as a result the income tax rate was reduced in four steps to 15% to make it directly competitive with the rate of income tax levied in the Channel Islands. It is, of course, impossible to say with any certainty what impact these reductions had, but it is fairly obvious that it was another step taken by the Manx authorities to make the Island more competitive. The basic elements required by a successful off-shore financial centre were available—political stability, financial stability achieved with low rates of taxation, physical and legal accessibility, good communications and the availability of professional expertise.

Banking

With continued decline in the tourist sector in the early 1970s the Government Treasury made a further attempt to stimulate compensatory growth in the financial sector. As a first step, the Treasury sought to encourage the banking sector to expand, and introduced in May 1973 a concession which allowed banks to pay gross interest on non-resident deposits and undertook not to pursue any income tax liabilities of non-residents which arose through earning interest on bank deposits. It was felt that the banking sector would need to be regulated if the Isle of Man was to develop into an attractive offshore financial centre and by exercising a control over banks which could pay gross interest to non-residents the quality of banking companies could be ensured. In the mid 1960s the Government had allowed fifteen companies to be registered with the word 'bank' in their title, although many of the companies had been formed with only £2 issued share capital; by 1973 the majority of those companies had not traded, whilst the other companies had not undertaken domestic banking operations on the Island. It was thought that these companies could cause considerable difficulty in the future if they were not brought under some form of control. One of the primary reasons for the Government subsequently introducing banking legislation was to enable it to regulate these fifteen companies as well as other existing and new banking companies which would be formed in the future.

The Banking Act 1975 was an act to provide for the licensing of banks and financial advertising, for the inspection of banks, for giving directions to banks and for the restriction on the use of the word 'bank'. It was envisaged that a licensing system would be created and that this would provide greater protection for depositors as well as protecting banks from some of their own over-enthusiastic colleagues. The policy succeeded in attracting additional banks into the Island from the dozen or so which had existed in the early 1970s, and the number had grown to over 40 by the end of the decade. Apart from the large British banks, there was little evidence of internationally-famous foreign banks in the Island, but the Government allowed a number of private local banks to be established during that period. A licensing system was operated by the Government, but the supervisory system was one which could be described as self-supervision relying on the integrity of the management of the banks. A boost to the Island's ambitions as an off-shore centre came in 1972 from an unlikely source—the British Government. In June 1972 it was decided to redraw the boundaries of the Scheduled Territories or the Sterling Area, the area in which sterling could be moved freely. The Island was one of the few remaining territories left within the area and the elimination of competition from many other low-tax areas gave an impetus to the expansion of financial business in the Island.

Although the number of banks rose and the deposits obviously increased as a result of that expansion, there was still one piece of ancient legislation, that is the Usury Act, which had an adverse effect on the growth of deposits. The Usury Act was really a throwback to the days of Henry VIII who

Table 15.1 *Isle of Man deposit base*

	Sterling £ millions	Non-Sterling £ millions	Total £ millions	% Change
14 July 1970 (1)	—	—	61	—
31 December 1974 (2)	—	—	171	—
31 December 1978	342	10	352	—
31 December 1979	511	19	530	51
31 December 1980	692	51	743	40
31 December 1981	912	120	1,032	39
31 December 1982	1,103	197	1,300	26
31 December 1983	1,336	263	1,599	23
31 December 1984	1,681	345	2,026	27
31 December 1985	2,161	390	2,551	26
31 December 1986	2,766	428	3,194	29
31 December 1987	3,220	446	3,666	15
31 December 1988	4,049	580	4,629	26

Sources: 1. a. Trends in money supply in UK (for branches of London Clearing Banks);
 b. Partial information on main banks registered in the Isle of Man;
 2. Bank of England special analysis, plus supplementary enquiries;
 Others: Isle of Man Treasury and Financial Supervision Commission.

introduced similar legislation in Britain which imposed an upper limit on the lawful rate of interest which could not be legally recovered. When market rates moved above the Usury Rate the finance industry had problems. As the Island was in the United Kingdom monetary area, bank deposits could simply be moved to the Channel Islands or the United Kingdom itself and earn a higher rate of interest, and it was not until 1979 that the Usury Act was repealed, although there had been no logical reason for retaining the Act for many years. Members of the Government had shown some naivety of monetary affairs in their long-held belief that the Usury Act kept interest rates low in the Isle of Man and therefore local authorities could borrow money more cheaply, a situation which was considered to be of great benefit to the Island. The fact that the Usury Act was in practice unworkable and that a rise in interest rates in the United Kingdom compelled the Isle of Man Government to raise the Usury Act limit was for many years not accepted as a reason for its repeal. Immediately after repeal of the Usury Act deposits leapt by 50% in one year. Subsequently the rate of growth fell to 40% in the second and third years and then stabilised at about 25% a year. Table 15.1 indicates the growth of deposits since the Usury Act was abolished in July 1979.

In the early 1970s, when the Treasury was attempting to create an environment in which the banking sector could expand, the initial policy proposal was to encourage the established leading banks but to discourage the small private banks. In the majority of low-tax offshore financial centres problems had been encountered in the banking sector usually through the over-extension of the small banking operation with resultant adverse publicity arising. Therefore the Manx Treasury's attitude was quite understandable in seeking to discourage the private banks and also to look towards the introduction of banking supervision.

It was not long, however, before members of the Government were subject to pressures to allow private banks to be established—especially if there were Manx shareholders—in order that the local investors could benefit from the expanding sector. Naturally it was suggested that local banks would be particularly interested in developing local industries and especially attuned to the needs of the local community. Also at that time the established banks in the Isle of Man were opposed to the idea that the Treasury should introduce banking supervision. They considered that they themselves were well regulated internally and there was no necessity for anyone to supervise them. The Treasury did not wish to antagonise the leading banks and therefore accepted that, although there would be a licensing system for banks and deposit takers in the Island, the Government would not introduce a supervisory system but the banks would be responsible for regulating their own affairs. The unfortunate consequences of the decisions were not felt until 1981 and 1982 when a small number of private banks and deposit-taking institutions developed considerable financial problems and collapsed. Although in international terms the organisations that collapsed were relatively small, the demise of the Savings and Investment Bank in particular generated considerable adverse publicity for the Island's financial sector and triggered off a sudden change in Government policy and a system of banking supervision was quickly introduced. A Financial Supervision Commission was set-up to maintain supervision over banks, deposit takers and associated activities in the financial sector; later an Insurance Authority was established. Despite the set-back in the banking sector in the early 1980s, bank deposits continued to rise on average by about 25% per year.

An embargo on the establishment of new banking companies was imposed by the Government during the period 1983–86, although it was understood that should a leading name wish to establish in the Island it would be considered. In the event there was no necessity for the Financial Supervision Commission to exercise its discretion, but once again action taken outside the Island rapidly changed the picture in 1987. In the Channel Islands, Jersey had held a census, and the result indicated that the population had risen more quickly than expected and faster than was considered desirable. With some speed the Jersey authorities acted to restrain the expansion of businesses and as a consequence the Isle of Man became a more attractive location. In 1987 a surge of new banking companies was established, all with highly reputable parentage, and demand for labour and accommodation rose in response.

Table 15.2 *Estimated number of persons employed: insurance, banking and finance*

	Number	Percentage of economically active persons
April 1971	760	3.4%
April 1981	1,515	5.8%
April 1986 (1)	2,150	7.9%
March 1989 (2)	2,879	9.5%

Note: (1) Treasury forecast (includes shipping management)
(2) Treasury census (includes shipping management)
Source: Census 1971, 1981.

Insurance

In 1974 the Treasury commissioned two independent reports from leading insurance brokers to assess the opportunities for developing the insurance industry in the Isle of Man. Both reports came to similar conclusions, that the Isle of Man could be developed as a centre for captive insurance (that is self-insurance usually by large companies or partnerships or associations of people or companies). The normal purpose behind the concept of captive insurance is to save on the cost of insuring particular risks, in other words any profits from the insurance activities of a captive will eventually be reflected in either lower premiums or the distribution of the profits of the captive back to its parent company. An alternative object in setting up a captive is to provide insurance cover for risks which the insurance market is unwilling or unable to provide. For example, in recent years professional indemnity cover has become very expensive and difficult to obtain, leading many large professional firms, such as accountants or architects, to establish captive insurance companies.

The captive insurance concept was being developed in some other low-tax areas, notably Bermuda and to a lesser extent Guernsey, and any proposals to develop similar activities in the Isle of Man would obviously have to be competitive. Normally such activities would be exempt from tax, but such a proposal was regarded within the Isle of Man Government as being too radical, and the Treasury was given the alternative of dropping the whole idea or accepting that such insurance companies would be exempt from tax on their underwriting profits but not on their investment income. The latter alternative was adopted, although there was no expectation that the Island would be successful in attracting insurance business.

History proved that the legislation in the Island was almost totally unattractive to the insurance industry and it was not until 1981 that the Treasury re-introduced the Exempt Insurance Companies Act, approved by the Island's Parliament, which enabled the insurance companies providing services to non-residents to be exempt from tax. In addition to providing the stimulus for the captive insurance companies, the legislation was designed to be attractive to life companies providing services entirely to non-residents. The legislation has been successful and numerous captive insurance companies have been attracted to the Island, including many of well-known parentage with substantial businesses in their own rights. Also there has been considerable success in attracting the life companies providing services for non-residents, mainly British nationals working and living abroad. The captive insurance companies provide relatively few job opportunities but do require the input of considerable skill and management expertise. The life companies are fairly demanding on labour resources, and were extremely useful to the Island in providing additional job opportunities during the recession of the early 1980s.

The Government formed an Insurance Authority to supervise the insurance industry.

By the middle of 1989 the Insurance Authority had licensed 64 captive insurance companies and nine life companies. Over 300 people were employed by the expanding life companies. The insurance companies had an issued share capital of £200 million and an annual premium income of £500 million. Services provided by the insurance brokers expanded with the development of the sector.

Companies controlled by non-residents

In off-shore financial centres there is a constant demand for companies which normally feature in financial planning for a wide range of purposes. The ease with which companies can be formed, legal requirements and costs relating to the companies are important factors taken into consideration by professional advisers.

A measure taken to stimulate the formation of non-resident companies was the introduction of the Company Registration Act 1974. A non-resident company is a company which is registered in the Island but trades outside the Island and the beneficial owners are not resident in the Island. In some ways it was unusual that this legislation should stimulate the formation of non-residents as the Act imposed an annual fee of £200 to be paid irrespective of the profits made by the company. At first it seems attractive, but prior to the introduction of this legislation no tax would have been paid by non-resident companies. However, the removal of uncertainty of the way the Government would act in future made the introduction of a flat rate of tax an attraction to those who saw advantage in having non-resident companies. Subsequently the annual fee was increased to £250 and in 1987 the fee was further increased to £450 and new legislation called the Non-Resident Company Duty Act replaced the old Company Registration Tax Act of 1974.

Table 15.3 *Income generated in insurance, banking and finance 1969–88 (£ thousands)*

	Income from employment and self-employment	Company profits	Total
1969–70	992	2,778	3,770
1970–71	1,201	3,487	4,688
1971–72	1,635	4,645	6,280
1972–73	2,599	5,707	8,306
1973–74	2,926	8,052	10,978
1974–75	5,110	13,136	18,246
1975–76	4,333	18,664	22,997
1976–77	4,538	22,367	26,905
1977–78	5,034	28,413	33,447
1978–79	6,370	30,768	37,138
1979–80	8,162	25,646	33,808
1980–81	10,612	26,597	37,209
1981–82	11,887	23,077	34,964
1982–83	12,833	23,447	36,280
1983–84	12,827	26,294	39,121
1984–85	13,144	26,697	39,841
1985–86	14,024	33,988	48,012
1986–87	15,763	49,303	65,066
1987–88	17,823	59,169	76,992

Source: Treasury.

In 1984 a further Act was introduced called the Income Tax (Exempt Companies) Act 1984 which allowed resident companies, which were undertaking certain activities outside the Island, and again were owned by non-resident beneficial owners, to pay a limited tax liability of £250. Initially the types of companies which benefited from this new legislation were those which are engaged in commodity trading, investment companies, shipping companies which do not operate from ports in the Isle of Man, shipping management companies and companies dealing in financial futures and trading options. The last two activities were not in the original list but were added at later dates. Considerable expansion of the scope of exemption from tax was agreed in 1989 and the vast majority of non-resident companies can acquire exempt status. The object of the legislation to attract non-resident business was to produce a spin-off in terms of income for the Island's financial sector and ultimately additional tax revenues for the Government. Differing views exist on the success or otherwise of this policy. In one respect it is an undoubted success because of the number of companies which have been formed in the Island, but on the other side of the coin is the attraction of activities which do not enhance the reputation of the Island. For example, companies have been formed to sell so-called university degrees, to provide diplomatic passports and, whilst the number of companies carrying out such activities is small, they are extremely difficult to control because often there are no local directors and no local premises other than a registered office which may be in an advocate's premises. The growth in the number of companies formed inevitably led to an expansion in companies which were set up purely to offer company management services. The

Table 15.4 *Number of companies registered and removed from register*

Year ending 31 December	Registered	Removed from register	Net increase	Remaining on register
1970	na	na	na	2,207
1971	na	na	na	2,810
1972	na	na	na	3,560
1973	na	na	na	5,150
1974	na	na	na	5,718
1975	na	na	na	6,260
1976	na	na	na	6,871
1977	1,490	235	1,255	8,126
1978	1,890	244	1,646	9,772
1979	2,522	399	2,123	11,895
1980	2,210	1,171	1,039	12,934
1981	2,201	848	1,353	14,287
1982	1,964	466	1,498	15,785
1983	2,074	1,338	736	16,521
1984	2,436	1,499	937	17,458
1985	3,088	1,889	1,199	18,657
1986	3,954	1,235	2,719	21,376
1987	4,230	2,173	2,057	23,433
1988	5,214	1,268	3,946	27,379

Note: companies registered include trading and non-trading companies;
na = not available.
Source: General Registry, Treasury.

Government made an attempt to control the activities of company management operators in 1982 but it was not successful and was abandoned in 1986. Some company management operations are thoroughly respectable and carried out by advocates and accountants, but there are certain fringe elements which tend to be responsible for the formation and sometimes management of companies which one would suggest were in the grey area of non-resident company operations. It still remains a problem, and probably always will, but the non-resident company does frequently form a part of a legitimate financial planning operation set-up in the Isle of Man.

Marine administration

The most recent and successful development in off-shore services has been in the area of marine administration. It has been possible to register ships in the Isle of Man for 200 years; however, only in the mid 1980s have ship owners and managers realised the potential of the Island as an economic and efficient management centre. This statement is true whether the ships are registered in the Island or elsewhere and whether they are sailing in home waters or worldwide. The Island does not operate a flag of convenience. It has always had its own shipping lines but increasingly the ships registered in the Island are servicing international trade rather than being restricted to the needs of the trade around the Irish Sea. The major advantage to ship owners registering their ships in the Island is the reduced sailing and staffing costs. Whilst the Island maintains international standards, these are low in terms of number of crew required, and more flexible as regards manning arrangements than those of the United Kingdom. Thus an Isle of Man ship is more competitive in international terms.

Ships which qualify to fly the Red Ensign may register in the Island, but only if they are owned or managed by a person or organisation which establishes a locally-registered company with its place of business in the Isle of Man. When the registry started with the Merchant Shipping Act 1786 there were no international agreements on standards among seafaring nations and it was not until the 1850s that some of the major seafaring nations considered that it was necessary to impose survey requirements and standards on their own ships. In more recent times the major seafaring nations decided that international minimum standards should be agreed and progressively these have been introduced—particularly during the last two decades—through the medium of the International Maritime Organisation with headquarters in London. As far as the Isle of Man was concerned these international conventions did not always apply, and ships registered in the Isle of Man could not be issued with international certificates. Therefore, the Isle of Man Government decided in consultation with the United Kingdom Government to take the necessary steps to make the Isle of Man a full maritime convention country and was operating fully on the principal conventions from 1 July 1986. In the Merchant Shipping (Registration) Act 1984, powers were taken to select ships which could be registered in the Isle of Man and within three years of the Act coming into force, 136 ships had been registered. A similar number of ships registered in other countries were also managed in the Island. Approximately 150 on-shore jobs had been created and employment found for 50 local seamen.

Conclusion

The development of the financial and com-

15.1 A sign of the times. A new office block in Douglas to cater for the growth of offshore services. Conversion of Clinch's brewery into shopping mews, bistro and wine-bar

mercial sector providing services to non-residents started rather late and initially with little enthusiasm. During the 1970s the Government Treasury made determined attempts to create the environment within which the sector could flourish and there was a rapid expansion of business.

In 1982 the collapse of banking companies overshadowed the sector, although a steady expansion of bank deposits was maintained. More importantly the banking defaults changed political thinking on the subject of supervision and a strong supervisory regime was imposed on the financial sector, re-creating public confidence.

A substantial contributor to national income, the financial sector has developed additional and diverse job opportunities both directly and indirectly in the construction, retailing and other service industries. Publicity has been generated for the Island, not all good, but the sector has been extremely valuable in balancing the decline in the tourist industry.

At the mid point in 1989 the prospects for the sector look good. Over the past three years, business people travelling by air to the Island exceeded those travelling by sea. In the same period, telephone calls to places outside the island doubled and rents of offices went up by 100%: indicators pointing to substantial expansion in the sector. Unlike most low-tax areas the Isle of Man has the great advantage that there is the physical space to accommodate an expansion of population if such a development was considered desirable (see Chapter 10 for a fuller discussion of this issue). A consultant's report, commissioned by the Isle of Man Government on *The Economic Implications of 1992*, published in August 1989, expressed the opinion:

> On balance, however, it appears reasonable to assume that over the next five years income generated by the I.O.M. financial services sector will increase in the range of between 10 per cent and 25 per cent per annum in real terms, with

15.2 Spurred by the success of the service sector, 1988–89 saw considerable activity in the property market, catering not only for the new New Residents but also for the more affluent sectors of the established population. This development is at Port-e-Chee near Douglas

15 per cent per annum being the 'central case' projection . . . The implication of these trends is that the share of National Income accounted for directly by the financial services sector will be approaching 50 per cent by the mid-1990s. (Peat Marwick McLintock et al., 1989).

CHAPTER 16

Tourism

Richard Prentice

Introduction

In 1887, the Jubilee Year of Queen Victoria's reign, 310,916 visitors landed at Douglas alone during the summer season. This was a record year for Victorian tourism on the Island, and indicates the importance of tourism to the Island a hundred years ago. The number of passengers landed at other ports is unknown, and so the true number of tourists is likely to have been greater still. Generally in the 1890s upwards of a quarter of a million passengers, mostly tourists, were landed at Douglas in the summer season of May to September (Table 16.1). The tourist accommodation industry had grown commensurately, and showed a wider spread across the Island than is now current. In particular, Ramsey rivalled Douglas as a resort of this period. In 1895 the *Isle of Man Times* guide listed 107 hotels and boarding houses in Douglas, but also many elsewhere, particularly in Ramsey (89 lodging and boarding houses, and nine hotels), Peel (49 lodging and boarding houses, and four hotels), Port Erin (52 boarding houses and hotels) and Port St Mary (40 boarding and lodging houses, and three hotels).

The boom era of tourism to the Island has long since gone, and the industry has suffered a protracted retrenchment and restructuring. The present chapter reviews the extent of this retrenchment, the response of the Manx Government and tourist industry and the contemporary nature of tourism on the Island.

Table 16.1 *Numbers of visitors landed at Douglas in the late Victorian era*

Year	Number May to September
1887	310,916
1888	267,908
1889	229,312
1890	260,786
1891	256,734
1892	258,835
1893	249,251
1894	251,003
1895	292,249
1896	305,525
1897	314,667

Source: Reports of the Board of Advertising for 1896 and 1897

Impacts of tourism on the Isle of Man

Tourism impacts are properly measured in economic, physical and cultural terms. On the Isle of Man these impacts are frequently assumed to be economic impacts alone and, further, are often equated with the economic conditions of the accommodation industry. In part, this reflects the past importance of the tourist accommodation industry on the Island, rather than its present importance. Economic statistics show both a comparative decline in the importance of the tourist accommodation sector, and also caution against over-emphasising the importance of this sector to the Island's economy. As recently as the financial year 1967–68 tourism was estimated to have contributed half of the

16.1 Port Soderick in 1898. Like a number of other small places, Port Soderick was a creation of the tourist industry, designed to give the Douglas day-trippers a different experience

national income of the Isle of Man (Tourist Industry Development Commission, 1970), although this figure may well have been an overstatement. In the financial year 1972–73 the tourist accommodation industry contributed £1.8 million to the Island's gross national income of £53.1 million, or 3.5%. In 1982–83 this sector contributed £2.7 million of the £189.7 million gross national income of the Island, or 1.5% (Webster and Dawson, 1985). The latter is not to deny that tourists spend money in other sectors, or that the tourist accommodation sector supports other sectors, such as wholesale distribution, by its purchases. Taking the tourist industry as a whole the Manx Treasury estimate that in 1972–73 the tourist industry generated 14% of the Island's income compared with the earlier estimate of half in 1967–68 by the Commission; in 1982–83 this proportion had fallen to 11% (Webster and Dawson, 1985). For the assessment the Manx Treasury included not only tourist accommodation, but also a proportion of public utilities, distributive services, catering services and miscellaneous services. Expressed in terms of employment, the tourist accommodation industry accounted for only 4% of employment at the 1981 census, and other catering and entertainments a further 4%. As the census is taken in the spring, these figures ignore temporary employment generated in the summer. However, these temporary jobs are not always filled by Manx residents and so the employment created is in part 'exported', often to young Irish females who come over to the Island to work for the summer season. The employment figures shown in Table 16.2 indicate that although there is a clear seasonality in the unemployment of Manx males engaged in the hotel and catering trades, the proportion of unemployed males engaged dir-

16.2 Many holiday-makers visited Douglas Head on Sundays to hear the open-air religious services and enjoy the view. Other attractions included a camera obscura and the short-lived Warwick's revolving tower seen on the right. Plans for a high level passenger suspension bridge never reached fruition

ectly in the tourist industry is small. Again, however, it should be remembered that other jobs can be created in the summer indirectly from tourism, and these effects are included in the bottom line of the table. In some years the effect is sizeable, but this may occur for reasons other than tourism. The census figures for employment in the tourist industry show a decline from 1971 to 1981, for in the former year 7% of employed persons worked in the sector. This is indicative of the Island's continuing restructuring away from tourism. This having been noted, the tourist industry is still of significant size on the Island to warrant particular economic attention.

The economic impact of tourists on the Isle of Man varies by season and by place. In the winter months friends and relatives are the main source of accommodation for visitors, but in the summer, hotels and guest houses predominate (Table 16.3). This is indicative of a different kind of tourist in the summer from their winter counterpart: summer tourists generally have no family links with the Island. The 1980s have seen little decline in the comparative importance of hotels and guest houses as sources of accommodation, despite the structural changes in the industry, and self-catering has not increased in importance, unlike in many 'traditional' resorts in England. Most period visitors stay in the Douglas or Onchan area, especially in the summer months. This is to be expected from the contemporary location of hotels and guest houses in this area. Even in the winter season, Douglas and Onchan are the main destinations; in this season this is a reflection both of the location of accommodation and of the Island's population for those visitors staying with relatives or friends. However, Ramsey and Port St Mary are also winter destinations for significant numbers of visitors.

Table 16.2 *Tourism and seasonal male unemployment*

	Unemployed males									
	March 1980 %	June 1980 %	March 1981 %	July 1981 %	March 1982 %	June 1982 %	April 1983 %	June 1983 %	April 1984 %	June 1984 %
Catering, cafe, and hotel occupation	6.3	3.5	7.1	7.4	6.3	5.5	9.0	6.0	9.3	6.9
Other occupations	93.7	96.5	92.9	92.6	93.7	94.5	91.0	94.0	90.7	93.1
Total number of males registered as unemployed	384	229	604	564	1,098	939	1,473	1,310	1,445	1,051
% decline on spring figure in total unemployed	*	40.4	*	6.6	*	14.5	*	11.1	*	27.3

*indicates not applicable
Source: Webster and Dawson (1985)

Tourist spending also varies by type of tourist and by season. The 'average' tourist to the Isle of Man in 1985 spent £21.89 per night on his/her holiday. Not all of this expenditure will have been on the Island, as the cost of travelling to the Island is included. If the latter sum is excluded, the 'average' tourist spent £14.39 per night. In contrast, the 'average' day tripper spent £23.01 on his or her trip for the day, or £9.24 excluding travel to and from the Island. For both period visitors and day trippers the largest single expenditure was the average cost of getting to and from the Island. As would be expected, this cost was comparatively greatest for day trippers, representing 60% of their expenditure on average. Seasonal variations in expenditure per night also exist. Because of the importance of the accommodation provided by friends and relatives in the winter, in this season accom-

Table 16.3 *Where visitors to the Isle of Man stay*

	Period visitors		January 1985 %	Staying tourists travelling by sea	
	1985 %	August 1985 %		August 1977 %	August 1981 %
Type of accommodation:					
Friends/relatives	18	15	67	10	14
Hotel/guest house	66	71	27	76	70
Self catering	10	12	1	7	12
Camp site	3	1	0	4	3
Other	2	1	5	3	1
Town/area:					
Douglas/Onchan	71	77	48	—	—
Port St Mary/South	15	12	19	—	—
Ramsey/North	10	8	27	—	—
Other	4	2	6	—	—

Sources: Isle of Man Passenger Survey, 1985; Webster and Dawson (1983).

modation expenditure per night is only on average a fifth of that in the summer. On a per night expenditure basis, the Island needs to attract summer visitors, rather than their winter counterparts, or seek to change the winter profile of visitors. The expansion of winter tourism may bring less than average income to the Island per tourist attracted, unless visitors can be attracted to hotels and guest houses.

Patterns for the total expenditure of the 'average' visitor are much the same. There is a marked difference between the expenditure of day trippers and period visitors, especially when fares to and from the Island are excluded from the comparison. Excluding fares to and from the Island, the period visitor on average spends twelve times the money on his or her visit than does the day tripper in August. Clearly, the Island needs to attract period visitors rather than day trippers. Winter visitors on average spend less than summer visitors, notably on accommodation. Interestingly, winter visitors tend to spend more on fares to the Island, at a time when fares are cheaper. This reflects the increasing importance of air travel in the winter, with the ferries not bringing the volume of tourists by sea as in the summer. Notably, even when the fares to and from the Island are excluded from the comparison, the 'average' period visitor spends less than half his/her expenditure on accommodation, even in August. This emphasises the indirect job creation effects of tourism. In particular, sizeable sums are spent on meals, snacks and on alcoholic drinks.

As noted at the outset of this discussion of impacts, economic impact is not the only form of impact which should be considered. Tourists can be intrusive, and may be unwanted by 'host communities'. What little work exists on this subject on the Isle of Man suggests that this is not the case. For example, the Tourist Trophy (TT) Motorcycle Races are both an annual and potentially disruptive event, especially in that roads are closed and large numbers of tourists are attracted to watch the races. Despite this potential disruption most Manx residents, even those living beside the course, do not generally object to the TT races; instead, adjustments are made to daily routines to minimise the disruption (Prentice, 1988). On the basis of the evidence of this one impact study it would be foolish to argue, however, that Manx residents demonstrate a strong objection to tourism and tourists.

Retrenchment of the past 25 years in the Manx tourism industry

The Isle of Man has not participated in the boom in British holiday-taking of the past 25 years. British residents took an estimated 34 million holidays in 1961, a holiday being defined in this case as a period of four or more nights away from home which was considered by the respondent to be a holiday. In 1985 the comparable figure was 49 million holidays (Central Statistical Office, 1979, 1987). These years also show a structural change in British holidays taken, which has not favoured the Isle of Man. In 1961 fewer than one in eight holidays were taken outside of the British Isles by residents of Great Britain; in 1985 the figure was an estimated one in three. In the latter year, a third of all foreign holidays taken by Britons were to Spain, or the Spanish holiday islands, indicating the popularity of a type of holiday which the Isle of Man cannot offer. In the twenty year period 1964–83 the number of passengers arriving on the Isle of Man during the tourist summer season varied year by year, sometimes by a substantial margin. While no catastrophic annual decline in 'ordinary' (that is, staying) passengers occurred in this period overall, a highpoint of 482,650 passengers in 1979 had fallen rapidly by 1982 to 348,016. Day trip passengers showed a peaking in number in the 1970s, and a collapse from 151,966 in 1979 to only 43,797 in 1983 (Table 16.4). Because of the annual variation in the passenger figures it is useful in the identification of longer term trends to divide these twenty years into five four year periods, and to produce an average figure for each period. The reworked figures are as follows:

	'Ordinary' passengers mean per season	Day passengers mean per season
1964–1967	359,315	95,711
1968–1971	387,096	109,821
1972–1975	394,431	128,500
1976–1979	412,504	133,484
1980–1983	386,684	76,348

From these average figures it is clear that the 1970s and particularly the latter 1970s were a recent highpoint in staying passenger arrivals in the main tourist season. The early 1980s were much the same as the period 1968–71 in terms of the numbers of staying passengers, who averaged around 386,500. Day-trippers showed a similar peak in the 1970s—and particularly in the latter 1970s—and then a marked fall, well below that of the 1960s numbers of passengers. More recently, specific official estimates of tourist arrivals for the 1980s show a very substantial and consistent decline in tourist numbers during the past decade, from the high point of half a million in 1979 to under 220,000 in 1988.

Decline has been most noticeable in the number of accommodation enterprises, which has shown a marked decline since the early 1960s. The number of hotels and guest houses enumerated by the Tourist Board in 1962 was 1,781; in 1985 this figure was 467. The number of bedrooms in these premises showed a decline, but not so marked, from 18,969 to 9,426 in the same period. This is indicative of the extensive amalgamation of neighbouring guest houses and private hotels into single enterprises which has occurred on the Island. Irrespective of this, the decline in the number of bedrooms in the period was still both substantial, representing a loss of half the number of bedrooms available to tourists in 1962, and continuous over the period. Self-catering accommodation saw an increase in number up to 1969, but since then

Table 16.4 *Passenger arrivals on the Isle of Man*

Year	Total	'Ordinary'	Day
1964	460,643	354,422	106,221
1965	462,124	355,261	106,383
1966	408,694	329,600	79,094
1967	488,642	397,975	90,667
1968	494,699	385,599	109,100
1969	532,808	401,407	131,401
1970	494,863	390,070	104,793
1971	465,297	371,308	93,989
1972	499,658	384,119	115,539
1973	529,224	403,884	125,340
1974	498,231	381,220	117,011
1975	564,611	408,499	156,112
1976	533,011	382,303	150,708
1977	485,428	373,526	112,902
1978	530,896	412,537	118,359
1979	634,616	482,650	151,966
1980	568,676	450,328	118,348
1981	456,643	379,602	77,041
1982	414,223	348,016	66,207
1983	412,585	368,788	43,797

Source: Webster and Dawson, 1983; 1985.

the number has intermittently fallen, although the average size has generally increased. In 1985 still only 15% of the total bedroom accommodation for tourists on the Island was in self-catering accommodation, despite the marked decline in the guest house sector. In the mid 1980s the Isle of Man accommodation industry was still predominantly hotel- and guest house-based, as it has been for the past one hundred years.

Characteristics of tourists visiting the Isle of Man

The restructuring experienced by the industry is mirrored in the changed average length of holiday taken by visitors on the Island over the past 20 years. The Island now caters much less for the traditional fortnight's holiday than it did even in the recent past. In 1969 the average (mean) length of stay of summer tourists on the Island was 11.1 nights (Tourism Industry Development Commission Report, 1970). In 1985 this figure had been reduced to 8.0 for August period visitors (Webster and Dawson, 1986). In the 1980s, therefore, the Isle of Man would have needed to attract 40% more tourists than in the 1960s in order to fill the same number of bed spaces. It is this factor, as well as the recent marked decline in the volume of tourists to the Island, which represents a substantial cause of the industry's retrenchment on the Island.

There has also been a notable change in the pattern of tourist origins to the Island in the 1980s, in part resultant of the higher rate of inflation in the Irish Republic, making holidays on the Island comparatively cheap for Irish residents, and the slower development of package holidays to other destinations by Irish tour operators. Between 1981 and 1985 the proportion of Irish tourists to the island in August doubled, and in August 1985 equalled a quarter of all tourists. The growth in the importance of Irish tourists has been matched by a decline in the proportion of tourists coming from north-west England in August. In 1981 more than four out of ten tourists came from the north-west in this month; by 1985 this proportion had been reduced to three out of ten. Northern Ireland has also increased in importance as a region of origin for tourists to the Island: in August 1985 one in seven tourists were from this Province. All other regions were of limited importance as origins of summer tourists, but not necessarily of day-trippers. The north of England stood out as an exceptional source of day-trippers in the summer of 1985, presumably reflecting the origins of tourists staying on the Lancashire coast, and principally those staying in Blackpool using the Fleetwood ferry

16.3 The cottages at Cregneash represent a vital resource both for 'heritage tourism' and for the education of visitors about Manx traditional life

16.4 A modern view of Lady Isabella, the huge waterwheel at Laxey. The wheel has been extensively refurbished and forms the centrepiece of an attractive wooded leisure area with picnic facilities and adventure walks

service. A further one in six day-trippers had their homes in north-west England and one in eight in Yorkshire or Humberside.

However, the annual pattern of visitors is different from that of the summer. In particular, Eire loses some of its importance, and the north-west of England stands out as the main source of period visitors. This said, in 1985 one in four period visitors either resided in the Irish Republic or Northern Ireland. The winter pattern of visitors emphasises links with the north-west of England; one in four period visitors in January 1985 came from this region. The south-east of England is a secondary source of visitors in the winter. However, these winter totals include a disproportionate number of business travellers, and so the Island's commercial links with London and the north-west may be distorting the winter tourist pattern.

The Island is a holiday destination for both non-manual and manual worker households. In 1985 nearly six out of ten period visitors were manual workers or their families, and these households represent the traditional volume market for tourism on the Island. The extent to which the passenger survey figures may be distorted by non-tourists is unlikely to be extensive in regard to social class for in the winter, when tourism is less important in terms of passengers carried, the proportion of non-manual workers is only slightly higher than in August. It is amongst day-trippers that this pattern differs. Three-quarters of all day-trippers in the summer of 1985 were manual workers or their families. This reflects the socio-economic profile of tourists holidaying on the Lancashire coast, and principally at Blackpool. Compared with 1977 the Island has lost, rather than gained, non-manual worker staying tourists, but for both years a characterisation of the Island as largely a manual worker holiday destination would be wrong. In 1977 a third of all staying tourists were from either higher or intermediate managerial, administrative or professional households, a proportion comparable with the holiday-taking of British tourists at the time.

Tourists visiting the Island include only a few retired persons, and this has been the pattern throughout the 1980s. Instead, the Island attracts both young and mature adults: in 1985, for example, nearly one in three staying tourists were aged 16 to 30 years, a notable increase on the situation of the early 1980s. Around four out of ten visitors to the Island were aged 31 to 60 years in 1985, a proportion which was notably higher in the winter. This difference may result from business travellers distorting the winter figure, rather than solely reflect a different tourist market in the winter. Notably, the Island is not substantially a tourist destination for households with children, although in August the proportion of children increases amongst the Island's visitors, to a quarter of staying tourists. The Tourist Board's promotion of 'events' to attract tourists can be expected to have altered the pattern of trip-making away from 'traditional' markets, but not extensively, as the objective of 'events' is to develop different markets, rather than to replace the traditional family holiday market. The importance of family groups as tourists remains. In August 1985 eight out of ten period visitors were travelling as families, as were seven out of ten day-trippers. The proportion of families compared with individuals changes in the winter, when most visitors

16.5 The Island's railways are still represented by the line from Douglas to Port Erin, now retained as a tourist attraction, but equally part of the Island's industrial archaeology

travel by themselves. The average size of group does show a decline from the beginning of the decade, but the proportion of families amongst visitors has remained much the same. Friends travelling together are particularly common in June, when they are attracted by the TT Races. In this month nearly three out of ten passengers surveyed leaving the Island in 1985 were travelling as a group of friends. Groups of friends are also disproportionately to be found amongst day-trippers to the Island. However, it is important not to overstate this: in the summer of 1985 day-trippers travelling with friends only amounted to one in five day-trip visitors.

In sum, it is clear that the characteristics of tourists to the Isle of Man vary by the time of year. The continuing importance of the traditional volume market of mainly manual workers and supervisory or clerical tourists, travelling as families with or without children, remains. The promotion of events may have changed the overall age and group profile of tourists, but this change has not been extensive. Instead of adjustments of this kind, the volume adjustment of the 1980s has been to attract more Irish visitors, to replace those lost from Great Britain. The Island is, however, vulnerable to the recent development of cheap air fares from Ireland to other destinations, and the long-term heightened importance of the Irish market cannot be presumed.

Transporting tourists to the Island

Despite the changing fortunes of Manx Airlines and the Isle of Man Steam Packet (discussed in Chapter 12), in the mid 1980s the ferries remained the volume carriers of tourists. In 1985, 228,821 staying tourists and day trippers were carried to the Island by ships in the summer season (May to September), compared with 42,105 by plane. In the months of July and August in excess of 125,000 tourists arrived by sea, compared with approximately 23,500 by air. In the early 1980s the proportion of tourists to other passengers carried changed, most notably for the airlines. As the passenger traffic won by the ferries has declined, the mix of passengers has increasingly moved towards tourists. Residents and business travellers have increasingly used the airline services. The dependence of the ferries on tourist traffic in the summer is clear from 1985 data: between May and September over eight out of ten ferry passengers were tourists. However, it should be noted that for the months of June to August tourists also represented in excess of half of the passengers carried by the airlines. However, the ferries are disproportionately dependent on tourists for their business throughout the year. For example, in January 1985, 42% of the passengers arriving by ferry were tourists, compared with 27% by air.

The dependence of the ferries on tourists can also be shown in the extent of extra scheduled sailings in the summer, compared with the winter season. In 1986, 131% more ferry services were scheduled from the United Kingdom in August than in January, that is, more than twice as many. In 1987, 174% more ferry services were scheduled for August than for January (Table 16.5). Table 16.5 also shows the recent cutbacks in ferry sailings resultant on the adjustment of the Steam Packet to the economic circumstances of the tourist trade of the 1980s: in August 1987 the number of scheduled sailings from Heysham was *fewer* than in January 1986, although the extent of this change may be exaggerated, because of changes in the capacity of the fleet to carry cars, which was previously a limiting factor.

The routes operated by sea and air in August 1987 are shown in Figure 16.1. Whilst it is true to say that the tourist to the Island has the widest range of departure points when travelling by air, both sets of routes also show concentration on certain routes. By air, the Isle of Man is clearly linked to Liverpool, Blackpool and Belfast and, secondly, to London and Manchester. By sea, the tourist not only has less overall choice of routes, but of these, more than four out of ten August

Table 16.5 *Scheduled sailings to Douglas from the United Kingdom*

Services from	August 1986	August 1987	January 1986	January 1987
Heysham	94	48	52	31
Belfast	12	11	0	0
Liverpool	9	14	0	0
Stranraer	5	4	0	0
Fleetwood	0	8	0	0
TOTAL	120	85	52	31

departures by sea are from Heysham. After Heysham, Dublin is now the second most important sea departure point, emphasising the important role of Irish residents as tourists to the Isle of Man in the later 1980s. Overall, three out of ten scheduled August sailings to the Island were from Ireland, North and South, in 1987.

Government departments and tourism

The Department of Tourism and Transport is the successor department of government to the former Tourist Board and is charged by the Tourist Act of 1975 to 'maintain, encourage, develop, protect and facilitate tourism in, to and from the Island to the best advantage of the Island' (Section 2 (1)). However, it should be noted that other branches of the Manx civil service also have tourism responsibilities, through the amenity resources they administer: these departments include the Department of Highways, Ports and Properties, the Department of Agriculture, Fisheries and Forestry, and the Manx Museum and National Trust. Tynwald has promoted tourism to the Island since the late Victorian era, establishing a committee, the Board of Advertising, in 1894. The activities of the Board of Advertising for 1896 and 1897 reveal the seriousness with which the task of advertising the Island was taken even when tourism boomed. To boost Easter visits, advertisements were placed in thirteen leading daily papers ' "calling" attention to the advantages of the island at this season of the year' (Report of the Board of Advertising for 1896). During the following summer season 803,000 four-page leaflets were inserted into the leading monthly magazines, and paragraph advertisements appeared in daily and weekly, and family and religious papers. The Board also reported placing full-page advertisements in publications issued by the principal railway companies. Railway companies were also circulated with posters, some framed and glazed for display in waiting rooms. Lantern slides of Manx scenery 'were presented to gentlemen giving public lectures and entertainments in the British Isles, upon their undertaking to exhibit them from time to time thereat' (ibid.). In 1897 'special and vigorous efforts were made' (Report of the Board of Advertising for 1887) to gain a share of the London market. General advertising continued as in the previous year, and in 1897, for example, 1,028,000 four-page leaflets appeared in the popular magazines.

The efforts of the Tourist Board to counter Mediterranean competition for holidaymakers since the 1960s had a long antecedence in promotional activity. Until recently, the Board was in practice primarily a promotional body, despite its legislative objective. For example, in 1980 the Board spent 43% of its budget directly on promotion and a further 30% on 'events', principally the TT races (Isle of Man Accounts). The Board, and subsequently the Department, is also charged, however, by statute, with the registration of tourist premises and the classification of these premises. Until the 1980s this was operated as a voluntary scheme; now all premises are registered and graded compulsorily. The other main functions of the Department are the promotion of Everymann package holidays and, since 1984, the encouragement of hotel modernisation by grants and loans. The confidential but much 'leaked' Poole Review of 1988 has recommended the separation of these functions from those of

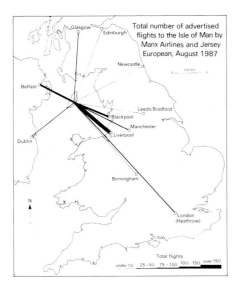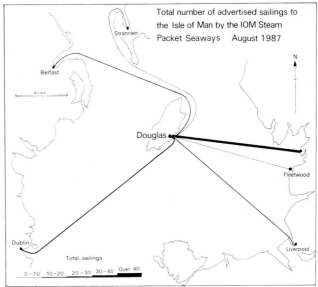

Fig. 16.1 Advertised flights and sailings to the Isle of Man, 1987

strategic planning for tourism. The Poole Review is reported to have recommended that strategic planning should be located in a new Economic Development Department in order to give a new direction to the industry. Tourism promotion is likely to become the task of a subordinate Tourism Agency.

The Isle of Man Tourist Industry Development Commission

An early response to competition from abroad was the creation of a Commission to advise on the industry. The Isle of Man Tourist Industry Development Commission was set up in 1968 to report on three topics, namely:

a. the then present state of the tourist industry of the Island, and to make recommendations for its extension and development;
b. schemes of financial assistance to the tourist industry from government, and to make recommendations for the modification or extension of such assistance; and
c. further actions to be taken by Tynwald to promote, develop and support the Island's tourist industry.

The objective was to gain outside advice on the industry: of the six members of the Commission two were from England, one from Northern Ireland and one from the Republic of Ireland. The Commission's report was published in 1970 (Tourist Industry Development Commission, 1970). The main recommendations of the Commission were four-fold:

a. financial assistance from Tynwald, additional to that already given, to the industry;
b. duty-free purchase facilities, principally on alcohol and tobacco;
c. developments; and
d. the extension of the season, both early and late.

Of these four recommendations, that on abrogation of the Common Purse arrangements with the United Kingdom has never to date been implemented. The impact of abrogation of this Agreement would have much wider impacts than on the sales of alcohol and tobacco alone. Likewise there is an important distinction between duty-free and duty-reduced goods. Duty-free would mean that the Government would receive no revenue at all: an unlikely circumstance to be effected. A more likely circumstance would be reductions on indirect taxes to enable price reductions generally on shopping goods, to encourage visitors to purchase in Manx shops (Tynwald Select Committee on the Common Purse, 1986). In this way Manx traders generally, and not just those involved with alcohol and tobacco, would benefit from abrogation.

The Commission was explicit in the need for Tynwald to support financially the modernisation of the accommodation industry, regarding the modernisation of accommodation 'a prime requirement' (para. 4.1.). Rehabilitation of the stock was considered a second-best alternative to rebuilding. The Commission was explicit in its long-term objective: 'In physical terms, whilst the paramount need is for new building, the Commission recognises that the Island will have to make do with existing hotels and guest houses for a number of years' (para. 4.3), and 'the solution must be in the demolition of blocks of obsolete boarding houses, and the erection, on the cleared sites of modern buildings' (para. 4(d)2). Whereas the Island has had some new hotel and chalet constructions, as recommended by the Commission, since this Report the renewal of the stock accommodation has *not* been largely by rebuilding, but instead by rehabilitation of the existing stock. Recognising that large-scale clearance and rebuilding was impossible in the short term the Commission also recommended the modernisation and merging of existing establishments. So much importance was placed by the Commission on mergers as a preliminary to the modernisation of the existing stock, that it recommended the Tourist Board to purchase two adjacent premises and convert them into one, modernise them, and use the refurbished premises as a 'show house'. The Commission was quite critical of the Manx system of loans and grants that had applied until the 1960s. The Commission recommended Bord Failte Eireann's policies as a basis for future Manx policies. The Commission also recommended the compulsory grading of all tourist premises, to replace voluntary registration which had been operated since 1962.

Development recommendations of the Commission included recommendations to expand the Island's yachting facilities, to maintain its golf courses, to improve provision for motorists so as to encourage 'mobile' (touring) holidays, and to develop angling as a tourist attraction. In particular, the Commission saw the need for 'urgent action to cope with wet weather' (para. 6(n)1). The recommendations were very much those of the era of the Commission: 'heritage', for example, the current promotional emphasis, gained only a scattered mention in the report. In the 1980s the distinctiveness of the Island has been promoted, but not its 'foreignness' as had been implied in the Commission's interpretation of heritage.

The extension of the season was recommended by developing special interest markets. Of the Commission's recommendations, those on extending the season have been most extensively followed by the promotion of 'events' to attract tourists with specific interests.

Restructuring of the accommodation industry in the past twenty-five years

The Manx accommodation industry is still predominantly run by family businesses, and the restructuring of the accommodation industry has taken other forms than changing the predominant family firm structure. Instead, it has taken the form of, firstly, amalgamations; secondly, the achievement of greater rates of bedroom occupancy than were achieved in the 1960s; and, thirdly, accommodation upgrading. The objective of hotel amalgamation is to increase the average number of bedrooms per hotel. In 1965 the average size of a Manx hotel or guest house was just under 12 bedrooms; in 1985 it was just over 20 bedrooms. The intervening years showed a consistent increase in average size. This process has resulted in a varied pattern of accommodation sizes on the Island (Table 16.6). Outside Douglas, Onchan, Ramsey, Port Erin and Port St Mary the average size of premises is markedly less than average. The second strategy of restructuring has been to seek to increase rates of bedroom occupancy. In this way the capital invested earns a greater return, for empty bedrooms are not earning a hotelier money. In 1965 there were approximately 19 staying visitors per bedroom available; in 1985 the number of visitors per bedroom was 27. If we assume that the average bedroom sleeps two visitors, this is still a low rate of occupancy, for with an average length of stay of seven nights, this means that in 1985 on average tourist bedrooms on the Island were only occupied for 14 weeks of the year. The early 1980s in fact saw a decline once again in bedroom occupancy, aafter a peak in occupancy in 1979. The industry is clearly yet to adjust fully to the decline in staying visitors experienced since this peak year (Table 16.4).

The third restructuring strategy has been to upgrade the facilities offered by the hotels and guest houses. This is not an easy task if the enterprises themselves are not profitable. Modernisation of the stock was a particular concern of the Tourist Industry Development Commission in its report of 1970, for Government action to encourage the modernisation of the stock had not been generally successful. In the 1960s modernisation had been subsidised by the Tourist Premises Improvement Act of 1961, as amended in 1963 and 1969. These Acts made loans available under attractive terms of interest. However, in the eight years ending in June 1969, only 339 applications had been made, of which 282 were approved (Tourist Industry Development Commission, 1970). In the 1960s 40% grants were also offered for new premises, or the improvement of existing premises, costing upwards of £50,000 and providing up to 200 bedrooms. These grants were clearly to encourage large schemes, and were made under the Improvement of Tourist Accommodation Scheme 1962, as amended by the Scheme of 1968. Only one grant was made under these schemes in the 1960s.

The 1970s saw a succession of legislation and regulations promoting the improvement of accommodation. The Tourist Premises (Compensation for Tenants' Improvement) Act 1970 gave a tenant of tourist premises the right of compensation for improvements made to the premises by himself or herself, thereby removing a constraint on investment by tenants. The improvements covered included amenities and lifts as well as structural alterations. The Tourist Premises Improvement Act 1974 provided a framework for schemes to be made to give further assistance towards the improvement of tourist premises. Amended by the Tourist Premises Improvement Act 1976, the law was changed again in 1977 by the repeal of the earlier legislation, and the enactment of the Tourist Premises (Provision and Improvement) Act 1977. This Act empowered the giving not only of finan-

Table 16.6 *Numbers of premises and average size of accommodation by area of the Island, 1983*

	Number of hotels guesthouses and self-catering accommodations	Average number of bedrooms
Towns:		
Douglas	508	17.8
Onchan	28	17.1
Ramsey	22	18.0
Peel	32	6.5
Castletown	18	8.2
Port Erin	57	17.6
Port St Mary	19	15.1
Laxey	28	3.4
Country areas:		
North	40	3.4
South	28	3.6
East	8	3.5
West	12	2.0

Source: Webster and Dawson (1985).

cial assistance, but also the guaranteeing of loans for the provision and improvement of tourist premises. The Act explicitly included 'the acquisition of premises adjoining existing tourist premises and the amalgamation of such premises into one tourist premises' (Section 1(a) (iii)) as a category of provision of premises. Some objectives of the Tourist Industry Development Commission Report of seven years previously can be seen in this Act. This Act remains the principal legislation for which Government assistance to the modernisation of the accommodation is effected.

Current financial support is set out in the Tourist Premises Development Scheme of 1984 as amended in 1985. Until 1984 this support was the responsibility not of the Tourist Board but of the Local Government Board, which remains the Board (now Department) responsible for private housing grants. In 1984 the more logical course of making the Tourist Board responsible for encouraging the improvement of tourist accommodation was effected. The 1984 Scheme requires a mix of criteria to be met by applicants which extend beyond the capital works themselves, as the Board (now Department) has been anxious to effect not only modernisation of the physical structures but also of the industry generally by its financial assistance. For example, the factors of relevance in assessing applications include:

- the standard of management;
- the provision of employment opportunities;
- the indication of a commitment to the industry;
- the standard of design and the quality of the completed project;
- the application of appropriate marketing techniques;
- the extension of the season; and
- the potential for the creation of new markets and activities.

Compared with the 1960s, the rates of assistance under the 1984 Scheme are much greater, and in part reflect both a belated response to the retrenchment of the industry, and a recognition of the financial help to the tourist accommodation elsewhere by governments. Major new developments, that is the provision of new high quality hotels and self catering accommodation, may qualify for up to 40% grants and loan assistance of up to 30%. Amalgamation works may attract loan assistance up to 100% of the value of the premises to be purchased, and a grant of up to 40% and a loan of up to 50% for alterations. The provision of en-suite facilities can qualify for a grant of up to 50% of the cost of provision, and a loan of up to 50%. The provision of passenger lifts likewise qualifies for generous grants and loans.

Despite the range of provisions available, the number of approved applications under the Tourism Premises Acts has not been great, except in the period immediately following the 1977 legislation. In total, in 1974–75 to 1983–84 the Government made grants totalling £1.37 million and loans of £4.96 million for the improvement of tourist premises. Clearly, the industry is modernising its accommodation slowly, even with recently generous grants and loans.

The quality of accommodation on the Isle of Man

Manx holidays are not uncompetitive in terms of price with holidays in the Mediterranean. However, package holidays abroad usually offer modern purpose-built accommodation, frequently with swimming pools, sports facilities and entertainment. Few Manx private hotels or guest houses can equal these facilities, despite the modernisation of the stock in the past thirty years. In 1987 the potential holiday-maker to the Island could find approximately 300 different accommodation enterprises listed in the main promotional brochures of the Department of Tourism and Transport. Nine out of ten of the private hotels listed were graded as either

Table 16.7 *Grading of accommodation enterprises advertised in main brochure by main types of accommodation, 1987*

	Keys					None/ ungraded	% of all accommodation advertised
	5 %	4 %	3 %	2 %	1 %		
Hotel	20	30	35	15	0	0	6
Private hotel	0	8	45	46	1	0	28
Guest house	1	6	33	42	15	3	27
Self catering	3	25	32	21	17	2	39

Note: the percentages summate across the table, except for the final column which summates vertically.

three or two key premises, as were three-quarters of the guest houses. Fewer than one out of ten of both of these popular types of accommodation was graded as having either an excellent or very high standard of decor, furnishings, service and catering (Table 16.7). The other popular type of accommodation, self-catering, showed a wider range of quality, with over a quarter of these premises or sites being graded at least of four key quality. Looked at from the extent of poorer quality accommodation, one in six of the advertised guest houses on the Island was of satisfactory or lower quality; as was a similar proportion of self-catering premises or sites. Only in the hotel sector were most of the premises graded at being of a higher standard, or better. The latter is the main sector of accommodation promoted for package holidays abroad by the British travel trade, and as such, the quality of Manx hotels compares favourably. However, hotels represent only a small minority of establishments as advertised in 1987. The Island's main types of accommodation are not of a comparable standard; giving the Island's accommodation sector a competitive disadvantage.

The quality of accommodation shows a marked regional bias on the Island. That in the east of the Island, in the area of Douglas and Onchan, is generally graded lower by the Department of Tourism and Transport than the accommodation elsewhere. Over a third of the advertised accommodation enterprises were graded as excellent, or of very high standard, or the camping equivalents, in the west, north and south of the Island, but less than one in ten of the enterprises in the east was graded comparably (Figure 16.2). In large part this reflects the good quality of self-catering accommodation outside the Douglas and Onchan urban area. In the east of the Island, one in ten self-catering accommodation enterprises was graded as excellent, or very high standard, or the camping equivalents, compared to one in four over the rest of the Island. Outside the Douglas area modern

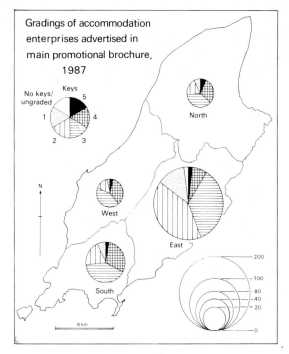

Fig. 16.2 The spatial distribution of different grades of tourist accommodation, 1987

purpose-built flats or refurbished cottages are common forms of self-catering accommodation. Whilst the Douglas area still dominates in accommodation provision the quality of this provision is, on the whole, inferior to that elsewhere on the Island. In the search for new markets the more affluent social groups, often taking several holidays, are an obvious target. The Isle of Man can offer scenery and relaxation, ideal for a supplementary holiday for the professional or executive. These groups will not be attracted unless the general standard of accommodation is high.

The Isle of Man Government has included questions on visitors' perceptions of the quality of accommodation in surveys of passengers from the Island. Table 16.8 is an attempt to set out the findings of these surveys in a composite form. Firstly, few visitors appear to think that their accommodation was poor; that is, that it failed to meet their minimum expectations. The accommodation on the Island is meeting existing visitors' minimum expectations, at least, therefore. However, an equal number of visitors in 1985 thought that their accommodation was only 'satisfactory' as thought it to be 'good'. This is hardly a strong recommendation of the general quality of accommodation by existing tourists, therefore. A quarter of respondents failed to answer this question, and this group is likely to include a disproportionate number of tourists who found their accommodation lacking in some way, but not such that an overall judgement could be made. Referring back to the 1977 figures, only a third of tourists thought their accommodation to have been 'very good value for money'. Taken together these surveys imply that many existing customers are not fully content with the quality of accommodation on the Island.

The University of Surrey's Tourism Strategy for the Island

The former Tourist Board commissioned the University of Surrey to develop a tourism strategy for the Island into the early 1990s. This report was published in late 1987 and identified the Manx tourist industry's central problem as the failure to adjust to new markets, 'The Isle of Man has failed to adapt its tourist facilities to satisfy emergent tourist markets in the face of a decline in its traditional long-holiday market' (Cooper et al., 1987,

Table 16.8 *Visitors' satisfaction with their accommodation*

	Period visitors 1985 %	August 1985 %	August 1985* %	Staying tourists travelling by sea August 1977 %	August 1981 %
Accommodation judged to be:					
Good	36	37	33	—	—
Satisfactory	34	35	40	—	—
Poor	6	10	12	—	—
No answer	24	18	15	—	—
Very good value for money	—	—	—	35	} 88
Good value for money	—	—	—	59	
Not good value for money	—	—	—	6	12

Note: * travelling by sea
Source: Isle of Man Passenger Survey, 1985; and Webster and Dawson (1983).

p. 1). The three main elements of the proposed strategy were as expected, namely:

- a strengthened role of the Tourist Board to coordinate and lead the industry;
- a marketing strategy to maintain and protect traditional markets, and to pioneer new markets principally to reduce seasonality; and
- a tourist development plan to produce identifiable and structured development priorities, to encourage both the upgrading of existing facilities and new developments.

The first element of the strategy called for the employment of tourism professionals, principally for development and research. A promotional campaign was recommended, to stress the importance of tourism to the Manx community, and a forum to act as a 'think-tank'. The report also recommended the commissioning of a study of the economic value of tourism to the Island. The Island's marketing was principally diagnosed as under-researched, particularly in terms of promotional effectiveness and different markets. In particular, the report recommended the identification of shorter holiday markets and of travellers with fewer seasonal constraints, and focusing of sales and promotional efforts on these markets. Also recommended was the development of the Island's business tourism sector, especially in terms of corporate meetings, selected conference trade and associated leisure packages. In particular the following developments were recommended:

- the continued upgrading of accommodation;
- new development of budget accommodation and self-catering accommodation;
- upgrading of existing sports and leisure facilities;
- improvement in the standards of visitor facilities;
- new development of water-based harbour/marina and themed heritage attractions;
- infrastructural improvements;
- in-service staff training; and
- upgrading of conference and exhibition facilities.

The consultants recommended a public sector lead 'given the scale of the task of developing the Isle of Man's tourist resources' (Cooper et al., 1987, p. 41). This lead was to embrace financial aid on a wider basis than accommodation upgrading alone. As such, the strategy was capital-intensive and mirrored developments in southern England. The tone of the report was that of the need both to respond to and to develop opportunities by a much more coordinated, informed and initiate industry. The longer-term value of the report may be to have generated discussion as to options for tourism on the Island as part of the subsequent Poole Review proposals to incorporate tourism strategy as an integral part of economic development planning on the Island.

Conclusion

The Isle of Man tourist industry has not benefited from the substantial increase in main holiday-making in its major market, Great Britain. New markets have been found to supplement the volume main holiday market, in particular 'events' and 'second' holidays, and to replace the lost British visitors with Irish holidaymakers, at least in the short term. However, these new markets have not been identified and developed in sufficient volume to arrest decline. Not only has the Island suffered from a change of taste in holidaying, towards the Mediterranean and sunshine, but it has also suffered from a shortening of the holiday period actually spent on the Island by tourists. The latter is indicative in part of the importance of extra or 'second' holidays to the industry. The Island has, however, been able to retain both a broad social class profile of tourists and a sizeable proportion of repeat visitors. It would seem that many tourists visiting the Island are not looking for the benefit of a Mediterranean holiday in a mass resort, but are instead

16.6 A night view of Douglas promenade. Generations of working class tourists found such romantic scenes irresistible

enjoying the scenic and heritage attributes of the Island. As such, the Island attracts a particular segment of the volume market. Efforts need to be directed to retaining this market segment, as well as developing new markets.

The tourist accommodation industry on the Island can no longer claim to represent the principal economic interest. Nevertheless, the importance of tourists to the Manx economy is not insignificant. The tourist accommodation industry has undergone a protracted restructuring to meet new market conditions; this restructuring has yet to be completed. The Island has not substantially rebuilt its tourist accommodation to meet modern market conditions. Instead piecemeal closures, amalgamations and modernisation by rehabilitation of existing buildings has occurred.

In that a holiday on the Isle of Man is competitive in price with many volume market destinations, and yet many potential tourists from north-west England go elsewhere, is not indicative of a prosperous future for the Island's tourist industry. In turn, low profit margins imply low rates of investment in modernisation of accommodation. Unless a major public initiative is forthcoming, the most likely scenario for the Manx tourist industry is that it will continue to restructure slowly, and progressively lose its market share of main holidays. The industry needs seriously to consider researching new markets and holiday decision-making. Likewise, the images of potential tourists of the Island need to be explored, as a basis for future promotion. It would seem essential that the Island's history of promoting itself vigorously as a holiday destination continues, and that travel agents or other sources of holiday information, such as major tourist information centres, should be actively encouraged to promote Manx holidays. This is a problem common to all of the British national tourist boards, and one particularly monitored by the Scottish Tourist Board. Overall, it would seem likely that the Island will need to spend progressively more on promoting itself as alternative heritage destinations are developed and promoted in the United Kingdom.

In summary, the future of the tourist industry on the Island is largely dependent on the ability of the industry and Manx Government to develop appropriate and effective strategies for the industry. The costs of attracting additional tourists are likely to rise, both as a result of modernising premises and attractions, and of promotion. As the University of Surrey report has made clear, in the competitive world of modern tourism, initiative, the will to change, and informed leadership of the industry are required. Equally, there is a need first to decide appropriate strategies to be achieved.

PART FOUR: THE ECONOMY

Conclusion

Each of the previous chapters has charted recent, and in some cases historic, changes in key sectors of the economy. Whilst the growth or decline of individual sectors has occurred at different points in time and for different specific reasons all change has taken place against the same background of trans-nationalisation. In the nineteenth century Manx farmers could direct their attentions to feeding a local population, the tourist industry focused upon attracting mill-workers from nearby industrial cities, and manufacturing rarely sought to export to the mainland let alone continental Europe. By the final decade of the twentieth century the economic and spatial context had radically altered. The Isle of Man is now part of a world economy which entrepreneurs ignore at their cost. The Island's tourist industry is competing with Spain, Greece and Turkey. The new financial sector has strong linkages with other offshore centres such as the Bahamas and the Cayman Islands. And Manx manufacturers are exporting items as diverse as shaving brushes and aircraft ejector seats to markets across the world. Changes in the way the Manx economy interacts with other economies thus provides us with an excellent example of trans-nationalisation and also of the economic and social changes which have come with this.

In parallel with the internationalisation of indigenous economies, the post-war period has also seen a trend towards the post-industrial society. Intrinsic to this is the shift from manufacturing saleable goods to the provision of personal, industrial and financial services. Not only has this had a profound effect upon the landscape in places such as the West Midlands of England but it has also offered differential opportunities to regions, economic sectors and social classes. In many ways the Isle of Man, again, exemplifies the trend towards post-industrial societies. As Chapter 15 makes clear, the Isle of Man is increasingly relying upon financial services for its economic existence. Manufacturing no longer consists of the locationally-tied traditional industries, but is increasingly made up of internationally footloose high technology industries which might, for example, be processing industrial diamonds. These new industries require less labour than their predecessors but they do demand skilled employees and environmentally attractive and easily accessible sites. However, whilst the Isle of Man does typify many aspects of a post-industrial economy, as in many other respects, it refuses to fit any stereotype exactly. Thus the Island already possessed a strong tourist industry with its associated invisible earnings. Yet at the very time when British tourism is booming and regions are striving to exploit their heritage commercially, the Isle of Man's tourist industry is already in serious decline.

If, as suggested, the Isle of Man is becoming a post-industrial economy integrated into a trans-national economic system, then indications of this should be seen in both the Island's balance of payments and in its sectoral structure. However, because there is no fiscal barrier between the Island and the United Kingdom it is extremely difficult to obtain an accurate estimate of the balance of payments. An exercise to establish the balance of payments position has been carried out by the Government in 1983 (see Table A.1), but there was relatively poor cooperation from many in the private sector and the results of the Treasury study on the balance of

Table A.1 *Isle of Man balance of payments estimates, 1983*

Balance (£)	UK	Elsewhere	Total
Visible: Exports/Imports of goods (f.o.b.)	(112,348,000)	(5,425,000)	(117,773,000)
Invisible Services: Transport and Travel	30,803,000	1,259,000	32,062,000
Financial Services, General Government and Other Services	(8,451,000)	5,951,000	(2,500,000)
Interest, Profit and Dividends: Private Sector, General Government and Public Corporations	25,225,000	7,938,000	33,163,000
All Other Payments/Receipts: Private Sector, General Government and Public Corporations	1,477,000	48,588,000	50,065,000
Invisible Balance:	49,054,000	63,736,000	112,790,000
Balance:	(63,294,000)	58,311,000	(4,983,000)

Note:
	UK	Elsewhere
Imports	173,857,000	12,128,000
Exports	61,509,000	7,703,000
Balance	112,348,800	5,425,000

Source: Treasury.

payments should be viewed with some caution. Because the Island has a small economy with few natural resources it is fairly obvious that there must be a substantial visible trade gap and that this is compensated for by a favourable balance on the invisible account due to the earnings from tourism and the financial sector and earnings from abroad in respect of investments.

Difficulties arise in estimating the various factors, for example many goods are brought in to the Island by ship in either vehicles or containers with little indication as to the type of goods being imported and no indication of value. Similar problems arise with exports, except that it is usually easier to identify the value because there is a relatively small number of companies exporting from the Island. Difficulties also arise in assessing the value of financial transactions because the Island is in the same monetary area as the United Kingdom and there is no exchange control and in addition the Island has over 21,000 registered companies. It is difficult to identify whether companies are acting in the domestic market or are effectively overseas companies.

The balance of payments factor cannot have the same relevance to the Island as it has to the United Kingdom since any imbalance is not reflected in the reserves of the Island's Government, but on the other hand it is a valuable indicator of economic strength. The Treasury survey took no account of capital movements which could be inward by both the new residents and new companies, but as the investment opportunities in the Island are somewhat restricted much of the inward movement of capital is matched by movement in the opposite direction. It is appreciated that new residents moving to the Island will almost certainly make a capital investment in property but somewhere along the chain the recipient of that investment will transfer the funds out of the Island for investment in the United Kingdom or some other market. During a period of expansion of the population and business activity there will undoubtedly be increased investment in the Island in terms of buildings and equipment. That increased investment will in turn generate imports of goods and services into the Island and the net position on the balance of payments will be far less than first impressions may imply.

A general opinion of the balance of payments can be formed by simply examining the state of the economy. A healthy prosperous economy will probably have a surplus on balance of payments, a declining economy is indicative of an outflow of capital, a falling level of consumer spending and an adverse balance of payments.

An indicator also available in the Isle of Man is the net income from abroad which has been rising rapidly, suggesting that assets held by the private sector have been growing and that the Island's economy is becoming increasingly international in its perspective.

Table A.2 *National Income (£'000s at factor cost)*

	1977/78	1987/88
Gross Domestic Product	112,711	268,058
Income From Abroad	8,478	65,217
Gross National Product	121,189	333,275

Source: Treasury.

Table A.3 comments on a different facet of the Island's economy, namely the relative contribution of its various sectors to income generation. The table shows the dominance of the service sector (particularly its financial component) and the relative weakness of agriculture and manufacturing. This is typical of a post-industrial economy.

Table A.3 *Income generated in basic sectors from Manx sources (£000's) (Factor Cost)*

	1977/78		1987/88	
	£	%	£	%
Manufacturing	14,508	12.3	42,314	14.0
Finance	33,447	28.4	76,992	25.4
Tourist Industry (1)	13,296	11.3	27,685	9.1
Construction	10,941	9.3	26,400	8.7
Agriculture and Fisheries	4,037	3.4	7,870	2.6
Public Administration	5,079	4.3	16,824	5.5
Other Services (2)	36,506	31.0	105,167	34.7
TOTAL	117,814		303,252	

Note: 1. Tourist accommodation, with proportions of public utilities, distributive services, catering and miscellaneous services.
2. Professional and scientific services, with proportions of the above sectors.
Source: Treasury.

CHAPTER 17

Geography of the Isle of Man into the 1990s
Danny McCarroll and Vaughan Robinson

This book began with a call for a return by academic geographers to an emphasis on landscape and place. To dealing with real places and their people rather than abstract concepts of space and process. Following this philosophy we have attempted to provide a contemporary, if necessarily eclectic, view of the physical and human geography of the Isle of Man, principally as a framework for teaching. In compiling the various sections, however, it has become clear that many opportunities remain for fruitful research.

For many years the solid geology of the Isle of Man has not received the attention it deserves. The most recent detailed work on the Manx Slate Series, for example, was completed in the mid 1960s and more recent micropalaeontological work is at variance with the established stratigraphy. Detailed study of the complex sedimentology and structure of these rocks would undoubtedly enhance understanding of the environment and palaeogeography of the closing Iapetus Ocean. The Poortown intrusive rocks would also repay investigation. The only published description, from early this century, was based on a single thin section. The Peel Sandstone and Carboniferous rocks of the south of the Island have seen a recent revival of interest. In both cases much remains to be learned, particularly concerning the complexities of the environments of deposition.

Work on the geomorphology of the Island has been dominated by the glacial and postglacial sediments and landforms. Early work on planation surfaces and strandlines has been called into question, but no detailed reinterpretation of the remnants of the preglacial landscape has yet been attempted. Elements of a preglacial drainage network, such as the Port Mooar Valley, clearly remain and the Manx uplands undoubtedly display surfaces of some sort. There is a need for a return to considering long-term landscape evolution, not simply in terms of Davisian cyclicity but in the light of an enhanced understanding of geomorphic processes.

The central problem of the Island's Quaternary history remains the environment of deposition of the 'foreign suite' of glacial deposits. The current debate centres over whether they represent subaerial or submarine deposition and has important implications for studies of sea level history, glacial isostacy and the interpretation of much of southern Britain's Quaternary landforms and deposits. Real advances in understanding the Manx evidence will, however, require a move away from the current inductive methodology of Quaternary sedimentology. Under the guise of objectivity, depositional sequences are first described and then interpreted in terms of some variation on each worker's preconceived general depositional model. It should be recognised that observation and description are not independent of theory. It would be more useful if Quaternary sedimentologists would state their theories explicitly and then go on to test them critically. Sedimentologists should also be aware of and address the implications of their theories for other branches of Quaternary science. Those arguing for a glaciomarine environment, for example, should attempt to explain the 'kame and kettle' topography of the west coast and the means by which the Great Irish Deer could have reached the Island so soon after deglaciation. Other promising areas for Quaternary research include the landforms and deposits of the south of the Island, which have been largely neglected, and the very thick deposits of the extreme north of the Island, which it may be possible to correlate with offshore sequences.

Lateglacial organic deposits continue to be exposed by coastal erosion and offer excellent opportunities for further study of the environment immediately following wastage of the last ice sheet. The Holocene deposits of Man have not received the attention they deserve. There is a wealth of knowledge to be gained, ranging from information on sea levels to understanding the environment of the early inhabitants. A basic requirement for a better understanding of Holocene conditions is more radiocarbon dates. At present there are insufficient to allow construction of local pollen assemblage zones which could be used for correlation. Perhaps the most promising site for future research is the Ballaugh Curraghs, which may yield a complete record of palaeoenvironmental conditions from the Lateglacial to the present day.

The Isle of Man supports a surprisingly wide range of natural habitats and associated flora and fauna. Several sites are of national or international importance. The Manx Museum and National Trust and voluntary bodies work hard to preserve this rich diversity. Although there have been important achievements in nature conservation, not least the continuing expansion of the colony of Manx Shearwaters on the Calf, once their great stronghold, some important sites have been lost. With increasing demand for development of rural areas other sites are certain to be threatened. If the diversity of scenery and wildlife which is the inheritance of the Manx people is to be preserved, then it is for the Manx Government to act. A first Wildlife and the Countryside Bill has reached the House of Keys and should afford some protection. Proper management of fragile environments, however, requires proper funding. There is a need for more research into Man's vulnerable ecosystems and of sufficient facilities for their protection.

There is no doubt that individual authors are better equipped to identify those topics in need of further research. However, even an editor can recognise certain omissions which might usefully be filled by the researcher interested in the human geography of the Island. That researcher enjoys a number of advantages over counterparts working in other localities. Because of the scale of the Island and its bureaucracy, officials are more accessible and more willing to help individual enquirers. Most of the contributors to the human part of this book have benefited from direct assistance by officials who by mainland standards are highly-placed. Few mainland researchers could hope to have direct access to officials at Ministerial level, for example. Secondly, because Manx society has jealously guarded its identity and its sense of place, it possesses relatively rich historical records, many of which are available to the general public. Added to this is the responsiveness and openness of the Manx Museum in Douglas, which again contrasts with the attitudes prevalent in some parts of the mainland. And lastly, the greatest resource available to any researcher of society is the people themselves. We have found Manx people to be exemplary. They are knowledgeable, interested and enormously cooperative. Set against these advantages are some obvious disadvantages for the researcher. These seem again to be a product of the Island's size but also the lack of a research tradition in certain areas. It is remarkable, for example, that mappable small-scale population data are still not available from the census some thirty years after these became commonplace on the mainland. Equally the absence of any small-scale population data prior to 1986 is a serious handicap to those interested in the changing social structure of the Island. In other fields, too, it is the inconsistency, absence or lack of continuity in data that forms the greatest barrier to research. The lack of any institute of higher education or central research unit is also a problem for some researchers. Knowledge is fragmented and often inaccessible to those unaware of which individuals or local bodies to approach. Perhaps proposals to create a government 'Think Tank' will obviate this difficulty.

Given the rapid and irrevocable changes currently being experienced by the Manx economy, perhaps the most important research task still outstanding is to document and analyse this change and its social and demographic concomitants. The twentieth century is undoubtedly a watershed for the Island and at the moment there are people still alive who can recall each successive phase of the Island's recent evolution. The Manx Folk Life Survey has gone a long way towards recording those ways of life and attitudes which prevailed in the early part of the century. Recorded oral histories held by the Manx Museum form an invaluable resource for researchers. It is, however, important that the Survey continues its work and does not restrict its definition of Folk Life too narrowly. The disappearing world of the traditional seafront boarding house is as much part of the Island's heritage as are the reminiscences of farmer-fishermen. We must not forget that today is tomorrow's history. Equally there is a need to document and understand the contemporary changes and strains brought about by the Island's new economic role. To what extent is the post-industrial society composed of discrete and non-communicating social groups? What are the opportunities for genuine social mobility on the Island for those born there? And what are the costs of the post-industrial society to those on its fringe or even excluded by it? It is a comment upon the rapidity of social change on the Island that some of these issues are already being voiced in public even though they were not captured by our contributors writing only twelve months ago. And finally perhaps in this context there is the broader research question of the internationalisation of post-industrial society. We have deliberately stressed the distinctiveness of the Island and its personality, but the Isle of Man is increasingly becoming part of a world economic system. Thought needs to be given to the paradox of encouraging distinctiveness whilst pursuing economic strategies that require homogenisation. And in the same vein there is a need for urgent research into the issue of the Single European Market in 1992 and the Island's response to it. The Isle of Man is at a particularly interesting and critical point in its history and it is vital that decisions are taken upon the basis of sound and detailed knowledge.

We hope that this book has provided some insight into the distinctive geography of the Isle of Man. Its beautiful and intriguing landscape, peculiar history and unique people. Changes are inevitable and in the next decade they are likely to be swift and perhaps irreversible. We sincerely hope that the Manx Government and people will continue to protect their environment and treasure their special sense of place. To those who would rather see the Island degenerate into just another small outlier of a uniform Europe, a few words of Manx:

Drogh ooir, Drogh yerrey as beggan grayse.

BIBLIOGRAPHY

Abell, P. H. (1984) *British Tramway Guide* (third edition), Feltham, P. H. Abell.

Alcock, L. (1970) 'Was there an Irish Sea culture province in the Dark Ages', in *The Irish Sea Province in Archaeology and History*, edited by D. Moore, Cardiff, Cambrian Archaeological Association.

Allen, D. E. (1978) 'The present day fauna and flora of Man as indicators of the date of the flandrian severance', in *Man and Environment in the Isle of Man*, edited by P. Davey, British Archaeological Reports, 54, Oxford.

────── (1985) *The Flora of the Isle of Man*, Douglas, Manx Museum.

Allen, J. R. L. (1974) 'Sedimentology of the Old Red Sandstone (Siluro-Devonian) in the Clee Hills area, Shropshire, England', *Sedimentary Geology*, **12**, pp. 73–167.

────── and Crowley, S. F. (1983) 'Lower Old Red Sandstone fluvial dispersal systems in the British Isles', *Transactions of the Royal Society of Edinburgh: Earth Sciences*, **74**, pp. 61–68.

Anderson, J. (1973) 'Ideology in geography', *Antipode*, **5**, pp. 1–6.

Anderton, R., Bridges, P. H., Leeder, M. R., and Sellwood, B. W. (1979) *A Dynamic Stratigraphy of the British Isles*, London, George Allen and Unwin.

Anon (1927) 'Ship building in Ramsey', *Journal of Manx Museum*, **3**, pp. 206 and 241.

Bawden, K. (1988) 'Industry in the Isle of Man', in *The Official Isle of Man Yearbook*, edited by T. Faragher, Douglas, Motor in Mann.

Bawden, T. A. (1971) 'The Manx paper making industry', *Proceedings, Isle of Man Natural History and Antiquarian Society*, **7**, pp. 449–77.

────── Garrad, L. S., Qualtrough, J. K., and Scatchard, W. J. (1972) *The Industrial Archaeology of the Isle of Man*, Douglas, Isle of Man Natural History and Antiquarian Society.

Berger, J. F. (1814) 'Mineralogical account of the Isle of Man', *Transactions of the Geological Society*, **2**, pp. 29–65.

Bersu, G. (1949) 'A promontory fort on the shore of Ramsey Bay, Isle of Man', *Antiquarian Journal*, **29**, pp. 62–79.

────── (1977) *Three Iron Age Round Houses in the Isle of Man*, Douglas, Manx Museum and National Trust.

────── and Wilson, D. M. (1966) *Three Viking Graves in the Isle of Man*, Society of Medieval Archaeology Monograph Series, 1.

Beresford, P. (1986) 'British ferries, that sinking feeling', *The Times*, 10 October.

Birch, J. W. (1958) 'Economic geography of the Isle of Man', *Geographical Journal*, **124**, pp. 494–510.

────── (1964) *The Isle of Man: A Study in Economic Geography*, Bristol, University of Bristol.

Bishop, W. W., and Coope, G. R. (1977) 'Stratigraphical and faunal evidence for Lateglacial and early Flandrian environments in south-west Scotland', in *Studies in the Scottish Lateglacial Environment*, edited by Gray, J. M., and Lowe, J. J., Oxford, Pergamon Press.

Board of Advertising (1896) *Third Annual Report*, Douglas, Isle of Man Government.

────── (1897) *Fourth Annual Report*, Douglas, Isle of Man Government.

Bolton, H. (1899) 'The palaeontology of the Manx slates of the Isle of Man', *Manchester Memoirs*, **43**, pp. 1–15.

Boulton, G. S., Jones, A. S., Clayton, K. M., and Kenning, M. J. (1977) 'A British ice-sheet model and patterns of glacial erosion and deposition in Britain', in *British Quaternary Studies*, edited by F. W. Shotton, Oxford, Oxford University Press.

────── and Jones, A. S. (1979) 'Stability of temperate ice caps and ice sheets resting on beds of deformable sediment', *Journal of Glaciology*, **24**, pp. 29–43.

Bowen, D. Q. (1973) 'The Pleistocene succession of the Irish Sea', *Proceedings of the Geologists Association*, **84**, pp. 249–73.

────── (1978) *Quaternary Geology*, Oxford, Pergamon Press.

Bowen, E. G. (1970) 'Britain and the British seas', in *The Irish Sea Province in Archaeology and History*, edited by D. Moore, Cardiff, Cambrian Archaeological Association.

────── (1972) *Britain and the Western Seaways*, London, Thames and Hudson.

Boyd, J. I. C. (1977) *The Isle of Man Railway* (fourth edition), Blandford, Oakwood Press.

Boyd-Dawkins, W. (1902) 'The red sandstone rocks of Peel, Isle of Man', *Geological Society of London*

Journal, **58**, pp. 633–46.
Bradley, R. (1978) *The Prehistoric Settlement of Britain*, London, Routledge and Kegan Paul.
Bradley, R. S. (1985) *Quaternary Palaeoclimatology*, Boston, Allen and Unwin.
Breeze, D. J. (1982) *The Northern Frontiers of Roman Britain*, London, Batsford.
Bregazzi, J. C. (1964) 'Douglas—past, present and future', *Proceedings, Isle of Man Natural History and Antiquarian Society*, **7**, pp. 26–34.
Broderick, G. (1978) *Chronica Regum Mannie et Insularum*, Belfast, Manx Museum and National Trust.
Brown, J. C. (1951) *Coast Erosion*, Report to the Isle of Man Harbour Commissioners, Douglas, Isle of Man Harbour Commissioners.
——— (1954) 'The story of Douglas harbour', *Proceedings, Isle of Man Natural History and Antiquarian Society*, **5**, pp. 350–57.
Brown, P. E., Miller, J. A., and Grasty, R. L. (1968) 'Isotopic ages of late Caledonian granitic intrusions in the British Isles', *Proceedings of the Yorkshire Geological Society*, **36**, pp. 261–76.
———, Soper, N. J., and York, D. (1965) 'Potassium-argon age pattern of the British caledonides', *Proceedings of the Yorkshire Geological Society*, **35**, pp. 103–38.
Brown, R. A. (1986) *Silloth and Douglas*, Douglas, Manx Transport Review.
Bruce, J. R. (1968) *Manx Archaeological Survey, 6th Report*, Douglas, Manx Museum and National Trust.
——— (1974) 'Chapel Hill, a prehistoric, early Christian and Viking site at Balladoole, Kirk Arbory, Isle of Man', *Proceedings, Isle of Man Natural History and Antiquarian Society*, **7**, pp. 632–65.
——— and Cubbon, W. (1930) 'Cronk yn How. An early Christian and Viking site, at Lezayre, Isle of Man', *Archaeologia Cambrensis*, **85**, pp. 267–308.
——— Megaw, B. R. S., and Megaw, E. M. (1947) 'A neolithic site at Ronaldsway, Isle of Man', *Proceedings of the Prehistoric Society*, **12**, pp. 139–60.

Caine, P. W. (1927) 'Brickmaking in the Isle of Man', *Isle of Man Examiner*, 14 January.
Central Statistical Office (1979) *Social Trends*, **9**, London, Her Majesty's Stationery Office.
——— (1987) *Social Trends*, **17**, London, Her Majesty's Stationary Office.
Chappell, C. (1986) *Island of Barbed Wire*, London, Corgi.
Charlesworth, J. K. (1957) *The Quaternary Era*, Vol. 2, London, Edward Arnold.
Colhoun, E. A. and Synge, F. M. (1980) 'The cirque moraines at Lough Nahanagan, County Wicklow, Ireland', *Proceedings of the Royal Irish Academy*, **80B**, pp. 25–45.
Committee of Tynwald (1897) *Report on Coast Erosion*, Minutes of a meeting, Douglas, Isle of Man Government.
——— (1924) *Report on Coast Erosion*, Minutes of a meeting, Douglas, Isle of Man Government.
Coope, G. R. (1977) 'Fossil coleopteron assemblages as sensitive indicators of climatic changes during the Devensian (last) cold stage', *Philosophical Transactions of the Royal Society of London*, **208B**, pp. 313–40.
——— and Brophy, J. A. (1972) 'Lateglacial environmental changes indicated by a coleopteron succession from North Wales', *Boreas*, **1**, pp. 97–142.
——— and Joachim, M. J. (1980) 'Lateglacial environmental changes interpreted from fossil coleoptera from St. Bees, Cumbria, north-west England', in *Studies in the Lateglacial of North-West Europe*, edited by J. J. Lowe, J. M. Gray, and J. E. Robinson, Oxford, Pergamon Press.
——— and Pennington, W. (1977) 'The Windermere Interstadial of the Late Devensian', *Philosophical Transactions of the Royal Society of London*, **280B**, pp. 337–39.
Cooper, C., Latham, J., and Westlake, J. (1987) *A Five Year Strategy for Tourism on the Isle of Man*, Guildford, University of Surrey.
Cornwell, J. D. (1972) 'A gravity survey of the Isle of Man', *Proceedings of the Yorkshire Geological Society*, **39**, pp. 93–106.
Corran, H. S. (1977) *The Isle of Man*, Newton Abbot, David and Charles.
Cowley, J. W. (1959) 'The Manx woollen industry through the centuries', *Proceedings, Isle of Man Natural History and Antiquarian Society*, **6**, pp. 39–45.
Cregeen, S. (1978) 'The Ballaharra excavations 1971', in *Man and Environment in the Isle of Man*, edited by P. Davey, British Archaeological Reports, 54, Oxford.
Crellin, J. (1938) 'Description of Kirk Michael in 1774', *Journal of the Manx Museum*, **4**, pp. 35–36.
Croll, J. (1864) 'On the physical cause of the change of climate during geological epochs', *Philosophical Magazine*, **28**, pp. 121–37.
Crowley, S. F. (1985) 'Lithostratigraphy of the Peel sandstones, Isle of Man', *Mercian Geologist*, **10**, pp. 73–76.
Cubbon, A. M. (1957) 'The ice age in the Isle of Man', *Proceedings, Isle of Man Natural History and Antiquarian Society*, **5**, pp. 499–512.
——— (1964) 'Clay Head cooking place sites, the excavation of a group of cairns', *Proceedings, Isle of Man Natural History and Antiquarian Society*, **6**, pp. 566–96.

――――― (1972–74) 'Clay Head cooking sites: an additional note on their dating', *Proceedings, Isle of Man Natural History and Antiquarian Society*, **8**, pp. 51–54.
Cumming, J. G. (1846) 'On the geology of the Isle of Man', *Quarterly Journal of the Geological Society of London*, **2**, pp. 317–48.
――――― (1847) 'On the geology of the Calf of Man', *Proceedings of the Geological Society*, **3**, pp. 179–85.
――――― (1848) *The Isle of Man its History, Physical, Ecclesiastical, Civil and Legendary*, London, John Van Voorst.
――――― (1853) 'On the superior limits of the glacial deposits of the Isle of Man', *Quarterly Journal of the Geological Society of London*, **10**, pp. 211–32.
Cunliffe, B. W. (1974) *Iron Age Communities in Britain*, London, Routledge & Kegan Paul.
Cuvier, G. (1812) *Recherches sur les Ossemens Fossiles de Quadrupèdes, ou l'on Rétablit les Caractères de Plusieurs Espèces d'Animaux que les Révolutions de Globe Paroissent Avoir Detruites*, 4, Paris, Deterville.

Dackombe, R. V., and Thomas, G. S. P. (1985) *Field Guide to the Quaternary of the Isle of Man*, Cambridge, Quaternary Research Association.
Davey, P. (1978) *Man and Environment in the Isle of Man*, British Archaeological Reports, 54, Oxford.
――――― (1978) 'Bronze age metalwork from the Isle of Man', in *Man and Environment in the Isle of Man*, edited by P. Davey, Oxford, British Archaeological Reports, 54, Oxford.
Darwin, C. (1842) 'Notes on the effects produced by the ancient glaciers of Caernarvonshire, and on the boulders transported by floating ice', *Philosophical Magazine*, **21**, pp. 180–88.
――――― (1848) 'On the transport of erratic blocks from a lower to a higher level', *Quarterly Journal of the Geological Society of London*, **4**, pp. 315–29.
Department of Local Government and the Environment (1987) *House Condition Survey 1984*, Douglas, Department of Local Government and the Environment.
Departments of Local Government and the Environment and Tourism and Transport (1987) *A Strategy for the Involvement of the Isle of Man Government in the Provision of Sports and Leisure Facilities in Douglas*, Douglas, Tynwald.
Dickson, C. A., Dickson, J. H. P., and Mitchell, G. F. (1970) 'The late Weichselian flora of the Isle of Man', *Philosophical Transactions of the Royal Society of London*, B, **258**, pp. 31–79.
Dickson, J. A. D. (1967) *The Structure and Sedimentation of the Carboniferous Rocks of the Castletown Area, Isle of Man*, London, University of London (unpublished doctoral thesis).

――――― Ford, T. D., and Swift, A. (1987) 'The stratigraphy of the carboniferous rocks around Castletown, Isle of Man', *Proceedings of the Yorkshire Geological Society*, **46**, pp. 203–29.
Downie, C., and Ford, T. D. (1966) 'Microfossils from the Manx slate series', *Proceedings of the Yorkshire Geological Society*, **35**, pp. 307–22.

Economic Adviser's Office (1986) *Isle of Man Passenger Survey 1985*, Douglas, Isle of Man Government.
Eliot-Hurst, M. E. (1985) 'Geography has neither existence nor future', in *The Future of Geography*, edited by R. J. Johnston, London, Methuen.
Erdtmann, G. (1925) 'Pollen statistics from the Curragh and Ballaugh, Isle of Man', *Proceedings of the Liverpool Geological Society*, **14**, pp. 158–63.
Eyles, C. H., and Eyles, N. (1984) 'Glaciomarine sediments of the Isle of Man as a key to Late Pleistocene stratigraphic investigations in the Irish Sea basin', *Geology*, **12**, pp. 359–64.
―――――, ―――――, and McCabe, M. A. (1985) 'Glaciomarine sediments of the Isle of Man as a key to Late Pleistocene stratigraphic investigations in the Irish Sea Basin: Reply', *Geology*, **13**, pp. 446–47.
Eyles, N. (1988) 'Sedimentation patterns on glaciated continental shelves—implications for onshore/offshore correlations (abstract of a paper presented at the QRA/MSG Meeting in Keyworth, January 1988)', *Quaternary Newsletter*, **54**, pp. 33–34.

Faragher, T. (1986) *The Official Isle of Man Year Book*, Douglas, Motor in Mann.
――――― (1988) *The Official Isle of Man Year Book*, Douglas, Motor in Mann.
Feltham, J. (1798) *A Tour Through the Isle of Mann*, Bath, R. Cruttwell.
Fletcher, M., and Reilly, P. (1988) 'Viking settlers in the Isle of Man: some simulation experiments', in *Computer and Quantitative Methods in Archaeology*, edited by C. L. N. Ruggles and S. P. Q. Rhatz, Oxford, British Archaeological Reports.
Fleure, H. J., and Dunlop, M. (1942) 'Glendarragh circle and alignments, the Braaid, Isle of Man', *Antiquaries Journal*, **11**, pp. 39–53.
――――― and Neely, G. J. H. (1936) 'Cashtal yn Ard, Isle of Man', *Antiquaries Journal*, **16**, pp. 373–95.
Ford, T. D. (1971) 'Slump structures in the Peel sandstone series, Isle of Man', *Proceedings, Isle of Man Natural History and Antiquarian Society*, **7**, pp. 440–48.
――――― (1987) 'Geological excursion guide, 4: the Isle of Man', *Geology Today*, **3**, pp. 64–69.
Freke, D. J. (1983) *Peel Castle Excavations: First Interim Report*, Douglas, St Patrick's Isle (IOM)

Archaeological Trust.

────── (1985a) *Peel Castle Excavations: Second Interim Report*, Douglas, St Patrick's Isle (IOM) Archaeological Trust.

────── (1985b) *Peel Castle Excavations: Third Interim Report*, Douglas, St Patrick's Isle (IOM) Archaeological Trust.

────── (1987) *Peel Castle Excavations: Fourth Interim Report*, Douglas, St Patrick's Isle (IOM) Archaeological Trust.

────── (Forthcoming) *Peel Castle Excavations 1982–1987: Final Report*, Liverpool, St Patrick's Isle (IOM) Archaeological Trust.

Gans, H. (1979) 'Symbolic enthnicity', *Ethnic and Racial Studies*, **2**, pp. 1–20.

Garrad, L. S. (1972) 'The wildlife of the Ayres', *Peregrine*, **4**, pp. 21–41.

────── (1978a) 'Evidence for the history of the vertebrate fauna of the Isle of Man', in *Man and Environment in the Isle of Man*, edited by P. Davey, British Archaeological Reports, 54, Oxford.

────── (1978b) 'Stone implements and their users in the Isle of Man', in *Man and Environment in the Isle of Man*, edited by P. Davey, British Archaeological Reports, 54, Oxford.

────── (1978c) 'Excavation of a lintel grave cemetery at Glentraugh, Santon, Isle of Man', in *Man and Environment in the Isle of Man*, edited by P. Davey, British Archaeological Reports, 54, Oxford.

────── (1985) *A History of Manx Gardens*, Douglas, Collector's Choice.

────── and Scatchard, W. J. (1973) 'Rope making in the Isle of Man', *The Local Historian*, **10**, pp. 332–33.

Gatrell, A. C. (1987) 'Population distribution on the Isle of Man: the 1986 census', *Geography*, **72**, p. 252–54.

Geikie, J. (1874–94) *The Great Ice Age*, London, W. Isbister (first edition, 1874), Dalby, Isbister and Co. (second edition, 1877), and Stanford (third edition, 1894).

Gelling, M. (1978) 'Norse and Gaelic in medieval Man: the place name element', in *Man and Environment in the Isle of Man*, edited by P. Davey, British Archaeological Reports, 54, Oxford.

────── (1990) 'The place name "Peel"', in *Excavations at Peel Castle 1982–87: Final Report*, edited by D. Freke, Liverpool, St Patrick's Isle (IOM) Archaeological Trust.

Gelling, P. S. (1952) 'Excavation of a promontory fort at Port Grenaugh, Santon, Isle of Man', *Proceedings, Isle of Man Natural History and Antiquarian Society*, **5**, p. 307–15.

────── (1957) 'Excavation of a promontory fort at Cass ny Hawin, Malew, Isle of Man', *Proceedings, Isle of Man Natural History and Antiquarian Society*, **6**, pp. 28–38.

────── (1958) 'Close ny Chollagh, an iron age fort at Scarlet, Isle of Man', *Proceedings, Prehistoric Society*, **24**, pp. 85–100.

────── (1962–63) 'Medieval shielings in the Isle of Man', *Medieval Archaeology*, **6–7**, pp. 156–72.

────── (1964) 'The Braaid site', *Journal of the Manx Museum*, **6**, pp. 201–05.

────── (1966) 'Excavation of a promontory fort on Ballanicholas, Kirk Marown, Isle of Man', *Proceedings, Isle of Man Natural History and Antiquarian Society*, **7**, pp. 181–91.

────── (1970a) 'The South Barrule hillfort reconsidered', *Journal of the Manx Museum*, **7**, pp. 145–47.

────── (1970b) 'A Norse homestead near Doarlish Cashen', *Medieval Archaeology*, **14**, p. 74.

────── (1977) 'Celtic continuity in the Isle of Man', in *Studies in Celtic Survival*, edited by L. Laing, Oxford, British Archaeological Reports.

────── (1978) 'The Iron Age', in *Man and Environment in the Isle of Man*, edited by P. Davey, Oxford, British Archaeological Reports.

Gent, H. (1983) 'Centralised storage in late prehistoric Britain', *Proceedings, Prehistoric Society*, **49**, pp. 243–67.

Gill, E. L. (1903) 'Keisley Limestone pebbles from the Isle of Man', *Quarterly Journal of the Geological Society of London*, **59**, pp. 307–10.

Gillot, J. E. (1954) 'Breccias in the Manx slate: their origin and stratigraphic relations', *Liverpool and Manchester Geological Journal*, **1**, pp. 370–80.

────── (1955) 'Metamorphism of the Manx Slates', *Geological Journal*, **92**, pp. 141–53.

────── (1956) 'Structural geology of the Manx Slates', *Geological Magazine*, **93**, pp. 301–13.

Glazer, D., and Moynihan, P. (1975) *Ethnicity: Theory and Experience*, Cambridge, Massachusetts, Harvard University Press.

Goodwyn, A. (1986) *Is This Any Way to Run a Shipping Line?*, Lancaster, Manx Electric Railway Society.

Gould, S. J. (1974) 'The origin and function of 'bizarre' structures: antler size and skull size in the 'Irish elk' *Megaloceras giganteus*', *Evolution*, **28**, pp. 191–220.

Gray, J. M. (1982) 'The last glaciers (Loch Lomond advance) in Snowdonia, North Wales', *Geological Journal*, **17**, pp. 111–33.

Gregory, J. M. (1920) 'The red rocks of a deep bore at the north end of the Isle of Man', *Transactions of the Institute of Mining Engineers of Scotland*, **59**, pp. 156–68.

Gresswell, R. K. (1966) *Report of Coastal Conditions Around Ayre, Isle of Man, with Special Reference to Advisability of Winning Gravel at and near the Point of Ayre*, Report to Messrs B. S. G. (Aggregates), Ramsey.

Haines-Young, R. H., and Petch, J. R. (1985) *Physical Geography: Its Nature and Methods*, London, Harper and Row.

Hallam, W. B. (1980) 'Fifteen decades of service to Man', *Sea Breezes*, **54**, pp. 417–19.

Harkness, R., and Nicholson, H. A. (1866) 'On the lower Silurian rocks of the Isle of Man', *Quarterly Journal of the Geological Society of London*, **22**, p. 488.

Harper, C. T. (1966) 'K-Ar ages of slates from the S caledonides of the British Isles', *Nature*, **212**, pp. 1339–41.

Harrison, A. (1968) 'Economic opportunities, 1700', *Journal of Manx Museum*, **7**, pp. 81–83.

Harrison, S. (editor) (1986) *One Hundred Years of Heritage. The Work of the Manx Museum and National Trust*, Douglas, Manx Museum and National Trust.

Haynes, J., Kiteley, R. J., Whatley, R. C. and Wilks, P. J. (1977) 'Microfaunas, microfloras and the environmental stratigraphy of the late glacial and holocene in Cardigan Bay', *Geological Journal*, **12**, pp. 129–58.

Hechter, M. (1975) *Internal Colonialism. The Celtic Fringe in British National Development*, London, Routledge & Kegan Paul.

Hendry, R. P. (1979) *Narrow Gauge Story*, Rugby, Hillside Publishing.

——— and Hendry, R. P. (1978) *Manx Electric Railway Album*, Rugby, Hillside Publishing.

——— and Hendry, R. P. (1980) *Manx Northern Railway*, Rugby, Hillside Publishing.

Hendy, J. (1980–86) 'Ferry scene', *Sea Breezes*, pp. 54–60.

Henshall, A. S. (1978) 'Manx megaliths again: an attempt at a structural analysis', in *Man and Environment in the Isle of Man*, edited by P. Davey, British Archaeological Reports, 54, Oxford.

Herbert, A. (1909) *The Isle of Man*, London, The Bodley Head.

Hibbert, S. (1825) 'Notice of the remains of an animal resembling the Scandinavian Elk, recently discovered in the Isle of Man, etc.', *Edinburgh Journal of Science*, **3**, p. 129.

Hobson, B. (1981) 'The igneous rocks of the Isle of Man', *Quarterly Journal of the Geological Society of London*, **47**, pp. 432–50.

Holgate, B. (1987) *Pagan Lady of Peel*, Douglas, St Patrick's Isle (IOM) Archaeological Trust.

Holt-Jensen, A. (1988) *Geography: History and Concepts: a Student's Guide*, second edition, London, Paul Chapman.

Home Affairs Board (1984) *Report on the Examination into the Policy and Financing of Manx Radio*, Douglas, Tynwald.

Horne, J. (1874) 'A sketch of the geology of the Isle of Man', *Transactions, Edinburgh Geology Society*, **2**, pp. 323–47.

Howie, G. W. (1929) 'Agriculture: present problems and some recent developments', *Journal of the Manx Museum*, **19**, pp. 147–48.

Huntley, B., and Birks, H. H. (1983) *An Atlas of the Past and Present Pollen Maps for Europe: 0–13000 years ago*, Cambridge, Cambridge University Press.

Industrial Advisory Council (1961) *First Report*, Douglas, Isle of Man Government.

Ineson, P. R., and Mitchell, J. G. (1979) 'K-Ar ages from ore deposits and related rocks of the Isle of Man', *Geological Magazine*, **116**, pp. 117–28.

Isle of Man Arts Council (1985) *Report for Period Ending 31 December 1985*, Douglas, Tynwald.

Isle of Man Board of Agriculture and Fisheries (1924–87) *Knockaloe Experimental Farm, Peel: Guides*, Douglas, Isle of Man Government.

——— (1975) *Report of the Commission of Enquiry into Agriculture and Horticulture*, Douglas, Isle of Man Government.

——— (1985) *Fisheries Statistics*, Douglas, Isle of Man Government.

——— (1986) *Annual Reports on Agricultural Returns*, Douglas, Isle of Man Government.

Isle of Man Department of Agriculture, Fisheries and Forestry (1987a) *Annual Report on Agricultural Returns*, Douglas, Isle of Man Government.

——— (1987b) *Statistical Returns*, Douglas, Isle of Man Government.

Isle of Man Department of Highways, Ports and Properties (1981–87) *Statistical Returns*, Douglas, Department of Highways, Ports and Properties.

Isle of Man Forestry, Mines and Lands Board (1984) *Report for the Year Ended 31 March 1984*, Douglas, Tynwald.

——— (1985) *Report for the Year Ended 31 March 1985*, Douglas, Tynwald.

——— (1986) *Report for the Year Ended 31 March 1986*, Douglas, Tynwald.

Isle of Man Government (1939) *Report of the Agricultural Commission Appointed by the Lieutenant Governor*, Douglas, Isle of Man Government.

——— (1968) *Interim Report of the Commission on the Imbalance of Population*, Douglas, Isle of Man Government.

——— (1970) *Tourist Premises (Compensation for Tenants' Improvements) Act 1970*, Chapter 26, Douglas, Isle of Man Government.

——— (1973) *Report of the Select Committee Appointed to Investigate the Effect of Population Changes on the Island Economy*, Douglas, Isle of Man Government.

——— (1974) *Tourist Premises Improvement Act 1974*, Chapter 24, Douglas, Isle of Man Government.

——— (1975) *Tourist Act 1975*, Chapter 19, Douglas, Isle of Man Government.

——— (1977a) *Tourist Premises (Provision and Improvement) Act 1977*, Chapter 8, Douglas, Isle of Man Government.

——— (1977b) *Tourist (General) Regulations 1977*, Circular 37/77, Douglas, Isle of Man Government.

——— (1980) *Report of the Select Committee of Tynwald on Population Growth and Immigration*, Douglas, Isle of Man Government.

——— (1982a) *Policy Planning Programme*, Douglas, Isle of Man Government.

——— (1982b) *Accounts*, Douglas, Isle of Man Government.

——— (1983) *Report of the Commission of Inquiry into the Fishing Industry*, Douglas, Isle of Man Government.

——— (1984) *Tourist Premises Development Scheme 1984*, Douglas, Isle of Man Government.

——— (1985a) *Second Report of the Standing Committee of Tynwald on Population Growth and the Control of Immigration*, Douglas, Isle of Man Government.

——— (1985b) *Local Government Act 1985*, Chapter 24, Douglas, Isle of Man Government.

——— (1985c) *House of Keys Select Committee Report on the Sealink-Steam Packet Merger*, Douglas, Isle of Man Government.

——— (1985d) *Tourist (General) (Amendment) Regulations 1985*, Douglas, Isle of Man Government.

——— (1985e) *Tourist Premises Development (Amendment) Scheme 1985*, Douglas, Isle of Man Government.

——— (1987) *The Development of a Prosperous and Caring Society*, Douglas, Isle of Man Government.

Isle of Man Health Service Advisory Council (1985) *Report for the Year Ended 31 March 1985*, Douglas, Isle of Man Government.

Isle of Man Local Government Board (1984) *Report to Tynwald on Housing*, Douglas, Isle of Man Local Government Board.

——— (1985) *Report for the Year Ended 31 March 1984*, Douglas, Isle of Man Local Government Board.

——— (1986) *Report for the Year Ended 31 March 1986*, Douglas, Isle of Man Local Government Board.

Isle of Man Steam Packet Co. Ltd. (1930) *Centenary Brochure*, Douglas, Isle of Man Steam Packet.

——— (1939) *Isle of Man for Holidays. Summer Services and Sea Excursions*, Douglas, Isle of Man Steam Packet.

——— (1960–1986) *Annual Reports and Accounts*, Douglas, Isle of Man Steam Packet.

Isle of Man Times (1895) *List Of Farmhouses and Country Lodgings, Boarding Houses and Hotels in the Isle of Man*, Douglas, J. Brown and Sons.

Isle of Man Tourist Board (1984) *Financial Aid for Tourist Accommodation*, Douglas, Isle of Man Tourist Board.

Isle of Man Weekly Times (1964) 'The story of the Laxey Glen flour mill', 6 March.

Jardine, W. G. (1975) 'Chronology of Holocene marine transgression and regression in south-western Scotland', *Boreas*, **4**, pp. 173–96.

Joachim, K. (1978) *Late Glacial Coleopteron Assemblages from the West Coast of the Isle of Man*, Birmingham, University of Birmingham (unpublished doctoral thesis).

Johnston, R. J. (1979) *Geography and Geographers*, London, Edward Arnold.

——— (1985a) 'Introduction: exploring the future of geography', in *The Future of Geography*, edited by R. J. Johnston, London, Methuen.

——— (1985b) 'To the ends of the earth', in *The Future of Geography*, edited by R. J. Johnston, London, Methuen.

Jolliffe, I. P. (1979) *Interim Report on Coastal Instability at Ballure Park, Isle of Man*, Report to the Isle of Man Harbour Board and Isle of Man Railways Board.

——— (1981) *An Investigation into Coastal Erosion Problems in the Isle of Man: Causes, Effects, and Remedial Strategies*, Report to the Isle of Man Harbour Board and Isle of Man Government.

Jones, D. H. (1966) 'Manx water mills', *Journal of Manx Museum*, **7**, pp. 11–17.

Jones, W. (1980) *The Nature and Movement of Beach Shingle around the Point of Ayre, Isle of Man*, London, Bedford College (unpublished BSc dissertation).

Kear, B. S. (1976) 'Soils of the Isle of Man', *Proceedings, Isle of Man Natural History and Antiquarian Society*, **8**, pp. 39–50.

Kendall, P. F. (1894) 'On the glacial geology of the Isle of Man', *Yn Lioar Manninagh*, **1**, pp. 397–437.

Kermode, P. M. C. (1930) 'Knock y Doonee', *Proceedings, Isle of Man Natural History and Antiquarian Society*, **3**, pp. 241–46.

——— and Herdman, W. A. (1914) *Manks Antiquities*, Liverpool, Liverpool University Press.

King, C. A. M. (1976) *Northern England*, London, Methuen.

Kinvig, R. H. (1955) 'Manx settlement in the U.S.A.', *Proceedings, Isle of Man Natural History and Antiquarian Society*, **5**, **4**, pp. 1–20.

——— (1975) *The Isle of Man: A Social, Cultural and Political History*, Liverpool, Liverpool University Press.

Kneale, M. (1987) 'Remember the Ramsey salt works', *Streetscene (IOM)*, **1**, **3**, pp. 18–24.

Kniveton, G. N. (1986) *Manx Aviation in War and Peace*, Douglas, Manx Experience.

Kurten, B. (1968) *Pleistocene mammals of Europe*, Chicago, Aldine Publishing Co.

Lacey, P. (1985) *The Independent Bus and Coach Operators of the Newbury Area 1919 to 1932*, Wokingham, P. Lacey.

Laing, L., and Laing, J. (1984–87) 'The Early Christian period settlement at Ronaldsway, Isle of Man: a reappraisal', *Proceedings, Isle of Man Natural History and Antiquarian Society*, **9**, pp. 389–416.

Lambden, W. T. (1953a) 'Early morning observations', *Bus and Coach*, September, p. 321.

——— (1953b) 'Small-sized one-man buses on town services', *Bus and Coach*, September, pp. 307–08.

——— (1958) 'Douglas Corporation's new vehicle. 26 seaters for one-man work', *Bus and Coach*, **30**, pp. 206–07.

——— (1964) *The Manx Transport Systems*, London, Omnibus Society.

Lamplugh, G. W. (1903) *The Geology of the Isle of Man*, Memoir of the Geological Survey of Great Britain, London.

Lewis, H. P. (1930) 'The Avonian succession in the south of the Isle of Man', *Quarterly Journal of the Geological Society of London*, **86**, pp. 234–88.

——— (1934) 'The occurrence of fossiliferous pebbles of Salopian age in the Peel sandstone (Isle of Man)', *Summary of Progress, Geological Survey of Great Britain*, **11**, pp. 91–109.

Longworth, I., and Cherry, J. (1986) *Archaeology in Britain since 1945*, London, British Museum.

Lowe, J. J., and Gray, J. M. (1980) 'The stratigraphic subdivision of the Lateglacial of North-West Europe: discussion', in *Studies in the Late Glacial of North-West Europe*, edited by J. J. Lowe, J. M. Gray, and J. E. Robinson, Oxford, Pergamon Press.

——— and Walker, M. J. C. (1984) *Reconstructing Quaternary Environments*, London, Longman.

McCabe, A. M. (1987) 'Quaternary deposits and glacial stratigraphy in Ireland', *Quaternary Science Reviews*, **6**, pp. 259–99.

——— Haynes, J. R., and Macmillan, N. (1986) 'Late Pleistocene tidewater glaciers and glaciomarine sequences from north County Mayo, Republic of Ireland', *Journal of Quaternary Science*, **1**, pp. 73–84.

——— Dardis, G. F. and Hanvey, P. M. (1987) 'Sedimentation at the margins of a Late Pleistocene ice-lobe terminating in shallow marine environments, Dundalk Bay, eastern Ireland', *Sedimentology*, **34**, pp. 473–93.

Mackay, and Schnellmann, (1963) *Mines and Minerals of the Isle of Man*, Douglas, Isle of Man Government.

MacKinder, H. J. (1902) *Britain and the British Seas*, London, Heinemann.

Manx Archaeological Survey (1911) *Third Report*, Douglas, Isle of Man Government.

Manx Journal of Agriculture (1934) **1**, old series, Douglas.

——— (1946), **1**, new series, Douglas.

Manx Museum and National Trust (1986) *Annual Report for the Year Ended 31 March 1985*, Douglas, Tynwald.

Mathieson, N. (1957a) 'Manx mines during the Atholl period', *Proceedings, Isle of Man Natural History and Antiquarian Society*, **5**, pp. 555–70.

——— (1957b) 'A forgotten factory. Port-e-Chee Cotton Mill 1772–80', *Journal of Manx Museum*, **6**, pp. 25–27 and 62–64.

Megaw, B. R. S. (1938) 'Manx megaliths and their ancestry', *Proceedings, Isle of Man Natural History and Antiquarian Society*, **4**, pp. 219–39.

——— (1938–40) 'The ancient village of Ronaldsway', *Journal of the Manx Museum*, **4**, pp. 181–82.

——— (1978) 'Norseman and native in the Kingdom of the Isles: a re-assessment of the Manx evidence', in *Man and Environment in the Isle of Man*, edited by P. Davey, British Archaeological Reports, 54, Oxford.

Megaw, E. L. (1978) 'The Manx Eary and its significance', in *Man and Environment in the Isle of Man*, edited by P. Davey, British Archaeological Reports, 54, Oxford.

Mitchell, G. F. (1958) 'A Lateglacial deposit near Ballaugh, Isle of Man', *New Phytologist*, **57**, pp. 256–63.

——— (1960) 'The pleistocene history of the Irish Sea', *Advancement of Science*, **17**, pp. 313–25.

——— (1965) 'The quaternary deposits of the Ballaugh and Kirk Michael districts', *Quarterly Journal Geological Society London*, **121**, pp. 358–81.

——— (1972) 'The pleistocene history of the Irish Sea: a second approximation', *Scientific Proceedings, Royal Dublin Society*, **4A**, pp. 181–99.

——— and Parkes, H. M. (1949) 'The giant deer in Ireland', *Proceedings, Royal Irish Academy*, **52B**, pp. 291–314.

Moffat, P. J. (1978) 'The Ronaldsway Culture: a review', in *Man and Environment in the Isle of Man*, edited by P. Davey, British Archaeological Reports, 54, Oxford.

Molyneux, S. G. (1979) 'New evidence for the age of the Manx group, Isle of Man', in *Caledonides of the British Isles Reviewed*, edited by A. L. Harris, C. H. Holland, and B. E. Leake, Special Publication 8, London, Geological Society of London.

Molyneux, T. (1697) 'A discourse concerning the large horns frequently found under ground in Ireland, concluding from them that the great American deer call'd a moose, was formerly common in that island: with remarks on some other things natural to that country', *Philosophical Transactions*, 1697, pp. 489–512.

Moorbath, S. (1962) 'Lead isotope abundance studies on mineral occurrences in the British Isles and their geological significance', *Philosophical Transactions of the Royal Society of London, A*, **254**, pp. 295–360.

Moore, A.W. (1900) *A History of the Isle of Man*, London, Unwin.

Moore, P. D. (1975) 'The origin of blanket mires', *Nature*, **256**, pp. 267–69.

Moore, T. H. (1974) 'Some social aspects of Castletown in the 18th and 19th century', *Proceedings, Isle of Man Natural History and Antiquarian Society*, **7**, p. 686–705.

MORI (1986) *Isle of Man New Business Survey*, London, MORI.

Morris, C. D. (1981) 'Keeill Vael, Druidale, Michael, Isle of Man. Survey and excavation 1979–80: interim report', *Archaeological Reports for 1980*, Durham, Universities of Durham and Newcastle upon Tyne.

────── (1983) 'The survey and excavations at Keeill Vael, Druidale, in their context', in *The Viking Age in the Isle of Man*, edited by C. Fell, P. Foote, G. Campbell and R. Thomson, London, University College London.

Neely, G. J. H. (1940) 'Excavations at Ronaldsway, Isle of Man', *Antiquaries Journal*, **20**, pp. 72–86.

Nockolds, S. R. (1931) 'On the Dhoon (IOM) granite: a study in contamination', *Mineralogical Magazine*, **22**, pp. 494–509.

Norris, S. (1904) *Douglas. The Naples of the North*, Douglas.

PA Consultants (1975) *Economic Survey of the Isle of Man, 1975*, London, PA Consultants.

Page, L. E. (1969) 'Diluvialism and its critics in Great Britain in the early nineteenth century', in *Towards a History of Geology*, edited by C. J. Schneer, Cambridge, Massachusetts, Massachusetts Institute of Technology Press.

Page, R. I. (1983) 'The Manx rune-stone', in *The Viking Age in the Isle of Man*, edited by C. Fell, P. Foote, G. Campbell and R. Thomson, London, University College London.

Pantin, H. M. (1975) 'Quaternary sediments of the north-eastern Irish Sea', *Quaternary Newsletter*, **17**, pp. 7–9.

────── (1977) 'Quaternary sediments of the Northern Irish Sea', in *The Quaternary History of the Irish Sea*, Geological Journal Special Issue 7, edited by C. Kidson, and M. J. Tooley.

────── (1978) *Quaternary Sediments from the North-East Irish Sea: Isle of Man to Cumbria*, Geological Survey of Great Britain, Bulletin, 64.

Pearson, F. K. (1970) *Isle of Man Tramways*, Newton Abbot, David and Charles.

Peat Marwick McLintock and Salford University Business Services Ltd. (1989) *The Economic Implications of 1992*, Consultants' report to Isle of Man Government.

Penny, L. F., Coope, G. R., and Catt, J. A. (1969) 'Age and insect fauna of the Dimlington silts, East Yorkshire', *Nature*, **224**, pp. 65–67.

Phillips, B. A. M. (1967) 'The Post-glacial raised shoreline around the northern plain, Isle of Man', *Northern Universities Geographical Journal*, **8**, pp. 43–50.

PSV Circle (1980) *Fleet History, Trams, Buses and Coaches of the Isle of Man*, PC6, London, PSV Circle and Omnibus Society.

Prentice, R. C. (1984) 'Housing in the Isle of Man', *Housing Review*, **33**, pp. 242–45.

────── (1985a) 'Rural bus services on the Isle of Man', *Swansea Geographer*, **22**, pp. 1–16.

────── (1985b) 'Buses on the Isle of Man—recent changes', *Eastern Counties Omnibus Society (Eastern Transport Collection), Terminus*, **191**, pp. 27–30.

────── (1986a) 'Directions for a housing policy in Wales', in *Housing and Homelessness in Wales*, edited by the Church in Wales, Penarth, Church in Wales Publications.

────── (1986b) 'Public attitudes to housing in Llanelli and the Isle of Man', *Housing Review*, **35**, pp. 186–88.

────── (1986c) *Rural Residents' Leisure Use of Manx National Glens*, St John's, Isle of Man Forestry, Mines and Lands Board.

────── (1988) 'The Manx TT races and residents' views: a case assessment of Doxey's 'irridex'', *Scottish Geographical Magazine*, **104**, pp. 155–60.

────── and Prentice, M. M. (1987) *Urban Residents' Leisure Use of Manx National Glens*, St John's, Department of Agriculture, Fisheries and Forestry.

Qualtrough, J. K. (1969) 'An introduction to Manx watermills', *Proceedings, Isle of Man Natural History and Antiquarian Society*, **7**, pp. 248–63.

Quayle, B. (1794) *A General View of the Agriculture of the Isle of Man*, London, Board of Agriculture.

Quayle, T. (1812) *A General View of the Agriculture of the Isle of Man*, London, Board of Agriculture.

Quilleash, M. (1964) 'The last of the Sulby mills', *Proceedings, Isle of Man Natural History and Antiquarian Society*, **6**, p. 503.

Rackam, O. (1980) *Ancient Woodlands*, London, Edward Arnold.

Radcliffe, J. W. (1971) 'Mines in Maughold', *Proceedings, Isle of Man Natural History and Antiquarian Society*, **7**, pp. 343–55.

Ralfe, P. G. (1924) 'A Manx lead mine in the thirteenth century', *Proceedings, Isle of Man Natural History and Antiquarian Society*, **2**, pp. 414–17.

Relph, E. (1976) *Place and Placelessness*, London, Pion.

Renshaw, J. C. (1976) *Coast Erosion in the North of the Isle of Man*, London, Bedford College (unpublished BSc dissertation).

Reynolds, P. J. (1987) *Ancient Farming*, Aylsborough, Shire Publications.

Robertson, D. (1794) *A Tour Through the Isle of Man*, London.

Robinson, V. (1987) 'Race, space and place: the geographer's contribution to the study of UK ethnic relations, 1957–1987', *New Community*, **14**, pp. 186–98.

Rothwell, A. E. (1919) 'Shipbuilding in the Isle of Man', *Manx Quarterly*, **22**, pp. 171–74.

Rowlands, B. M. (1971) 'Radiocarbon evidence of the age of an Irish Sea glaciation in the Vale of Clwyd', *Nature*, **230**, pp. 9–11.

Russell, G. (1978) 'The structure and vegetation history of the Manx hill peats', in *Man and Environment in the Isle of Man*, edited by P. Davey, British Archaeological Reports, 54, Oxford.

Shotton, F. W., and Williams, R. E. G. (1971) 'Birmingham University radiocarbon dates VI', *Radiocarbon*, **13**, pp. 141–56.

——— (1973) 'Birmingham University radiocarbon dates VII', *Radiocarbon*, **15**, pp. 451–68.

Simpson, A. (1961) *The stratigraphy and tectonics of the Manx slate series, Isle of Man*, London, University of London (unpublished doctoral thesis).

——— (1963) 'The stratigraphy and tectonics of the Manx Slate Series, Isle of Man', *Quarterly Journal of the Geological Society of London*, **119**, pp. 367–400.

——— (1964a) 'The metamorphism of the Manx Slate Series, Isle of Man', *Geological Magazine*, **101**, pp. 20–36.

——— (1964b) 'Deformed acid intrusions in the Manx Slate Series, Isle of Man', *Geological Journal*, **4**, pp. 189–206.

——— (1965) 'The syntectonic Foxdale-Archallagan granite and its metamorphic aureole', *Geological Journal*, **4**, pp. 415–34.

Sissons, J. B. (1976) *The Geomorphology of the British Isles: Scotland*, London, Methuen.

——— (1980) 'The Loch Lomond advance in the Lake District, northern England', *Transactions, Royal Society Edinburgh: Earth Sciences*, **71**, pp. 13–27.

Sizer, C. A. (1962) 'Remains of the Irish giant deer in Leicestershire', *Leicester Literary and Philosophical Society*, **56**, pp. 1–4.

Sizer, J., and Kelly, A. (1987) 'One third of housing in Isle of Man unsatisfactory', *Environmental Health News*, **2**, 23, p. 3.

Smart, R. (1986) 'The flints', in *The Peel Castle Excavations: the Half Moon Battery*, edited by R. H. White, Douglas, St Patrick's Isle (IOM) Archaeological Trust.

Smith, A. G. (1984) 'Newferry and the Boreal-Atlantic transition', *New Phytologist*, **98**, pp. 35–55.

——— Grigson, C., Hillman, G., and Tooley, M. J. (1981) 'The Neolithic', in *The Environment in British Pre-History*, edited by I. G. Simmons and M. J. Tooley, London, Duckworth.

Smith, B. (1927) 'On the carboniferous limestone series of the northern part of the Isle of Man', *Geological Survey of Great Britain, Summary of Progress for 1926*, pp. 108–18.

——— (1930) 'Borings through the glacial drifts of the northern plain of the Isle of Man, *Geological Survey of Great Britain, Summary of Progress*, **3**, pp. 14–23.

Smith, M. G., and Pilcher, J. R. (1973) 'Radiocarbon dates and vegetational history of the British Isles', *New Phytologist*, **72**, pp. 903–14.

Stenning, E. H. (1928) 'Discovery of a deposit of bones in the raised beach at Castletown', *Proceedings, Isle of Man Natural History and Antiquarian Society*, **3**, pp. 234–38.

——— (1958) *Portrait of the Isle of Man*, London, Robert Hale.

Stephens, N., Creighton, J. R., and Hannon, M. A. (1975) 'The late pleistocene period in north-eastern Ireland: an assessment', *Irish Geographer*, **8**, pp. 1–23.

——— and McCabe, A. M. (1971) 'Late pleistocene ice movements and patterns of late and post-glacial shorelines on the coast of Ulster, Ireland', in *The Quaternary History of the Irish Sea*, Geological Journal Special Issue 7, edited by C. Kidson and M. J. Tooley.

Stuart, A. J. (1977) 'The vertebrates of the last cold stage in Britain and Ireland', *Philosophical Transactions, Royal Society London*, **B280**, pp. 295–312.

Synge, F. M. (1977) 'Records of sea levels during the Late Devensian', *Philosophical Transactions, Royal Society of London*, **B280**, pp. 211–28.

——— (1979) 'Quaternary glaciation in Ireland', *Quaternary Research Association (GB) News*, **28**, pp. 1–18.

Taylor, P. (1985) 'The value of a geographical perspective', in *The Future of Geography*, edited by R. J. Johnston, London, Methuen.

Temple, P. H. (1956) *Some Aspects of the Geomorphology of the Isle of Man*, Liverpool, University of Liverpool (unpublished master's thesis).

Thomas, G. S., and Gershuny, J. I. (1985) *Countryside Recreation in Scottish Daily Life*, Scottish Leisure Survey report, 3, Perth, Countryside Commission for Scotland.

Thomas, G. S. P. (1973) Letter to Harbour Board, 21 May 1973 (unpublished).

—— (1976) *The Quaternary Stratigraphy of the Isle of Man*, Cardiff, University of Wales (unpublished doctoral thesis).

—— (1977) 'The Quaternary of the Isle of Man', in *The Quaternary History of the Irish Sea*, Geological Journal Special Issue, 7, edited by C. Kidson and M. J. Tooley.

—— (1984) 'On the glacio-dynamic structure of the Bride Moraine, Isle of Man', *Boreas*, **13**, pp. 355–64.

—— (1985) 'The Quaternary of the northern Irish Sea basin', in *The Geomorphology of Northwest England*, edited by R. H. Johnson, Manchester, Manchester University Press.

—— Connaughton, M., and Dackombe, R. V. (1985) 'Facies variation in a Late Pleistocene supraglacial outwash sandur from the Isle of Man', *Geological Journal*, **20**, pp. 193–213.

—— and Dackombe, R. V. (1985) 'Comments on glaciomarine sediments of the Isle of Man as a key to late pleistocene stratigraphic investigations in the Irish Sea Basin', *Geology*, **13**, pp. 445–46.

Tiddeman, R. H. (1872) 'On the evidence for the ice sheet in north Lancashire and adjacent parts of Yorkshire and Westmoreland', *Quarterly Journal of the Geological Society*, **28**, pp. 471–89.

Tinsley, H., and Grigson, C. (1981) 'The bronze age', in *The Environment in British Prehistory*, edited by I. G. Simmons and M. J. Tooley, London, Duckworth.

Tooley, M. J. (1978) 'Flandrian sea-level changes and vegetational history of the Isle of Man: a review', in *Man and Environment in the Isle of Man*, edited by P. Davey, British Archaeological Reports, 54, Oxford.

Tourist Industry Development Commission (1970) *Report*, Douglas, Isle of Man Government.

Tourist Industry Liaison Committee (1982) *Report and Recommendations Concerning Abrogation of the Common Purse Agreement*, Douglas, Tourist Industry Liaison Committee.

Tynwald Report (1982) *Report to Tynwald on the Jolliffe Report of 1981*, Douglas, Isle of Man Government.

Tynwald Select Committee on the Common Purse (1986) The Customs and Exise Agreement 1979, Douglas, Isle of Man Government.

Walker, M. J. C. (1980) 'Lateglacial history of the Brecon Beacons, South Wales', *Nature*, **287**, pp. 133–35.

Wallace, (1972) 'A breeding bird census of the Ayres 1965', *Peregrine*, **4**, pp. 18–21.

Ward, J. C. (1880) 'Notes on the geology of the Isle of Man', *Geological Magazine*, **7**, pp. 1–9.

Watts, W. A. (1977) 'The Late Devensian vegetation of Ireland', *Philosophical Transactions Royal Society London*, **B280**, pp. 273–93.

Watson, E. (1971) 'Remains of pingos in Wales and the Isle of Man', *Geological Journal*, **7**, pp. 381–87.

Webster, J. H. (1988) *Isle of Man Digest of Economic and Social Statistics 1988*, Douglas, Isle of Man Government.

—— and Dawson, W. (1984) *Isle of Man Digest of Economic and Social Statistics 1985*, Douglas, Isle of Man Government.

Williams, B. A. (1960) *Report of the Isle of Man Industrial Survey*, Douglas, Isle of Man Government.

Williamson, K. (1938–39) 'Paper making in Man', *Journal of the Manx Museum*, **4**, pp. 41–42, 126–29, and 147.

Wingfield, R. T. R. (1987) 'Giant sand waves and relict periglacial features on the sea bed west of Anglesey', *Proceedings, Geologists Association*, **98**, pp. 400–04.

Woodman, P. (1978) 'A re-appraisal of the Manx mesolithic', in *Man and Environment in the Isle of Man*, edited by P. Davey, British Archaeological Reports, 54, Oxford.

Woods, G. (1811) *An account of the past and present state of the Isle of Man—including a sketch of its mineralogy*, London.

Wright, J. (1902) 'The foraminifera of the Pleistocene clay, Shellag', *Yn Lioar Manninagh*, **3**, pp. 627–29.

—— and Reade, T. M. (1906) 'The Pleistocene clays and sands of the Isle of Man', *Proceedings, Liverpool Geological Society*, **17**, pp. 103–17.

Wright J. E., Hull, J. M., McQuillin, R., and Arnold, S. E. (1971) *Irish Sea Investigations*, Institute of Geological Sciences report 71/19.

Wright, M. D. (1980–82) 'Excavation at Peel Castle, 1947', *Proceedings, Isle of Man Natural History and Antiquarian Society*, **9**, pp. 21–57.

GEOGRAPHICAL INDEX

Andreas 71, 73, 183, 188, 223, 226
Arbory 73, 76, 146, 155
Archallagan 217
Ayres 16, 65, 66–68, 84, 85, 86–88, 171, 173

Ballabont 19
Ballacagen 72, 73
Ballacain 86
Ballacoarey 224, 226
Ballachurry 85, 86
Ballacorkish 222
Balladoole 33, 38, 70, 76
Ballaghenie 68
Ballaglas 189
Ballakinnay 68, 85, 87, 90
Ballaleigh 14
Ballamooar 85
Ballanard 224
Ballanorris 73
Ballasalla 14, 17, 34, 35, 47, 189, 195, 224, 227
Ballaugh 17, 26, 44, 56, 59, 60, 62, 65, 71, 76, 82, 90, 91, 139, 146, 153, 155
Ballaugh Curraghs 16, 17, 57, 59, 65, 66, 74, 76, 84, 85, 90–98, 173, 189, 207, 272
Ballawillin 224
Ballure 41, 43, 44, 49, 50, 57, 82, 83, 189
Bay Fine 76
Bay ny Carrikey 57, 70
Beary Mountain 10
Beinn-y-Phott 10, 69, 72, 189
Billown 33, 35, 38, 85, 224
Bishops Court Glen 171
Block Eary 191
Blue Point 16, 57, 68, 87
Braaid 14, 15
Bradda Head 28, 221, 222
Bradda Hill 11, 13
Braddan 71, 74, 146, 167, 229
Braust 73, 86
Bride 16, 71, 80, 87, 146, 153, 155, 198, 226
Bride Hills 16, 42, 45, 48, 52, 65, 66, 68

Calf of Man 10, 13, 76, 85, 86, 89–90, 173, 221, 272
Carrick y Voddey 76
Cashtal yn Ard 71
Cass-ny-Hawin 32, 70, 85
Castletown 8, 17, 18, 20, 22, 34, 44, 76, 88, 136–38, 142, 143, 153, 155, 166, 174, 176, 177, 181–83, 189, 192, 195, 215, 224, 228
Castleward 73
Central valley 15–16, 48, 85, 93
Chapel Hill 34, 70, 76
Clagh Ouyr 10, 12
Close e Kewin 85
Close ny Chollagh 33–36, 73, 76
Close y Garey 75
Colby 17
Colden 69, 85
Cranstal 79, 80, 81, 82
Creggans 17
Cregneish 5, 24, 173, 176, 254
Crogga River 14, 15
Cronk Darragh 73
Cronk ny Arrey Laa 10
Cronk ny Shiannagh 76
Cronk Sumark 73
Cronk Urleigh 14
Cronkbourne 141, 142, 167
Cross Welkin Hill 34
Curragh Aspic 94
Curragh Beg 93
Curragh e Cowle 85, 90
Curraghglass 16, 93

Dalby 14, 85
Dalby Mountain 10
Derbyhaven 32, 33, 88, 189, 215, 224
Dhoo-Glass River 14, 134
Dhoon 19, 21, 28, 85
Dhoon stream 12, 16
Dog Mills 43, 52, 80, 81, 82
Douglas 4, 10, 13, 14, 15, 38, 57, 84, 89, 100, 134, 136–44, 153, 155, 158, 161, 163, 165, 166, 173, 174–83, 186, 189–92, 194, 195, 202, 215, 217, 224, 228, 229, 245, 248, 250, 256, 261, 263, 266
Douglas Bay 15
Dreswick Harbour 32
Druidale 72

Eary 86
Eary Cushlin 173
Eary ny Shiannagh 76
Ellan a Voddij 76

Fishers Hill 77
Fleshwick Bay 13

GEOGRAPHICAL INDEX 285

Foxdale 19, 21, 23, 28, 38, 140, 146, 161, 186, 194, 195, 221, 222, 223, 237
Foxdale River 14

Gansey 80
German 71, 138, 146, 153, 155
Ginger Hall 70
Glen Auldyn 85
Glen Ballyre 56, 58, 63
Glen Dhoo 17, 44, 59, 60
Glen Helen 172, 192
Glenfaba 224
Glen Maye 14, 85, 172, 223
Glen Mooar 44, 49, 50, 59, 77, 81, 82, 224
Glen Roy 85
Glen Rushen 224
Glen Trunk 78
Glen Wyllin 54, 56, 58, 59, 70, 82
Glenchass 221
Gob ny Calla 85
Gob ny Rona 90
Gob ny Ushtey 10
Greeba 16, 93, 94, 217
Greeba Mountain 10
Groudle Glen 13, 85, 172, 189, 194
Guilcagh 76, 86, 94

Jurby 56, 57, 71, 73, 78, 146, 153, 155, 166, 183, 192, 198, 235
Jurby Head 16, 50, 82
Jurby Ridge 16, 55

Kallow Point 34, 35
Keristal 14, 15
Kerrowdhoo Reservoir 13
Kerrowkneale 73
Killane River 16, 17, 90, 91, 94
King Orry's Grave 71
Kiondroghad 73
Kionlough Glen 81, 82
Kionslieu 86
Kirk Michael 14, 43, 44, 47, 49, 50, 54, 56, 60, 62, 78, 79, 81, 195
Knockaloe 143, 211, 212, 214

Laggagh Mooar 90, 93
Langness 17, 22, 31, 32, 33, 85, 86, 88, 173
Laxey 12, 13, 23, 38, 99, 100, 140, 146, 155, 189, 193, 194, 215, 221–23, 227, 231, 237
Laxey Glen 10, 44, 59, 226, 231
Lezayre 94, 145, 146, 168, 189
Lheim ny Chynnee 76
Lhen River 16, 17, 48, 59, 73, 76, 82, 94
Lonan 71, 73, 140
Lough Cranstal 16, 42, 57, 65–68, 71, 72, 85, 93
Loughan Ruy 74
Loughcroute 94

Malew 10, 13, 73, 146
Marown 171, 146, 188
Maughold 71, 73, 76, 85, 90, 146, 153, 155, 168, 221, 223, 224
Maughold Head 12, 28, 59, 73, 88, 173
Meayl Hill see Mull Hill
Michael Parish 78, 153, 155, 165, 166
Morragh 224
Mount Karrin 10, 72
Mull Hill (or Meayl Hill) 10, 13, 26, 71
Mullagh Ouyr 10, 69, 72
Mullenlowne 74

Nappin Ponds 85
Narradale 85
Neb River 14, 15
Niarbyl 57
North Barrule 10, 12, 59, 85
Northern Plain 16–17, 18, 20, 37, 40, 57, 60

Oakhill 14, 15
Oatland 28
Onchan 71, 146, 153, 155, 232, 250, 261, 263
Orrisdale 78
Orrisdale Head 50, 53, 54, 70

Patrick 71, 153
Peel 8, 10, 14, 15, 18, 22, 24, 26, 30, 99, 136, 137, 139, 140, 143, 146, 153, 155, 166, 182, 188–92, 195, 198, 205, 215, 217, 228–30, 248
Perwick 76, 85
Phurt 16, 37, 42, 50, 57, 65–68, 71
Point of Ayre 37, 38, 40–43, 48–50, 57, 68, 77, 81–83, 87, 189, 224
Poortown 19, 28, 38, 271
Port Cornaa 185
Port Erin 17, 85, 89, 139, 143, 153, 155, 168, 192, 195, 211, 215, 217, 256, 261
Port Grenaugh 13, 14, 17
Port-e-Chee 188, 228, 229, 246
Port e Vullen 12
Port Mooar Valley 11, 12, 88, 271
Port St Mary 34, 35, 76, 80, 89, 143, 168, 182, 189, 207, 221, 224, 228, 229, 250, 261
Port Soderick 14, 249
Poyll Vaaish 17, 33, 34, 70, 77, 88, 224
Pulrose 189, 190, 191

Ramsey 12, 39, 41, 59, 81–83, 93, 100, 136, 137, 143, 153, 155, 166, 168, 174, 176, 177, 178, 180–83, 188–95, 198, 215, 224, 228, 231, 248, 250, 261
Regaby Beg 224
Rhendhoo 73, 86
Ronaldsway 14, 15, 17, 70, 73, 181–85, 217, 231, 233, 234
Rue Point 81, 82, 87
Rushen 76, 146, 155

St Johns 10, 14, 15, 16, 73, 75, 99, 193, 218, 227
St Judes 224
Sandwick Bay 32
Santon 71, 146
Santon Burn 14
Sartfield 82, 94
Scarlett Point 23, 33–36, 38, 85, 86, 88–89
Shellag Point 16, 43, 47, 50, 52, 81
Silverdale 172
Skeirrup 85
Slieau Curn 14
Slieau Dhoo 6, 10
Slieau Freoghane 10
Slieau Managh 72
Slieau Meayll 13
Slieau Ruy 28
Slieau Whallion 10
Snaefell 10, 12, 59, 69, 85, 189, 196, 223, 230

South Barrule 10, 29, 41, 43, 69, 72, 85, 94, 189, 217, 224
Spanish Head 173
Stoney Mountain 29
Strandhall 70, 71, 76
Sulby 17, 90, 189, 191, 226, 227
Sulby Glen 17, 59
Sulby River 16, 17, 59

Thistle Head 85
Traie Fogog Bay 31
Traie ny Halsall 85
Tromode 141, 212, 227, 228
Turkeyland 32, 33, 38

Union Mills 16

West Craig 224

SUBJECT INDEX

Agriculture 99, 109, 110, 114, 117, 121, 134, 135, 136, 140, 145, 204, 207–15
Air transport 182–85, 257–58
Alluvial fans 17, 31, 32, 42, 44, 46, 56, 57, 59, 61
Alluvium 15, 16, 17
Atholl, Dukes of 74, 122, 126, 127, 136, 137, 138, 208, 220, 221

Balance of payments 269–70
Banking 239–41
Beach: raised (postglacial) 16, 17, 47, 57, 66, 67
Birds 85–91
Bishop of Sodor and Man 116, 125
Bishop Wilson 122
Bishops Court 115, 116
Boards of Government 78, 80, 84, 129, 130, 168, 234, 258–59
Brick-making 39, 45, 219, 220, 224–25
Bride moraine 59, 67
Brine extraction 82, 224
Bronze Age 69, 72, 73, 99, 105–06
Bus services 194, 196–200
Butterflies (*Lepidoptera*) 88, 97

Castle Mona 138, 140
Castle Rushen 75, 115, 117, 118, 119, 120
Celts 23, 108, 109
Channels: abandoned 17; dry 13, 16, 45; meltwater 13, 48, 57, 59
Christian, William 122, 126
Christianity 73, 105, 108, 109, 111–15
Chronicles of the Kings of Man and the Isles 74, 113
Church 114, 115, 116, 124, 135, 221
Coal 22, 37, 74, 99, 182, 189, 194, 219, 224
Coleoptera (beetles) 58, 61–65, 66, 96
Conservation 84–94, 171
Constitution 99, 123–32
Cost of living 156, 191–92
Crosses 108–11, 113, 114

Darraghs (bog oak) 71
Derby, Earls of (see also Stanley family) 73, 119–22, 125, 126, 130
Disease 135, 137, 139
Douglas Head suspension bridge 141
Drainage: diversions 14, 15, 59; ice-marginal 14
Drumlins 17, 43, 47

Education 142, 170
Elderly population 144, 153, 168
Electoral system 129
Emigration 136, 137, 139, 140, 143, 144–47, 151, 156, 205, 230
Employment structure 154, 217, 233, 235, 242, 245, 250
Engineering industry 228, 231, 233, 235
Erosion: coastal 9, 66, 77–83
Executive council 130

Fertility 140, 142, 144, 150, 153
Financial sector 100, 153, 184, 206, 238–47, 268–69
Fishing 99, 134, 135, 140, 142, 145, 204, 208, 215–17
Flax 134
Flint tools 68, 69, 70, 71, 72, 99, 103–05
Folk Life Survey 173–74, 272
Forestry 84, 171–72, 217–18
Fossils 25, 30, 31, 32, 34, 41, 47, 60, 69, 76
Freeport 233–34

Gas 182, 189, 190, 191
Geology 18–54, 88
Godred Crovan 74, 113, 114, 117
Glacial: drift deposits 8, 10, 14, 15, 16, 17, 18, 37, 40–45, 57, 77, 99, 271; erratics 17, 41, 43, 45, 46, 47, 48, 73; landforms 8, 10, 15, 16, 43, 44, 45, 47, 48, 52, 54, 57, 271; meltwater 17, 48, 57; striations 10, 17, 41, 43, 45, 46, 47, 48, 59
Glaciation 8, 13, 40–54, 87, 271
Glens 14, 100, 171–73, 176, 194, 218
Godred Crovan 74, 113, 114, 117
Granites 8, 18, 20, 21, 23, 25, 28–30, 38, 41, 46, 59

Health care 168–69
Hillforts (see also Promontory forts) 106, 113
Holocene (Flandrian) 9, 40, 52, 56, 60, 63–69, 77, 272
House of Keys 84, 125, 126, 127, 128, 129, 130, 137
Housing 134, 135, 137, 138, 140, 151, 163–68
Housing Improvement Society 141

Immigration 100, 107, 136, 137, 147, 150, 151, 152, 168, 237
Income: personal 163
Industrial policy 230–35
Insurance sector 241–42
Internment camps 143
Ireland 141, 155, 249, 254, 257

Iron Age 73, 99, 106–10, 112
Isle of Man Development Plan 84, 85
Isle of Man Natural History and Antiquarian Society 75

Keeill 78, 108–10, 111, 114
Kettle holes 42, 43, 47, 50, 54, 55, 56, 59, 60, 61, 62, 63

Lady Isabella wheel (Laxey) 99, 189, 221, 228, 255
Lagoonal deposits 16, 42, 43, 57, 67, 68
Lake Andreas 16, 43, 45, 48, 59, 66, 90
Land tenure 117, 121, 134, 135
Lateglacial 9, 40, 55–63, 74–76, 272
Legislative Council 128
Leisure provision 170–76
Lieutenant Governor 127, 128, 129, 130, 131, 144, 231
Lime kiln 78, 189, 224
Limestone 8, 17, 18, 20, 32–38, 189, 219, 224
Lintel grave 78, 109, 111, 112
Loch, Henry 127
Loch Lomond Stadial 40, 56, 57, 59, 61, 63, 65

MacDermot Commission 128
Magnus Barefoot 113, 117
Magnus of Norway 74
Mammals 65, 74–76
Manx Electric Railway 83, 100, 189, 194–95
Manx Museum 49, 75, 76, 84, 89, 91, 93, 108, 169, 172, 173, 176, 258, 272, 273; other museums 176
Manx National Nature Reserves 84, 173
Manx National Trust 86, 87, 89, 172, 173
Manx Nature Conservation Trust 84, 94, 173
Manx Radio 175
Manx Shearwater 76, 89, 272
Manx slate 8, 12, 13, 16, 17, 24–28, 29, 38, 88, 89, 223, 224, 271
Manx steam railways 192–95
Marine administration 244–45
Mec Vannin 129, 149
Medieval 69, 74, 93, 109, 111–14, 117, 118, 135, 220, 221, 227
Megalithic tombs 103, 104, 106
Megaloceras giganteus (Great Deer or Giant Irish Elk) 9, 57, 74–75, 271
Mesolithic 63, 65, 68, 69–71, 72, 76, 99, 103
Millennium 123, 125
Milling 135, 161, 189, 219, 220, 226
Mining 38, 88, 99, 139, 140, 142, 146, 194, 220, 221–23, 230
Model village, Cronkbourne 141, 142, 167
Mortality 139, 140, 144, 150, 153, 168
Moths 57, 97, 98

Neolithic 63, 65, 66, 67, 68, 69, 71–72, 76, 99, 103–05, 106
New Residents 147, 149, 150, 151, 152, 155, 158, 205, 206
Norse Period 73, 99, 109, 110–15, 118, 119; Norse settlement 89
North American Manx Association 146, 147, 148

Orchids 87, 94
Ores 8, 23, 37, 38, 73, 99, 106, 219, 221–23, 230

Paper industry 220, 229
Peat 10, 15, 16, 55, 65, 66, 69, 72, 74, 99, 105, 134; peat-cutting (turbaries) 72, 93, 94, 189, 219
Peel Castle 103, 109, 111, 112, 116, 118, 119, 121, 122
Peel sandstone 8, 14, 18, 22, 30–32, 38, 224, 271
Periglacial 12, 44, 55, 56, 57, 59
Pingo 59, 60, 61, 62, 65, 74
Plants 85, 87–94
Pleistocene 12, 49
Pollen record 58, 60–69, 71, 72, 94, 272
Port facilities 138, 140, 161, 177, 181–82
Postglacial: 55–76, 90; gravels 16; raised beach 16
Power generation 162, 188–92
Preglacial: drainage 13; marine platform 37; shorelines 16; valleys 10, 59, 88
Promontory forts 73, 76, 106, 107, 112

Quarrying 223–24
Quarterlands 110

Radiocarbon dating 9, 56, 65, 68, 69, 72, 73, 105, 272
Railways 161, 189, 192–95, 220, 222
Relations with EEC 131, 215, 216, 273
Relations with UK 115, 118, 119–22, 126, 130, 131, 240
Revestment 122, 126, 131, 204, 220
Roads 161, 195–96, 220
Roman influence 106–08
Ronaldsway Airport 182–84, 216
Round houses 72, 105–07, 109
Rushen Abbey 74, 116, 117, 118, 120, 122, 221

St German's Cathedral 114, 115
Saliferous marls 8, 18, 22, 37, 38, 189; extraction 82, 224
Sand and Gravel 13, 14, 39, 43, 47, 52, 53, 226
Sand dunes 16, 67, 87
Sea transport 107, 177–82
Seals 87, 89, 90
Sheading 73, 99, 114, 128
Sheep 90, 212, 213
Ship-building 220, 228
Shorelines: preglacial 16
Smuggling 121, 126, 137, 204
Snaefell Mountain Railway 100, 193, 194
Society for the Preservation of the Manx Countryside 84
Soil 207

Solifluction 10, 44, 50
Stanley family (see also Derby, Earls of) 15, 119–22, 125, 126, 135, 136, 137
Steam Packet Co. 140, 177–81, 257

Taxation 116, 122, 126, 127, 147, 156, 202, 205, 220, 238, 239, 242, 243, 246
Telecommunications 162, 185–88
Terraces: alluvial 10, 16, 59; altiplanation 12; lake 14; solifluction 10, 44, 59
Textile industry 226–27, 228–29, 231, 233
Tithes 116, 145, 205
Tourism 84, 100, 138, 140, 143, 183, 184, 185, 195, 205, 206, 220, 229, 248–67, 268–69
Tourist accommodation 260–64
Treen 109
Tynwald 78, 79, 83, 114, 123, 124, 128, 129, 130, 131, 149, 168, 195

Unemployment 143, 144, 151, 169, 205, 230, 233, 249, 251

Victorian era 3, 99, 140–42, 166, 177, 192–94, 221–29, 248
Vikings (see also Norse Period): 105, 110–13, 123, 124, 125; boat burials 109, 111; burial mounds 111
Volcanic rocks 8, 18, 23, 25, 34–36, 88

Water power 189, 219, 237
Welfare services 150, 151, 168–70
Wildlife Park, Curraghs 91, 94
Wildwood (Manx native) 64, 65, 70, 71, 72–74, 88, 99
Windermere (Lateglacial) Interstadial 56, 59, 65, 75
Woollen mills 134, 161
World War Two, impact of 143, 178, 183, 202, 205